Current Topics in Clinical Virology

Edited by Peter Morgan-Capner

Public Health Laboratory Service

Public Health Laboratory Service
61 Colindale Avenue
London NW9 5DF

ISBN 0 901144 30 4

Typeset and printed by The Laverham Press, Salisbury, Wiltshire

Preface

Some five years ago the Public Health Laboratory Service (PHLS) published a book of virological reviews edited by Philip Mortimer. This volume attempts to follow the success of *Public Health Virology – 12 Reports* by bringing together 14 reviews of areas of interest to those having responsibilities in diagnostic virology. Strictly speaking, some chapters discuss infective agents which are not viruses, such as *Coxiella burnetii*, *Mycoplasma pneumoniae* and *Chlamydia trachomatis*, but they find a home in this volume because their diagnosis is usually made in the virology laboratory.

Some viruses of major clinical and public health importance are not covered in this volume. Notable by their absence are human immunodeficiency virus and hepatitis B, except where they appear in the chapter on blood transfusion microbiology. Reviews of these viruses are common, however, and my objective was to obtain authoritative discussion of topics which less frequently attract the attention of reviewers.

After suggesting to Philip Mortimer that he and the Public Health Laboratory Service should commission a second volume of essays, I found myself having that responsibility. I only hope that one of the readers of this volume feels stimulated to edit a third in the series. There is no shortage of suitable topics, as exemplified by recent descriptions of further 'new' viruses, namely hepatitis C and E. When the authors are as helpful as those who have contributed to this volume, it is a rewarding endeavour.

Peter Morgan-Capner

Contributors

Dr J Barbara   North London Blood Transfusion Centre

Dr EO Caul   Bristol Public Health Laboratory

Dr JE Cradock-Watson   Manchester Public Health Laboratory

Professor DH Crawford   London School of Hygiene and Tropical Medicine

Dr J Dadswell   Reading Public Health Laboratory

Dr S Gardner   Virus Reference Laboratory, Central Public Health Laboratory

Professor PD Griffiths   Royal Free Hospital, London

Dr RJC Hart   Exeter Public Health Laboratory

JM Hawkins   UK National External Quality Assessment Scheme for Microbiology, Quality Assurance Laboratory, Central Public Health Laboratory

Dr NA Hotchin   London School of Hygiene and Tropical Medicine

Dr A King   Central Veterinary Laboratory, Ministry of Agriculture, Fisheries and Food, Weybridge

Dr G Lloyd   PHLS Centre for Applied Microbiology and Research, Porton Down, Wiltshire

Professor ND Noah   King's College Hospital, London

Dr T Riordan   Exeter Public Health Laboratory

Dr J Sellwood   Reading Public Health Laboratory

Dr A Webster   Royal Free Hospital, London

Dr P White   Norwich Public Health Laboratory

Dr T Wreghitt   Cambridge Public Health Laboratory

Contents

Contents

Chapter 1

Varicella–zoster virus infection during pregnancy

JE Cradock-Watson

Chickenpox, one of the most common and contagious infectious diseases of childhood, is also one of the mildest. The disease is seldom severe and complications are rare. For some patients, however, the disease has special hazards: these patients include children with leukaemia, patients whose immunity is depressed by drugs or disease, pregnant women and neonates whose mothers get chickenpox shortly before or after delivery. I shall review the risks for pregnant women and neonates and, as far as possible, reassess them in the light of work at present being carried out in the Public Health Laboratory Service (PHLS).

Pathogenesis of postnatal chickenpox

The pathogenesis of chickenpox is still not entirely clear. The current view, outlined by Grose [1] and Plotkin [2], is based largely on the mousepox model of Fenner [3], studies of simian varicella [4,5] and reports of fatal infection in normal and immunocompromised children [6,7]. Following implantation at some unknown site in the respiratory tract, varicella-zoster (VZ) virus passes to the regional lymph nodes. From about the fourth to the sixth day the virus spreads by primary viraemia to replication sites in the peripheral lymph nodes, the lung, liver, bone marrow and suprarenal glands. This presumptive primary viraemia has not been demonstrated in man, but its occurrence is supported by rare observations of spots during the prodromal period. It is at this stage that immunoglobulin could influence the outcome. Visceral involvement may also follow the secondary viraemia, which leads to infection of the skin and mucous membranes. This viraemia is associated with mononuclear cells and has been detected from five days before, to one day after the onset of the rash in patients with normal immune function [8–10]. Being cell-borne, it is unlikely to be affected by passively administered antibody. The patient is infectious from 1–2

1

days before the rash until the time when the last vesicles are dry. This is normally not more than five days after the onset, but in immuno-compromised patients the viraemia, period of infectivity and clinical course may all be prolonged [11–13]. There is still uncertainty about the mode of transmission and the site from which this occurs. Vesicular fluid contains viable virus but dry crusts are not infectious and virus isolation from the respiratory tract is rare. Infection is easily achieved by close family contact, but studies of hospital outbreaks have shown that airborne transmission can also occur over greater distances, especially when assisted by differences in air pressure [14,15]. Fomites are prob-ably not important. The incubation period is 10–21 days, commonly 14–16 days, but can be shorter in immunocompromised patients and in neonates infected transplacentally.

IgG, IgM and usually IgA antibodies appear 2–5 days after the onset of the rash and rapidly increase in titre. IgM and IgA reach a peak in 2–3 weeks and then decline over a period of months. IgM eventually disappears. IgG reaches a peak in 3–4 weeks and then declines slowly, but persists indefinitely at a lower titre in virtually all individuals. Resting levels cover a wide range, however, and IgG is sometimes difficult to detect by even the most sensitive methods.

Epidemiological background

Epidemiological data come mainly from the United States of America, where chickenpox has been nationally notifiable (but seriously under-reported) since 1972. However, the pattern in the USA probably applies to all developed temperate regions. The incidence is seasonal, reaching a peak from March to May and falling to its lowest level during the late summer and autumn [16,17]. Chickenpox in the 5–9 year age group accounts for the majority of cases [16,18,19]. There is also a high incidence in children aged 1–4 years, and recent data from the Royal College of General Practitioners suggest that the proportion of cases in this age group increased in England and Wales between 1982 and 1985 [20]. In the home, where close contact can occur with a small number of susceptible individuals, the disease is transmitted easily, and secondary attack rates of 61% and 87% for normal children [19,21] and 89% for seronegative women [22] have been reported. In a school containing both susceptible and immune individuals the disease spreads more slowly and attack rates of this order may be achieved only after two waves of infection [23]. Immunity is normally lifelong, but subclinical [24] and even clinical [25,26] reinfection have been reported. The latter is largely confined to vaccinees, who may be reinfected with wild virus, and immunocompromised persons in whom reinfection may be indistinguishable from generalised reactivation.

These epidemiological features ensure that over 90% of adults are immune [27–29]. Studies of the age-specific prevalence of antibody in

England [30], West Germany [22,31] and Sweden [32] provide a similar picture.

Prevention of chickenpox with immune globulin

Early work in the USA by Ross [19] showed that normal human immunoglobulin failed to prevent chickenpox in immunocompetent children and adults but reduced the number of spots. A further trial was therefore conducted with immunoglobulin prepared from patients convalescing from herpes zoster (zoster immune globulin, ZIG). Brunell *et al* [33] compared ZIG with normal immunoglobulin and showed conclusively that 2ml of intramuscular ZIG prevented chickenpox in normal children when given within 72h of exposure. A smaller dose (0.5ml) prevented chickenpox in only 1 of 4 children, but dramatically reduced the number of spots in the other three [34]. It was later observed, however, that 2ml ZIG failed to prevent chickenpox in a child with leukaemia. Further studies on leukaemic and immunocompromised children were therefore carried out with larger doses of ZIG, but, for ethical reasons, no controls. These children were not completely protected, but had clinical attack rates ranging from 15%–67% [35–38]. The observed attack rates were strongly influenced by closeness of contact, and may have been lowered in some studies by the inadvertent inclusion of seropositive subjects; they can only be compared with the reported attack rates of 61%–87% in normal children and 50%–100% in immunocompromised children [39,40].

Infection in ZIG-treated immunocompromised children was usually mild, and sometimes subclinical. Complications were fewer in subjects who received high-titre rather than low-titre batches [38]. Despite the number of uncontrolled variables, these trials give the strong impression that ZIG, even if it has little effect on the incidence of infection, reduces the severity of illness in immunocompromised patients and increases the proportion of asymptomatic infections.

In the USA the limited supply of ZIG from patients with zoster was later augmented by preparing varicella-zoster immune globulin (VZIG) from normal donors with high titres of complement-fixing antibody. In the United Kingdom the anti-varicella-zoster immunoglobulin in current use is also prepared in this way and will be referred to as VZIG [41]. It replaces the earlier anti-chickenpox immunoglobulin (ZIG) prepared from patients convalescing from zoster. Guidelines published by the Department of Health (DoH) recommend giving VZIG to contacts in the following groups: children with leukaemia, immunosuppressed patients, recipients of bone marrow transplants, seronegative pregnant women, and neonates whose mothers develop chickenpox during the last six days of pregnancy or soon after delivery [42]. In the USA the Immunization Practices Advisory Committee (ACIP) makes broadly similar recommendations, but recommends VZIG for neonates whose

3

mothers develop chickenpox within five days before, and two days after, delivery [43]. The ACIP also recommends giving VZIG to neonates weighing less than 1,000 grams or of less than 28 weeks' gestation, regardless of the timing of the mother's rash, on the grounds that such immature infants may lack maternal antibody.

Incidence of chickenpox in pregnancy

Thanks to the high prevalence of immunity in adults, chickenpox is uncommon in pregnancy. Sever and White [44], in a survey in the USA between 1958 and 1964, noted 20 cases of chickenpox in 30,000 pregnancies – slightly less than the incidence of rubella. In Great Britain from 1950 to 1952, the Manson study of virus infections in pregnancy included 353 cases of chickenpox and 654 cases of rubella [45]. If, as is likely, many of the clinically diagnosed cases of rubella in these two surveys actually had other causes, then the numbers of cases of genuine rubella and chickenpox were probably similar. Since selective vaccination has reduced rubella susceptibility in pregnant women in the UK from 15%–20% to about 3% [46,47], chickenpox during pregnancy is now almost certainly more common than rubella. This disparity will increase if mass vaccination of boys and girls interrupts the circulation of rubella virus. The actual incidence of chickenpox during pregnancy, however, is not known.

Complications of chickenpox for the pregnant woman

Pregnant women are at risk from the more serious complications of chickenpox such as encephalitis and pneumonia which, for unknown reasons, occur predominantly in adults [48]. Encephalitis is rare, and there is no evidence that its incidence or severity is increased by pregnancy. Some degree of pulmonary involvement, however, is not uncommon in adults. Weber and Pellechia [49] reported that in a male military establishment in which all cases of chickenpox were admitted to hospital, routine radiographs showed signs of pneumonitis in 18 out of 110 cases (16%). However, only two patients had corresponding symptoms and signs and the authors concluded that varicella pneumonitis was usually a mild self-limiting disorder. Mermelstein and Freireich [50] also noted lack of correlation between radiological and clinical signs and warned that some cases could be unpredictably fulminating and life-threatening.

Retrospective surveys of case reports have given the impression that varicella pneumonia is particularly common and severe in pregnancy. Harris and Rhoades [51] reported seven deaths in 17 pregnant women with pneumonia (41%), and an increased incidence of abortion and prematurity. Young and Gershon [52], reviewing these and five additional cases, reported ten deaths (45%), whereas Triebwasser *et al* [53], reviewing 236 cases (60% males) in non-pregnant adults, reported a fatality rate of only 11%. These figures, particularly for pregnant patients, are almost certainly biased by selective reporting of serious cases admitted to hospital. Siegel *et al* [54], who studied 150 women with chickenpox in pregnancy, reported only one death from pneumonia. Paryani and Arvin [55], in a prospective study of 43 cases of chickenpox at all stages of pregnancy, reported clinical pneumonia in four women (9%), of whom two required ventilatory support, one went into premature labour and one died. The existing information does not allow accurate assessment of the risk. Some reviewers have concluded that there is no evidence that the illness is more severe or that complications are more common in pregnant women than in other adults with chickenpox [52,56]. There is no doubt, however, that varicella pneumonia can be a fulminating and even fatal condition, and some physicians with recent experience of severe cases recommend that all women with chickenpox in pregnancy should be admitted to hospital where their condition can be monitored and they can be treated promptly with acyclovir if necessary [57,58].

Although pneumonia is associated with abortion and premature delivery, there is no evidence that uncomplicated chickenpox during pregnancy causes these complications [44,54]. It should be remembered, however, that fetal losses in early pregnancy are probably underreported.

Should ZIG be given to susceptible pregnant women in contact with chickenpox? Enders reported that ZIG given within 10 days of contact reduced the clinical attack rate from 89% to 41% [22]. The benefit was greatest when ZIG was given within 72h. In the study of chickenpox in pregnancy at present being conducted by the PHLS Communicable Disease Surveillance Centre (CDSC), ZIG or VZIG has been given to 44 seronegative pregnant contacts, of whom 21 (48%) developed chickenpox (seven severely) and 11 (25%) had subclinical infections. A delay in the administration of VZIG of up to 10 days did not affect the results. There were no controls in the CDSC study, but the proportion of asymptomatic infections suggests that VZIG reduces severity and should be given if supplies permit.

Risks to the fetus and neonate of chickenpox during pregnancy

The main hazards consist of intrauterine and perinatal infection, with three possible clinical expressions: first, maternal chickenpox during the first five months of pregnancy may be followed by the characteristic pattern of abnormalities known as the 'congenital varicella syndrome' or 'varicella embryopathy'; secondly, chickenpox during the second and third trimesters may lead to the appearance of herpes zoster in an otherwise healthy child; thirdly, maternal chickenpox just before or

5

after delivery may cause severe and even fatal disease in the neonate. In addition, some fetal infections are probably asymptomatic.

VARICELLA EMBRYOPATHY

In 1947 Laforet and Lynch [59] described abnormalities in a newborn infant whose mother had chickenpox when eight weeks pregnant. The defects, which the authors attributed to fetal infection, included pigmented areas in the skin of the left leg, atrophy and paralysis of the right leg, bilateral cortical atrophy, bilateral optic atrophy and chorioretinitis, and dysfunction of the sphincters controlling the anus and bladder. After a stormy six months the infant was lost to follow-up. Srabstein *et al* [60], reviewing four previous reports and adding a fifth case of their own, noted striking similarities and postulated the existence of a 'congenital varicella syndrome' caused by chickenpox in early pregnancy. More than 30 infants with major abnormalities following chickenpox during the first 20 weeks of pregnancy have now been reported by numerous authors who have increasingly come to regard the association as causal rather than coincidental. The abnormalities are diverse but often include certain characteristic defects, which are now generally attributed to fetal infection. They have been described in detail in recent reviews [61–63] and can be summarised as follows:

1   Areas of skin loss or scarring, usually unilateral and segmental in distribution.
2   Hypoplasia of bone and muscle in an upper or lower limb, often with rudimentary digits. The affected limb, which may be partially paralysed, is usually on the same side as the skin lesion, and corresponds to the same segments. Hypoplasia of the upper limb can include the scapula and clavicle.
3   Abnormalities of the nervous system, including cortical and cerebellar atrophy, microcephaly, Horner's syndrome, sphincter dysfunction affecting the bowel and bladder, convulsions and nystagmus. Necropsy reveals extensive damage to the brain and spinal cord.
4   Abnormalities of the eye (not necessarily unilateral), including microphthalmia, chorioretinitis, optic atrophy and cataracts. Ocular defects are not always evident at birth but may develop later, perhaps indicating progressive damage.
5   Malformations of the gastrointestinal and genitourinary tracts.
6   General handicaps such as fetal growth retardation, prematurity, failure to thrive, psychomotor retardation and repeated infections.
7   Death in infancy in many cases, often due to recurrent aspiration pneumonia.

HERPES ZOSTER IN CHILDHOOD

Most cases of chickenpox occur at an age when the immune system is virtually mature. The ensuing period of latency in the sensory ganglia lasts for many years. Herpes zoster occurs when cell-mediated immunity, weakened by age, drugs or disease, fails to curtail reactivation. Several studies have shown that chickenpox during the first year of life predisposes to zoster in childhood [64–66]. In addition, Baba *et al* [65] reported that the incidence of childhood zoster was five times greater after chickenpox at 0–2 months than after chickenpox at 2–11 months. Baba suggests that the period of latency tends to become progressively shorter as varicella occurs earlier in life because the immature immune system is less efficient at limiting both the initial infection and subsequent reactivation.

Not all cases of zoster in childhood have a history of early postnatal chickenpox. Brunell *et al* [67,68] described six children in whom zoster appeared at ages between 3.5 months and 3.5 years. None had previously had chickenpox, but their mothers had had the illness when 3–7 months pregnant. The association between chickenpox in pregnancy and zoster in childhood has been noted by many other authors and there are now at least 25 such cases in the literature. Brunell suggests that these infants were infected *in utero* at a time when their immune systems were too immature to maintain latency.

In general, case reports of childhood zoster, whether resulting from pre- or post-natal infection, support the hypothesis that the earlier the child gets chickenpox, the earlier it is likely to get zoster.

PATHOGENESIS OF VARICELLA EMBRYOPATHY

The clinical manifestations of intrauterine infection include embryopathy and the development of zoster in infants who were normal at birth. Higa [63] analysed case reports of 27 infants in the former group and 25 in the latter. Almost all the mothers of the 27 malformed infants had chickenpox at 8–20 weeks, whereas the majority of the women whose infants developed zoster had the disease at later stages of pregnancy. Higa notes that early postnatal chickenpox predisposes to early zoster and concludes that the principal manifestations of varicella embryopathy are caused not by primary infection of the fetus but by *in utero* reactivation of virus whose period of latency in the fetal nervous system has been shortened by immunological immaturity. The neuromuscular defects can be attributed to denervation, and the cortical and ocular abnormalities to encephalitis. Grose [69] takes a similar view and points out that the most severe defects of the limbs tend to follow infection during the period of intense musculoskeletal and neurological development at 6–12 weeks of gestation. Chickenpox during the second trimester tends to cause mainly ocular defects. Both these reviewers therefore regard the syndrome as being one of 'congenital zoster' rather than 'congenital varicella'.

INCIDENCE OF VARICELLA EMBRYOPATHY

For many reasons the incidence of embryopathy is difficult to determine. First, the syndrome is ill defined. Many of the abnormalities, such as skin scars and associated hypoplasia of a limb, are characteristic; others, such as ocular defects and the more general handicaps of growth and development, can have other causes, known or unknown. Secondly, compared with rubella, the syndrome is rare and less clearly associated with a particular stage of pregnancy. It follows that even a large controlled survey, if assessed purely clinically, may not reveal any increase in abnormalities. Moreover, recruitment to a survey is prospective only from the time of reporting to the epidemiologist; a bias may be introduced by the patient or her doctor if, for example, they report only severe infections at certain stages of pregnancy. Thirdly, technical limitations have frustrated laboratory diagnosis: virus has never been isolated, herpes-type particles have been demonstrated only twice [70,71] and serological evidence has been obtained from only six neonates – four with specific IgM at birth [71–74] and two with IgG antibody during the second and fourth years of life [75,76].

Controlled prospective studies of viral infections in pregnancy have not revealed any excess of embryopathy after clinically-diagnosed chickenpox. Siegel [77], who studied 135 live births after maternal chickenpox, found four infants with abnormalities – an incidence of 3% compared with 3.4% in controls. One infant had cataracts and another had microcephaly after maternal infection at eight and nine weeks, respectively. In retrospect, these abnormalities could have been due to varicella. In the Manson study [45], after deducting four abortions, there were 288 live births and five stillbirths (one with hydrocephalus) after chickenpox. The live births included six abnormal infants: five were born after chickenpox during the third trimester and included three cases of congenital heart disease but no defects characteristic of varicella. One infant with mental retardation had been exposed to maternal chickenpox at six weeks. The overall incidence of defects was only 2.1% in the varicella group and 2.3% in the controls. Hill *et al* [78] studied 24 cases of chickenpox in pregnancy and found no abnormalities in 23 live births.

Two prospective studies confined to chickenpox in pregnancy have yielded five suspected cases. Paryani and Arvin [55] reported 41 women whose pregnancies went to term after chickenpox at different stages: 11 in the first trimester, 11 in the second and 19 in the third. One infant, at risk during the first trimester, had cutaneous and bony defects of the right leg, bilateral chorioretinitis, cortical atrophy and abnormalities of the left kidney and ureter. These defects were considered to be characteristic of varicella embryopathy, but specific IgM was not detected and tests for lymphocyte proliferation in the presence of antigen gave negative results. Enders [22] reported 170 pregnant women with confirmed chickenpox. Their infections were not uniformly distributed throughout pregnancy, but 85 (50%) occurred during the first trimester and 131 (77%) during the first five months. The outcome of pregnancy was known for 120 women, of whom 107 produced liveborn infants. Defects characteristic of varicella embryopathy occurred in four infants, of whom three were exposed during the second or third months and one at 18–19 weeks. Two of these infants survived and two died. The overall incidence of varicella embryopathy in liveborn infants was therefore 3.7%, but becomes nearly 5% if the denominator is restricted to cases occurring during the first 20 weeks of pregnancy. The incidence of embryopathy in the prospective study being conducted by the PHLS is at present one in 64 for the first 20 weeks of pregnancy. Adding this case to the four reported by Enders gives an incidence for the first 20 weeks of about 3.5% in liveborn infants. The diagnosis in these five cases is now supported by tests for specific IgM and IgG antibodies carried out in the Manchester Public Health Laboratory (PHL).

Herpes zoster in pregnancy
There are at least eight reports of infants with defects suggestive of congenital infection (microcephaly, microphthalmia, cataract, retardation) being born to mothers with a history of zoster during the first four months of pregnancy [79,80]. However, in no case was there any laboratory evidence of intrauterine infection with VZ virus. Two of the mothers had also had rubella in early pregnancy, which could have accounted for the abnormalities. The association between gestational zoster and fetal abnormality in the other cases could have been coincidental. Paryani and Arvin [55] studied 14 women with confirmed zoster at different stages of pregnancy but found no clinical evidence of infection in any of the infants and no serological evidence of infection in any of 12 who were tested. The PHLS survey now includes 40 women who have gone to term after confirmed zoster at 5–35 weeks of pregnancy. None of the infants has shown any signs of embryopathy and no specific IgM has been detected in any of 36 infants who have been tested. Viraemia may occasionally occur in zoster, but no convincing case of fetal infection has yet been reported.

Neonatal chickenpox
Chickenpox during the first few months of life is uncommon, partly because of lack of exposure and partly because of the presence of IgG antibody derived from the mother. The newborn child of a non-immune mother has no maternal IgG and may contract the disease if the mother develops the rash around the time of delivery. Chickenpox appearing at

9

the age of ten days or more could result from postnatal contact or airborne transmission, but a rash appearing earlier is more likely to be due to intrauterine infection.

The mechanism of intrauterine infection is not known and may not necessarily be the same in every case. Chickenpox can involve the mucous membranes and internal organs as well as the skin. Infants born at the time of the mother's rash could therefore be infected in the birth canal during delivery, and infants born later might be infected trans-amniotically. The majority, however, are thought to be infected by the haematogenous route, via the placenta, as a result of maternal viraemia. There are few published histological studies of the placenta in cases of neonatal chickenpox but there are two reports describing placentitis in cases in which a macerated fetus was expelled one month after maternal infection. In one case, in which the mother had chickenpox at four months of gestation, the villi were degenerate and necrotic, with granu-lomatous foci containing epithelioid cells [81]. In the second case, in which the mother had chickenpox at eight months, the villi showed infiltration with plasma cells, fibrosis and calcification [82]. The infant's rash follows the mother's rash by an interval, usually 7–17 days, which reflects the time taken for infection to cross the placenta and establish viraemia in the infant. This interval, which it is convenient to call the incubation period of neonatal infection, is more variable, and often shorter, than the incubation period of ordinary chickenpox, presumably because the initial stage of replication is circumvented. It is sometimes less than seven days and occasionally as little as one day, which suggests that fetal infection can be initiated by primary as well as by secondary maternal viraemia.

Between 1878 and 1968, 47 cases of congenitally-acquired varicella in neonates aged up to 10 days were described in different countries, mostly as single case reports. They were tabulated at intervals by successive authors who added new cases of their own [83–87]. An incubation period of 1–18 days was followed by a variable clinical course. Some infants had only a few spots with little or no constitutional upset; others had a severe illness with a prolonged course leading to pneumonia which was sometimes fatal. Involvement of the central nervous system was uncommon. Six infants (14%) died during the second week of life, apparently as a direct result of the infection. Necropsies on five infants showed generalised visceral dissemination involving the lungs, spleen, heart, pancreas, kidneys and suprarenal glands. Histological examination revealed focal areas of haemorrhage and necrosis in many internal organs, particularly the lung, with evidence of viral infection in the form of intranuclear inclusions [85,88–91]. In one case VZ virus was isolated from the lung as well as from the skin [90].

In 1974, Meyers [92] reviewed the severity and outcome in 36 previously published cases, plus an additional five reported to the Centers for Disease Control during the preceding two years. Four neonates died within two weeks of birth. All the deaths, and most of the severe infections, occurred among 19 neonates in whom the rash appeared 5–10 days after birth, giving a fatality rate of 21% in this group. In 22 infants in whom the rash appeared before the age of five days there were no deaths.

An explanation for the greater severity in infants aged 5–10 days was proposed by Brunell [93], who measured the titres of complement-fixing antibody in mothers and infants at birth in seven cases in which the mothers developed chickenpox 3–76 days before delivery. Three infants born three or four days after the maternal rash had no antibody at birth despite high titres in their mothers: all three got chickenpox. One infant born five days after the mother's rash had a low titre of antibody, and three infants born six or more days after the mother's rash all had high titres similar to those in their mothers. Brunell considered the antibody in these last four cases to be maternal in origin because it was detectable by complement-fixation and none of the infants got chickenpox. The absence of antibody in neonates born less than five days after the mother's rash was due to the time required for IgG antibody to appear in the mother and cross the placenta. The greater severity of chickenpox in infants aged 5–10 days could therefore have been due to the fact that the majority were born less than five days after the mother's rash and consequently lacked maternal antibody. The milder illnesses in infants aged less than five days generally resulted from maternal chickenpox more than five days before delivery and could have been modified by IgG derived from the mothers.

The crucial factor affecting severity thus appeared to be the interval between the mother's rash and delivery. Gershon [94], analysing 50 published cases from this point of view in 1975, reported seven deaths among 23 infants (30%) when the mother's rash occurred less than five days before delivery, but no deaths among 27 infants when the mother's rash occurred earlier. The period of high risk was later extended to include maternal chickenpox up to two days after delivery on the grounds that maternal viraemia might cause transplacental infection during this time [95]. Severe disease has also been reported in infants aged 11–28 days who acquired chickenpox, presumably postnatally, from their mothers or other close contacts. Rubin *et al* [96], analysing 14 such cases, reported that the disease was severe in five infants, three of whom died.

The widely quoted fatality rate of 30% continues to cause anxiety. However, the rarity of death from neonatal chickenpox in the UK before the introduction of ZIG suggests that the risk was overestimated in the past, probably because of selective reporting. During the 20 years from 1954 to 1973, only 21 deaths were reported to the Office of

Population Censuses and Surveys (OPCS). Nevertheless, severe and even fatal cases do occur and VZIG is therefore recommended in the UK for neonates whose mothers develop chickenpox during the last six days of pregnancy or soon after delivery, and for seronegative neonates exposed to chickenpox in the home [42].

The risk of severe disease, although diminishing after the neonatal period, probably remains high for several months. Preblud *et al* [97] reported that the risk of death for infants with chickenpox during the first year of life was four times that for children aged 1–14 years, but unfortunately the ages of infants developing fatal infection were not known. In a recent survey of chickenpox in England and Wales from 1967 to 1985, Joseph and Noah [20] reported that half the deaths during the first year of life occurred during the first month.

Problems of laboratory diagnosis

The outcome of chickenpox at different stages of pregnancy can be assessed only by a prospective clinical and virological study. Hitherto, however, serological investigation has been limited by the available techniques. The complement-fixation test, although satisfactory for confirming the diagnosis of chickenpox in adults, does not reliably measure IgG antibody in neonates. The neutralisation test [98] is laborious and requires a suitable cell-free antigen. Tests involving indirect haemagglutination [99] or immune adherence [100] are vulnerable to cross reactions and tend to give variable results in different laboratories. None of these methods can detect specific IgM antibody in unfractionated serum. A major improvement came with the introduction of immunofluorescence (IF). This was originally described as a fluorescent antibody test for membrane antigens (FAMA), in which suspended cells infected with VZ virus were treated first with patient's serum and then with fluorescein-labelled antibody to human IgG [101]. It is also possible to use fixed monolayers of infected cells on coverslips [102], with the advantage that previously prepared coverslips can be stored at −20°C and then removed for staining as required. IgG antibody can be titrated by testing serial dilutions of serum. IF has the advantage that it is possible to observe the site of the reaction and to compare infected and uninfected cells, but it requires lengthy preparatory work followed by subjective interpretation.

Although IF is still used by some workers, there is an increasing tendency to replace it by indirect (antiglobulin) forms of radioimmunoassay (RIA) [103,104] or enzyme-linked immunoassay (ELISA) [105–107], using a semipurified antigen prepared from disrupted cells. RIA needs radioactive reagents. ELISA, which is of comparable sensitivity, avoids this disadvantage and may eventually replace IF for screening purposes. It can give a quantitative result with sufficient accuracy from a single serum dilution.

IF and ELISA can be adapted to detect specific IgM by using labelled antibody to IgM in the final stage of the test. However, this procedure is not ideal because false positive results may occur if IgM with anti-IgG activity is present, and false negatives if IgG displaces IgM from combination with the antigen. The preferred method, for this and other viruses, is now antibody capture [108]. In the Manchester PHL, this takes the form of an IgM antibody capture radioimmunoassay (MACRIA) in which sensitivity and specificity are ensured by the use of a monoclonal antibody labelled with 125-I [109]. An enzyme label can also be used. MACRIA gives a quantitative result from a single dilution, but it measures the proportion of the total IgM that is specific rather than the specific IgM concentration. As long as the total IgM is sufficient to saturate the detection system, the concentration of specific IgM is largely irrelevant. This feature, far from being a disadvantage, makes the capture technique particularly suitable for neonates, in whom the proportion of specific IgM may be high even if the total IgM is within normal limits. In the ELISA (IgG) and MACRIA (IgM) tests now in use in Manchester, the antibody content in arbitrary linear units is estimated from a single dilution by comparing the unknown serum with a range of previously prepared standards in every batch of tests. One unit is defined as the minimum amount of antibody that can be distinguished from the range of activity in negative sera (3SD).

These serological developments are a great help in investigating the problems associated with chickenpox in pregnancy. For example, pregnant contacts and neonates at risk from maternal varicella can be screened for IgG antibody; subsequent infection in the mother or neonate, with or without symptoms, can be confirmed or excluded by studying the IgG and IgM responses. Infants with signs of embryopathy can be tested for the customary markers of congenital infection, namely IgM antibody at birth and IgG after the age of one year. Below, we give some of the results obtained with these methods in a joint PHLS study being conducted by CDSC and the Manchester PHL.

SPECIFIC IMMUNOGLOBULINS IN INFANTS WITH SUSPECTED VARICELLA EMBRYOPATHY

Between 1973 and 1989, the Manchester PHL received 30 sera from 20 infants (14 in the UK and six in West Germany) who were born with apparent embryopathy after maternal chickenpox at 7–26 weeks of gestation. Full descriptions of six of these cases have been published elsewhere [22,71,110]. Table 1 gives the serological results and brief clinical summaries, although in some cases few details were available. IgM antibody was detected in 11 of 18 infants at ages up to 197 days, and persistent IgG in 7 of 8 at ages 10–36 months. In all, 16 infants had serological evidence of congenital infection after maternal varicella at 8–26 weeks of gestation. One infant (case 7), who had neither IgM nor

Table 1 Antibodies, abnormalities and clinical outcome in 20 infants with suspected varicella embryopathy

Case no (ref)	Maternal varicella	IgM units (age)	IgG units (age)	Abnormalities	Further outcome
1	7wk	<1 (1d)	nt	Microcephaly, hypoplasia of limbs	Died 45min after birth
2	8wk	5.3 (birth) <1 (17d)	nt	Hypoplasia of L limbs, impaired anal reflex, neurogenic bladder	
3 [110]	11wk	<1 (53d)	nt	Rudimentary digits L hand, depigmented skin scars L arm, R leg, scalp, L Horner's syndrome	Recurrent pneumonia Died aged 3 months
4 [71]	12wk	>40 (birth)	nt	Vesicles on scalp, skin loss on trunk L4,5, fixed flexion of fingers both hands, bilateral microphthalmia and chorioretinitis, absent R kidney, thymic hypoplasia, cerebral atrophy	Died 36h after birth EM: herpes particles in vesicle fluid
5	12wk	nt	1.1 (35m)	Twin. Hypoplasia and scarring L arm. Other twin had persistent IgG antibody but no symptoms	
6	13wk	1.3 (birth)	nt	Multiple congenital abnormalities	Died soon after birth
7	13wk	<1 (128d)	<1 (9m) >40 (12m)	Skin scar L T4, with hypoplasia of underlying ribs, neurogenic bladder	Zoster, serologically confirmed, at 1 year
8	13wk	15.3 (119d)	nt	Congenital skin defect R frontal area, R facial palsy, progressive hydrocephalus	Pneumonia Died aged 4 months
9 [22]	12wk	<1 (52d)	5.1 (13m)	Suspected varicella embryopathy	

14

10 [22]	14wk	<1 (birth)	10.9 (10m)	Hypoplasia L arm, skin scars L axilla, bilateral microphthalmia	
11	15wk	2.2 (2d)	4.3 (3y)	Hypoplasia R arm, limited extension at elbow, R middle finger in fixed flexion	
12	16wk	1.9 (99d)	3.5 (30m)	Bilateral microphthalmia, L retinal dysplasia	
13	16wk	<1 (birth)	1.3 (11m)	Small area of skin loss between R ear and clavicle, requiring graft	
14	17wk	3.1 (birth)	nt	Scarring on legs and head, not segmental, also small erythematous spots all over body	
15	18wk	nt	4.5 (13m)	Chorioretinitis	
16 [22]	18wk	>40 (2d)	nt	Suspected varicella embryopathy	Died aged 3 weeks
17	20wk	<1 (1d)	nt	Skin scars and some vesicles	
18 [22]	24wk	7.8 (13d)	nt	Suspected varicella embryopathy	Died
19	17wk	26 (56d) 16 (91d) 7.1 (197d)	nt nt nt	Dysplasia of limbs, skin scars and vesicles, microphthalmia, CNS symptoms	Died aged 9 months
20	17wk	16 (131d)	nt	Small scars on limbs, retinal scars, incoordination of breathing and swallowing, subdural effusion, retardation	

nt = not tested

15

persistent IgG, was nevertheless probably infected since he developed zoster at the age of one year; two weeks later his IgM and IgG titres were 20 and >40 units, respectively. The first available specimen from this infant was taken at the age of 128 days – too late, perhaps, for IgM. Three infants (cases 1, 3, 17) were IgM-negative but could not be retested: two of them died before the age of three months and one was lost to follow-up. Eight of the 20 infants are known to have died.

These results show that IgM antibody can be detected in many cases of suspected varicella embryopathy. They give no idea of the duration of the fetal IgM response, which probably varies widely, nor do they indicate the incidence of abnormality. Nevertheless, they do suggest that IgM tests could be used to investigate the frequency of fetal infection, with or without symptoms, after maternal chickenpox at successive stages of pregnancy. The potential value of testing for persistent IgG antibody is less certain, because two infants (cases 5, 13) had only weakly positive results which were difficult to interpret, and one (case 7) had no detectable antibody before developing zoster. We have found, however, that many infants with confirmed neonatal chickenpox, although producing good IgG and IgM responses at the time, have no detectable IgG when retested 1–2 years later, so lack of persistent IgG in infants who were exposed to chickenpox *in utero* does not necessarily exclude congenital infection. It seems that in many infants the virus provides only a very weak antigenic stimulus during the period of latency.

Outcome in neonates after perinatal maternal varicella

Since 1979, all pregnant women and neonates for whom VZIG was requested from CDSC have been followed up prospectively. These patients include 201 women with chickenpox during the last 28 days of pregnancy, and 80 with chickenpox 1–28 days after delivery. VZIG (100mg up to the end of 1985 and 250mg thereafter) was issued for all neonates whose mothers had chickenpox during the last week of pregnancy or after delivery. IgG antibody was present, usually in high titre, in all neonates whose mothers had chickenpox more than seven days before delivery, but in none who were born less than three days after the onset of the mother's rash. At intermediate stages progressively fewer neonates were born with antibody: their titres were comparable with those achieved by giving VZIG but were generally lower than those in cord sera from women with no recent history of chickenpox or zoster. One infant, whose mother had chickenpox five days before delivery, died 36 hours after birth with respiratory distress but with no evidence of chickenpox. There were no other deaths. The outcome in the remaining 280 infants is summarised in Table 2. In all, 169 (60%) were infected: 134 (48%) had chickenpox and 35 (13%) had asymptomatic infections diagnosed from the presence of specific IgM. Chickenpox was

Table 2 Outcome in 280 infants whose mothers had chickenpox between 28 days before and 28 days after delivery

Onset of mother's rash in days before (−) or after (+) delivery	No of infants followed up	Total no infected (%)	No with asymptomatic infection (%)	Total no with chickenpox rash (%)	No with severe chickenpox	Incubation period, range in days (median)
−28 to −15	30	17 (57)	12 (40)	5* (17)	0	<9–21 (12)
−14 to −8	45	21 (47)	9 (20)	12† (27)	1	<4–24 (13)
−7 to 0	125	88 (70)	10 (8)	78‡ (62)	15	2–25 (13)
+1 to +7	51	31 (61)	4 (8)	27 (53)	3	8–23 (13)
+8 to +14	20	11 (55)	0	11 (55)	0	
+15 to +28	9	1 (11)	0	1 (11)	0	
Total	280	169 (60)	35 (13)	134 (48)	19 (7)	<9–25 (13)

*4 neonates had spots at birth
†6 neonates had spots at birth
‡6 neonates had spots at birth

severe in only 19 infants, of whom 18 were infected during the two-week period centred on delivery and 16 during the high-risk period from four days before to two days after delivery. In the others, the illness was mild, with few spots and little or no constitutional upset.

Two main conclusions can be drawn from this study. First, antibody acquired passively from the mother does not always prevent infection in the neonate. For example, when the mother had chickenpox during the last 20 days of pregnancy, 30 neonates developed chickenpox despite the presence of antibody at birth: the majority were presumably infected transplacentally before acquiring antibody. In addition, when 36 newborn children of immune mothers (who had had chickenpox in the past) were exposed to siblings with varicella in the home, three had asymptomatic infections and three had clinical chickenpox despite the presence of antibody at the time of contact. Failure of maternal antibody to protect neonates exposed to chickenpox has also been reported by others [111–113]. VZIG, therefore, cannot be expected to protect neonates completely. In the PHLS study it can have done little to reduce the incidence of neonatal chickenpox, which reached a peak of 62% during the week before delivery and remained high for two weeks after birth. In a similar study in Sweden, Hanngren reported a clinical attack rate of 51% in 41 ZIG-treated neonates exposed during the high-risk period and concluded that ZIG had no effect on the incidence of infection [114].

Secondly, the majority of neonatal infections were mild, and gave no cause for concern. How far can this generally favourable outcome be attributed to VZIG? Hanngren reported no deaths, and in the PHLS study the only infant who died had no evidence of VZ virus infection. It is tempting to attribute the low mortality to VZIG, especially as the dose was relatively greater than that given to older children. However, as we have seen, the previously estimated fatality rate of 30% is not a valid basis for comparison since it was derived from case reports scattered widely in time and place. During the 13 years from 1974 to 1986, when the use of VZIG in the UK became increasingly common, eight neonatal deaths were reported to OPCS. This small reduction from 21 deaths during the preceding 20 years cannot be attributed to VZIG alone, particularly as at least three of the deaths were in VZIG-treated infants [115]; it may reflect other factors such as better clinical care and the use of acyclovir. In the PHLS study, any effect of VZIG on severity was difficult to identify, not just because of lack of controls, but because some neonates also received maternal antibody and, in recent years, acyclovir as well. The effect of VZIG is therefore uncertain, but the general mildness of most cases of neonatal chickenpox in the PHLS and Swedish studies suggests that VZIG does reduce severity and should therefore be given to neonates whose mothers develop chickenpox during the last seven days of pregnancy or the first 14 days after delivery. However, a better method of prophylaxis is needed.

Most of the severe cases in the PHLS study were infected between four days before and two days after delivery, confirming that this is indeed a potentially hazardous time. The increased risk associated with this period has been attributed to transplacental infection, lack of maternal antibody and immaturity of the cellular immune response. If we assume that eight days is the minimum incubation period of infection acquired by the conventional route, then shorter intervals must indicate intrauterine infection. In the PHLS study, there were 22 incubation periods of 2–7 days, presumably indicating transplacental infection, but there was no evidence that these cases were especially severe. Lack of maternal antibody is a factor that also applies to infections acquired after the high-risk period. Cellular immunity is difficult to assess but is presumably immature throughout the perinatal period. Each of these three factors may contribute to severity, but none can be identified as the sole determinant.

Conclusion

Chickenpox is uncommon during pregnancy because at least 90% of women are immune; in the UK the incidence at the present time is probably about the same as for rubella. Chickenpox is more severe in adults than in children, and more likely to be complicated by pneumonia which can be fulminating and even fatal. Seronegative pregnant women in close contact with the disease should therefore be given VZIG, if supplies permit, and observed carefully with a view to treatment with acyclovir.

Chickenpox during the first five months of pregnancy occasionally causes embryopathy with a wide range of defects including skin scars, hypoplasia of the limbs, abnormalities of the eye and CNS, visceral malformations and general handicaps of growth and development. The incidence of embryopathy after chickenpox during the first 20 weeks of pregnancy is approximately 3%–5%, and the diagnosis can be confirmed in many cases by detecting IgM antibody at birth. The nature and distribution of the lesions suggests that they are caused by reactivation *in utero* rather than by the primary infection and that the syndrome should be regarded as one of congenital zoster rather than congenital varicella. The occurrence of zoster in otherwise normal infants who were exposed to maternal chickenpox *in utero* supports this interpretation and suggests that many fetal infections may be asymptomatic. IgM tests could be used to investigate the frequency of fetal infection, with or without symptoms, after chickenpox at successive stages of pregnancy. Herpes zoster during pregnancy does not present a significant risk to the fetus.

Maternal chickenpox just before or after delivery can cause neonatal varicella. This was previously thought to be a life-threatening condition, with a fatality rate of 30% for infants whose mothers developed chicken-pox less than five days before delivery. In the PHLS survey, 177 women had chickenpox during the two-week period centred on delivery. The only infant who died had no evidence of VZ virus infection. Of the surviving infants, 119 (68%) were infected and 105 (60%) had clinical chickenpox, but most cases were mild and only 18 were severe. VZIG was given to all infants exposed during this period but its effect was uncertain, partly because of lack of controls and partly because earlier estimates of the fatality rate were biased by selective reporting. However, VZIG probably does reduce severity and should therefore be given to neonates whose mothers develop chickenpox during the last seven days of pregnancy or the first 14 days after delivery. Neonates whose mothers have chickenpox more than seven days before delivery are born with high titres of maternal IgG antibody and do not require VZIG.

Acknowledgements
I thank Dr Elizabeth Miller for allowing me to include results from the PHLS study of VZ virus infections in pregnancy, Professor Gisela Enders for allowing me to test sera from infants in her study of chickenpox in pregnancy, Dr MO Savage and Dr HJ Heggarty for allowing me to include results from their published cases of varicella embryopathy, and Mrs Margaret Ridehalgh for technical assistance. Table 2 is reproduced by permission of the *Lancet*.

References
1   Grose C. Variation on a theme by Fenner: the pathogenesis of chickenpox. *Pediatrics* 1981, **68**, 735–7.
2   Plotkin SA. Clinical and pathogenetic aspects of varicella-zoster. *Postgrad Med J* 1985, **61 (suppl 4)**, 7–14.
3   Fenner F. The pathogenesis of the acute exanthems: an interpretation based on experimental investigations with mousepox (infectious ectromelia of mice). *Lancet* 1948, **ii**, 915–20.
4   Wenner HA, Abel D, Barrick S, Seshumurty P. Clinical and pathogenic studies of Medical Lake macaque virus infections in cynomolgus monkeys (simian varicella). *J Infect Dis* 1977, **135**, 611–22.
5   Wenner HA, Barrick S, Abel D, Seshumurty P. The pathogenesis of simian varicella virus in cynomolgus monkeys. *Proc Soc Exp Biol Med* 1975, **150**, 318–23.
6   Cheatham WJ, Weller TH, Dolan TF, Dower JC. Varicella: report of two fatal cases with necropsy, virus isolation, and serologic studies. *Am J Pathol* 1956, **32**, 1015–35.
7   Miliauskas JR, Webber BL. Disseminated varicella at autopsy in children with cancer. *Cancer* 1984, **53**, 1518–23.
8   Ozaki T, Ichikawa T, Matsui Y, *et al*. Viremic phase in

nonimmunocompromised children with varicella. *J Pediatr* 1984, **104**, 85–7.

9   Asano Y, Itakura N, Hiroishi Y, *et al*. Viremia is present in nonimmunocompromised children with varicella. *J Pediatr* 1985, **106**, 69–71.

10 Asano Y, Itakura N, Hiroishi Y, *et al*. Viral replication and immunologic responses in children naturally infected with varicella-zoster virus and in varicella vaccine recipients. *J Infect Dis* 1985, **152**, 863–8.

11 Feldman S, Epp E. Detection of viremia during incubation of varicella. *J Pediatr* 1979, **94**, 746–8.

12 Myers MG. Viremia caused by varicella-zoster virus: associated with malignant progressive varicella. *J Infect Dis* 1979, **140**, 229–33.

13 Feldman S. Varicella zoster infection of the fetus, neonate, and immunocompromised child. *Adv Pediatr Infect Dis* 1986, **1**, 99–115.

14 Leclair JM, Zaia JA, Levin MJ, Congdon RG, Goldmann DA. Airborne transmission of chickenpox in a hospital. *N Eng J Med* 1980, **302**, 450–3.

15 Gustafson TL, Lavely GB, Brawner ER, Hutcheson RH, Wright PF, Schaffner W. An outbreak of airborne nosocomial varicella. *Pediatrics* 1982, **70**, 550–6.

16 Preblud SR, D'Angelo LJ. Chickenpox in the United States, 1972–1977. *J Infect Dis* 1979, **140**, 257–60.

17 Preblud SR, Orenstein WA, Bart KJ. Varicella: clinical manifestations, epidemiology and health impact on children. *Pediatr Infect Dis* 1984, **3**, 505–9.

18 Preblud SR. Age-specific risks of varicella complications. *Pediatrics* 1981, **68**, 14–7.

19 Ross AH. Modification of chickenpox in family contacts by administration of gamma globulin. *N Eng J Med* 1962, **267**, 369–76.

20 Joseph CA, Noah ND. Epidemiology of chickenpox in England and Wales, 1967–85. *Br Med J* 1988, **296**, 673–6.

21 Hope-Simpson RE. Infectiousness of communicable diseases in the household (measles, chickenpox, and mumps). *Lancet* 1952, **ii**, 549–54.

22 Enders G. Management of varicella-zoster contact and infection in pregnancy using a standardized varicella-zoster ELISA test. *Postgrad Med J* 1985, **61 (suppl 4)**, 23–30.

23 Gershon AA, Krugman S. Seroepidemiologic survey of varicella: value of specific fluorescent antibody test. *Pediatrics* 1975, **56**, 1005–5.

24 Arvin AM, Koropchak CM, Wittek AE. Immunologic evidence of reinfection with varicella-zoster virus. *J Infect Dis* 1983, **148**, 200–5.

25 Gershon AA, Steinberg SP, Gelb L. Clinical reinfection with

varicella-zoster virus. *J Infect Dis* 1984, **149**, 137–42.

26 Kibirige MS, Heney D, Bailey CC. Immunisation in the immuno-compromised child. *Arch Dis Child* 1988, **63**, 679–80.

27 Gershon AA, Raker R, Steinberg S, Topf-Olstein B, Drusin LM. Antibody to varicella-zoster virus in parturient women and their offspring during the first year of life. *Pediatrics* 1976, **58**, 692–6.

28 Gershon AA, Steinberg SP. Antibody responses to varicella-zoster virus and the role of antibody in host defense. *Am J Med Sci* 1981, **282**, 12–7.

29 Muench R, Nassim C, Niku S, Sullivan-Bolyai JZ. Sero-epidemiology of varicella. *J Infect Dis* 1986, **153**, 153–5.

30 Heath RB. Herpesvirus infections: varicella-zoster. In: Zuckerman AJ, Banatvala JE, Pattison JR (eds). Principles and practice of clinical virology. Chichester: Wiley, 1987, 51–73.

31 Enders G. Varicella-zoster virus infection in pregnancy. *Prog Med Virol* 1984, **29**, 166–96.

32 Grandien M, Appelgren P, Espmark A, Hanngren K. Determination of varicella immunity by the indirect immunofluorescence test in urgent clinical situations. *Scand J Infect Dis* 1976, **8**, 65–9.

33 Brunell PA, Ross A, Miller LH, Kuo B. Prevention of varicella by zoster immune globulin. *New Engl J Med* 1969, **280**, 1191–4.

34 Brunell PA, Ross A, Miller L, Cohen M, Schmerler A. Zoster immune globulin. *Pediatr Res* 1970, **4**, 462.

35 Brunell PA, Gershon AA, Hughes WT, Riley HD, Smith J. Prevention of varicella in high risk children: a collaborative study. *Pediatrics* 1972, **50**, 718–22.

36 Gershon AA, Steinberg S, Brunell PA. Zoster immune globulin: a further assessment. *New Engl J Med* 1974, **290**, 243–5.

37 Judelsohn RG, Meyers JD, Ellis RJ, Thomas EK. Efficacy of zoster immune globulin. *Pediatrics* 1974, **53**, 476–80.

38 Orenstein WA, Heymann DL, Ellis RJ, *et al.* Prophylaxis of varicella in high-risk children: dose-response effect of zoster immune globulin. *J Pediatr* 1981, **98**, 368–73.

39 Meyers JD, Witte JJ. Zoster-immune globulin in high-risk children. *J Infect Dis* 1974, **129**, 616–8.

40 Geiser CF, Bishop Y, Myers M, Jaffe N, Yankee R. Prophylaxis of varicella in children with neoplastic disease: comparative results with zoster immune plasma and gamma globulin. *Cancer* 1975, **35**, 1027–30.

41 Hejjas M, Salker R, Barbara JAJ. Screening of blood donors for high titre antibody to herpes varicella-zoster. *Vox Sang* 1980, **39**, 335–8.

42 Department of Health and Social Security, Welsh Office, Scottish Home and Health Department. Immunisation against infectious disease. Joint Committee on Vaccination and Immunisation. Lon-

don: Her Majesty's Stationery Office, 1988, 115–7.

43 Immunization Practices Advisory Committee. Varicella-zoster immune globulin for the prevention of chickenpox. *MMWR* 1984, **33**, 84–90, 95–100.

44 Sever J, White LR. Intrauterine viral infections. *Ann Rev Med* 1968, **19**, 471–86.

45 Manson MM, Logan WPD, Loy RM. Rubella and other virus infections during pregnancy. London: Her Majesty's Stationery Office, 1960. (Ministry of Health Reports on Public Health and Medical Subjects no 101.)

46 Miller CL, Miller E, Waight PA. Rubella susceptibility and the continuing risk of infection in pregnancy. *Br Med J* 1987, **294**, 1277–8.

47 Noah ND, Fowle SE. Immunity to rubella in women of childbearing age in the United Kingdom. *Br Med J* 1988, **297**, 1301–4.

48 Guess HA, Broughton DD, Melton LJ, Kurland LT. Population-based studies of varicella complications. *Pediatrics* 1986, **78**, 723–7.

49 Weber DM, Pellechia JA. Varicella pneumonia. Study of prevalence in adult men. *J Am Med Ass* 1965, **192**, 572–3.

50 Mermelstein RH, Freireich AW. Varicella pneumonia. *Ann Int Med* 1961, **55**, 456–63.

51 Harris RE, Rhoades ER. Varicella pneumonia complicating pregnancy. Report of a case and review of literature. *Obstet Gynecol* 1965, **25**, 734–40.

52 Young NA, Gershon AA. Chickenpox, measles and mumps. In: Remington JS, Klein JO (eds). Infectious diseases of the fetus and newborn infant, 2nd edn. Philadelphia: Saunders, 1983, 375–427.

53 Triebwasser JH, Harris RE, Bryant RE, Rhoades ER. Varicella pneumonia in adults. Report of seven cases and a review of literature. *Medicine* 1967, **46**, 409–23.

54 Siegel M, Fuerst HT, Peress NS. Comparative fetal mortality in maternal virus diseases. A prospective study on rubella, measles, mumps, chicken pox and hepatitis. *N Engl J Med* 1966, **274**, 768–71.

55 Paryani SG, Arvin AM. Intrauterine infection with varicella-zoster virus after maternal varicella. *N Engl J Med* 1986, **314**, 1542–6.

56 Brunell PA. Varicella-zoster infections in pregnancy. *J Am Med Ass* 1967, **199**, 315–7.

57 Boyd K, Walker E. Use of acyclovir to treat chickenpox in pregnancy. *Br Med J* 1988, **296**, 393–4.

58 White RG. Chickenpox in pregnancy. *Br Med J* 1988, **296**, 864.

59 Laforet EG, Lynch CL Jr. Multiple congenital defects following maternal varicella: report of a case. *New Engl J Med* 1947, **236**, 534–7.

60 Srabstein JC, Morris N, Larke RPB, deSa DJ, Castelino BB, Sum

E. Is there a congenital varicella syndrome? *J Pediatr* 1974, **84**, 239–43.

61   Brunell PA. Fetal and neonatal varicella-zoster infections. *Semin Perinatol* 1983, **7**, 47–56.

62   Alkalay AL, Pomerance JJ, Rimoin DL. Fetal varicella syndrome. *J Pediatr* 1987, **111**, 320–3.

63   Higa K, Dan K, Manabe H. Varicella-zoster virus infections during pregnancy: hypothesis concerning the mechanisms of congenital malformations. *Obstet Gynecol* 1987, **69**, 214–22.

64   Latif R, Shope TC. Herpes zoster in normal and immuno-compromised children. *Am J Dis Child* 1983, **137**, 801–2.

65   Baba K, Yabuuchi H, Takahashi M, Ogra P. Increased incidence of herpes zoster in normal children infected with varicella zoster virus during infancy: community-based follow-up study. *J Pediatr* 1986, **108**, 372–7.

66   Guess HA, Broughton DD, Melton LJ, Kurland LT. Epidemiology of herpes zoster in children and adolescents: a population-based study. *Pediatrics* 1985, **76**, 512–7.

67   Brunell PA, Miller LH, Lovejoy F. Zoster in children. *Am J Dis Child* 1968, **115**, 432–7.

68   Brunell PA, Kotchmar GS. Zoster in infancy: failure to maintain virus latency following intrauterine infection. *J Pediatr* 1981, **98**, 71–3.

69   Grose C. Varicella zoster virus: pathogenesis of the human disease, the virus and virus replication, and the major viral glycoproteins and proteins. In: Hyman RW, ed. Natural history of varicella-zoster virus. Boca Raton, Fl: CRC Press, 1987, 1–65.

70   Dodion-Fransen J, Dekegel D, Thiry L. Congenital varicella-zoster infection related to maternal disease in early pregnancy. *Scand J Infect Dis* 1973, **5**, 149–53.

71   Essex-Cater A, Heggarty H. Fatal congenital varicella syndrome. *J Infect* 1983, **7**, 77–8.

72   Brice JEH. Congenital varicella resulting from infection during second trimester of pregnancy. *Arch Dis Child* 1976, **51**, 474–6.

73   Hajdi G, Meszner Z, Nyerges G, Buky B, Simon M. Congenital varicella syndrome. *Infection* 1986, **14**, 177–80.

74   Cuthbertson G, Weiner CP, Giller RH, Grose C. Prenatal diagnosis of second-trimester congenital varicella syndrome by virus-specific immunoglobulin M. *J Pediatr* 1987, **111**, 592–5.

75   Frey HM, Bialkin G, Gershon AA. Congenital varicella: a case report of a serologically proved long-term survivor. *Pediatrics* 1977, **59**, 110–2.

76   Kotchmar GS, Grose C, Brunell PA. Complete spectrum of the varicella congenital defects syndrome in 5-year old child. *Pediatr Infect Dis* 1984, **3**, 142–5.

77 Siegel M. Congenital malformations following chickenpox, measles, mumps, and hepatitis. Results of a cohort study. *J Am Med Ass* 1973, **226**, 1521–4.

78 Hill AB, Doll R, Galloway TMcL, Hughes JPW. Virus diseases in pregnancy and congenital defects. *Brit J Prev Soc Med* 1958, **12**, 1–7.

79 Webster MH, Smith CS. Congenital abnormalities and maternal herpes zoster. *Br Med J* 1977, **ii**, 1193.

80 Brazin SA, Simkovich JW, Johnson WT. Herpes zoster during pregnancy. *Obstet Gynecol* 1979, **53**, 175–81.

81 Garcia AGP. Fetal infection in chickenpox and alastrim, with histopathologic study of the placenta. *Pediatrics* 1963, **32**, 895–901.

82 Hanshaw JB, Dudgeon JA, Marshall WC. Viral diseases of the fetus and newborn. 2nd ed. Philadelphia: Saunders, 1985.

83 Odessky L, Newman B, Wein GB. Congenital varicella. *New York J Med* 1954, **54**, 2849–52.

84 Freud P. Congenital varicella. *Am J Dis Child* 1958, **96**, 730–3.

85 Pearson HE. Parturition varicella-zoster. *Obstet Gynecol* 1964, **23**, 21–7.

86 Raine DN. Varicella infection contracted in utero: sex incidence and incubation period. *Am J Obstet Gynecol* 1966, **94**, 1144–5.

87 Hyatt HW. Neonatal varicella. Report of a case in a 5-day-old infant and review of the literature. *J Nat Med Ass* 1967, **59**, 32–4.

88 Oppenheimer EH. Congenital chickenpox with disseminated visceral lesions. *Bull Johns Hopkins Hosp* 1944, **74**, 240–7.

89 Lucchesi PF, LaBoccetta AC, Peale AR. Varicella neonatorum. *Am J Dis Child* 1947, **73**, 44–54.

90 Ehrlich RM, Turner JAP, Clarke M. Neonatal varicella: a case report with isolation of the virus. *J Pediatr* 1958, **53**, 139–47.

91 Steen J, Pedersen RV. Varicella in a newborn girl. *J Oslo City Hosp* 1959, **9**, 36–45.

92 Meyers JD. Congenital varicella in term infants: risk reconsidered. *J Infect Dis* 1974, **129**, 215–7.

93 Brunell PA. Placental transfer of varicella-zoster antibody. *Pediatrics* 1966, **38**, 1034–8.

94 Gershon AA. Varicella in mother and infant: problems old and new. In: Krugman S, Gershon AA, eds. Infections of the fetus and newborn. New York: Alan R Liss, 1975, 79–95. (Progress in clinical and biological research, vol 3.)

95 Preblud SR, Zaia JA, Nieberg PI, Hinman AR, Levin MJ. Management of pregnant patient exposed to varicella. *J Pediatr* 1979, **95**, 334–5.

96 Rubin L, Leggiadro R, Elie MT, Lipsitz P. Disseminated varicella in a neonate: Implications for immunoprophylaxis of neonates postnatally exposed to varicella. *Pediatr Infect Dis* 1986, **5**, 100–2.

97   Preblud SR, Bregman DJ, Vernon LL. Deaths from varicella in infants. *Pediatr Infect Dis* 1985, **4**, 503–7.
98   Caunt AE, Shaw DG. Neutralization tests with varicella-zoster virus. *J Hyg* 1969, **67**, 343–52.
99   Furukawa T, Plotkin SA. Indirect hemagglutination tests for varicella-zoster infection. *Infect Immun* 1972, **5**, 835–9.
100 Gershon AA, Kalter ZG, Steinberg S. Detection of antibody to varicella-zoster virus by immune adherence hemagglutination. *Proc Soc Exp Biol Med* 1976, **151**, 762–5.
101 Williams V, Gershon A, Brunell PA. Serologic response to varicella-zoster membrane antigens measured by indirect immunofluorescence. *J Infect Dis* 1974, **130**, 669–72.
102 Cradock-Watson JE, Ridehalgh MKS, Bourne MS. Specific immunoglobulin responses after varicella and herpes zoster. *J Hyg* 1979, **82**, 319–36.
103 Arvin AM, Koropchak CM. Immunoglobulins M and G to varicella-zoster virus measured by solid-phase radioimmunoassay: antibody responses to varicella and herpes zoster infections. *J Clin Microbiol* 1980, **12**, 367–74.
104 Campbell-Benzie A, Kangro HO, Heath RB. The development and evaluation of a solid-phase radioimmunoassay (RIA) procedure for the determination of susceptibility to varicella. *J Virol Methods* 1981, **2**, 149–58.
105 Forghani B, Schmidt NJ, Dennis J. Antibody assays for varicella-zoster virus: comparison of enzyme immunoassay with neutralization, immune adherence hemagglutination, and complement fixation. *J Clin Microbiol* 1978, **8**, 545–52.
106 Shanley J, Myers M, Edmond B, Steele R. Enzyme-linked immunosorbent assay for detection of antibody to varicella-zoster virus. *J Clin Microbiol* 1982, **15**, 208–11.
107 Shehab Z, Brunell PA. Enzyme-linked immunosorbent assay for susceptibility to varicella. *J Infect Dis* 1983, **148**, 472–6.
108 Brown DWG. Viral diagnosis by antibody capture assay. In: Mortimer PP, ed. Public Health Virology: 12 reports. London: Public Health Laboratory Service, 1986, 92–108.
109 Kangro HO, Ward A, Argent S, Heath RB, Cradock-Watson JE, Ridehalgh MKS. Detection of specific IgM in varicella and herpes zoster by antibody-capture radioimmunoassay. *Epedimiol Infect* 1988, **101**, 187–95.
110 Savage MO, Moosa A, Gordon RR. Maternal varicella infection as a cause of fetal malformations. *Lancet* 1973, **i**, 352–4.
111 Readett MD, McGibbon C. Neonatal varicella. *Lancet* 1961, **i**, 644–5.
112 Baba K, Yabuuchi H, Takahashi M, Ogra P. Immunologic and epidemiologic aspects of varicella infection acquired during infan-

cy and early childhood. *J Pediatr* 1982, **100**, 881–5.

113 Gustafson TL, Shehab Z, Brunell PA. Outbreak of varicella in a newborn intensive care nursery. *Am J Dis Child* 1984, **138**, 548–50.

114 Hanngren K, Grandien M, Granstrom G. Effect of zoster immunoglobulin for varicella prophylaxis in the newborn. *Scand J Infect Dis* 1985, **17**, 343–7.

115 Holland P, Isaacs D, Moxon ER. Fatal neonatal varicella infection. *Lancet* 1986, **ii**, 1156.

Human viruses and water

J Sellwood and JV Dadswell

With increasing public awareness of environmental issues, there is a realisation that potable water supplies are potentially vulnerable to pollution. The upsurge in recreational activities associated with surface waters, both inland and marine, has also caused the quality of such waters to be questioned. Among the many possible pollutants, the presence of human viruses must be considered. A recent review of water-associated illness has detailed instances of probable viral gastro-enteritis resulting from contamination of potable water and from recreational activities associated with surface waters [1]. A survey of bathers using certain UK marine waters has also suggested a possible connection between bathing and probable viral gastroenteritis [2]. Moreover, shellfish-associated gastroenteritis is now well recognised [3].

The human viruses present in natural waters are almost entirely derived from sewage, thus reflecting those present in human excreta that are sufficiently hardy to withstand adverse environmental conditions. These include enteroviruses, hepatitis A and non-A, non-B viruses, adenoviruses, reoviruses, rotaviruses and 'gastroenteritis viruses' such as Norwalk-like, calici- and astroviruses.

Viruses belonging to these groups may therefore be found in raw sewage and subsequently appear, after works treatment, in both sludge and final effluent. Most inland sewage treatment works discharge their effluent directly into a watercourse where it becomes greatly diluted. Nevertheless, surface water subsequently abstracted downstream is likely to contain some virus which must be prevented, by appropriate treatment, from entering a potable supply. To a lesser extent, the virus content of sewage sludge also contributes a virus load to the environment when it is disposed of, usually by spreading on agricultural land or by dumping at sea.

Enteric viruses, disseminated as outlined above, may also penetrate into the ground; natural filtration processes ensure that water subsequently abstracted from the water-bearing rock (an aquifer) by, for example, borehole will be largely virus-free.

Coastal towns in the UK discharge their sewage directly into the sea with minimal pre-treatment. This constitutes a major environmental load of enteric virus, to which is added the virus remaining in rivers flowing into the sea.

It is fortunate that, with environmental exposure, natural processes combine to ensure viral decay; these are little understood but include temperature, sunlight, and the activity of bacteria, protozoa and other organisms.

The various aspects of this virus–water association are now considered in more detail.

The water environment

SEWAGE

The virus content of sewage will reflect the viruses circulating in the community at any one time. The organic content is present in a large volume of liquid, industrial and domestic. The types and numbers of viruses identified from sewage will depend on the method of detection. In London, for example, 9×10^3 plaque-forming units (pfu) per litre have been found using an assay which detects those viruses that grow well in monkey kidney cells under agar and form plaques, ie poliovirus and coxsackievirus B [4]. The full range of viruses found in faeces is described on p. 38. Those detected in the plaque assay are an underestimate of the range of types actually present.

Enteroviruses in sewage have been detected in greatest number and types in the summer and autumn, mirroring the seasonality of this group of viruses in the community [5]. These viruses may survive for days or weeks if protected within faecal material at low temperature [6]. Stability in the environment varies among enterovirus serotypes, poliovirus being one of the more easily denatured. Other virus groups may be more resistant, especially reovirus and rotavirus.

The treatment of raw sewage varies in the UK from area to area. Coastal authorities do not pre-treat sewage before maceration and discharge to the sea through short or long sea outfalls. Inland authorities must use sewage treatment works to separate the solid and liquid phases and treat each appropriately before disposal. The raw sewage entering a treatment works passes through a coarse mesh, which traps larger debris, then is retained in settling tanks for several hours. During this time heavy material is deposited on the tank floor and this sludge is then scraped away to be treated as described later. The upper liquid portion is pumped out from the surface to be treated as effluent. Several sedimentation stages take place during the complete treatment process, where more heavy material is removed to be treated as sludge. None of the procedures involved remove viruses specifically, but many viruses at this stage will be adsorbed on solids and so will be treated or disposed of with the sludge.

EFFLUENT

At sewage treatment works this liquid phase of sewage is processed by utilising the action of aerobic microorganisms on the organic content. In the activated sludge process, air or oxygen is bubbled vigorously through the liquid. An alternative method is the percolating filtration process which involves the liquid trickling through deep beds of clinker coated with a well-established microecosystem. The clinker is surrounded by a biofilm of bacteria, algae, protozoa; near the surface of the bed where oxygen is most plentiful, nematodes and small crustacea occur. All these organisms use the organic content of the effluent as a food source. Viruses are most likely to be contained in clumps of debris so may be ingested, remain in particles suspended in the effluent, or be deposited when heavy material is again allowed to settle out.

The virus content of raw sewage can be reduced by between 90% [4] and 26% at the final effluent stage, though the actual numbers remain high: 5×10^{11}pfu/litre in a Midlands effluent [7]. As mentioned previously, these detection rates include the enteroviruses that plaque; the other virus types are not so easily detected.

Disposal is usually to the local river system. In the USA there is some chemical disinfection of the effluent before discharge but actual reduction of the virus content is not proven [8,9]. Concern has been voiced about the possible dangers of chlorine residues present in this type of treated effluent. In many countries, including the UK, the effluent/river water may be used for drinking water abstraction or for irrigation. Effluent used for irrigation may percolate for some distance through soil, with the potential to contaminate water sources [10].

SLUDGE

This heavy organic material is allowed to settle out at various stages during sewage treatment. A common method of treatment is for this solid phase to undergo mesophilic digestion (without the addition of external heat), during which the sludge is kept in large anaerobic tanks for 15–30 days at 30–35°C [11]. The ammonia produced will denature a proportion of viruses [12], though the common enterovirus serotypes have been recovered after full digestion. Detection of viruses from sludge is a difficult and probably inefficient process, due to the toxicity of the material for tissue culture [13]. Reliable figures of virus quantity before and after digestion are therefore not available.

The disposal of sludge takes many forms. Liquid sludge (5% solids content) can be applied to grassland or ploughed into soil for its nutrient content. Virus persistence on grass and in soil depends on absorption to solids, low temperature, soil moisture and virus type. Virus has been recovered 21 weeks after sludge application to land in Denmark [14]. Liquid sludge can also be de-watered to a dry mass for use as landfill in rubbish tips. This desiccation denatures the viruses. The alternative method of sludge disposal is to discharge it to the sea. The fate of the viruses involved is discussed on p. 33.

SURFACE WATER

Most rivers in the UK receive effluent from sewage treatment works. All virus groups present in sewage will therefore be present in river water. The virus content of different rivers has been reported as between 647pfu and 2pfu per litre [7,15]. Of the major water sources in Wales (including rivers and reservoirs), 61% were found to contain enteroviruses [15]. Some of these viruses will remain in the water as it travels downstream but a proportion will become solids-associated and so will persist in river sediment.

River water and effluent are used for irrigation on crops that will be processed and cooked. They must not be used for vegetables or fruit that will be eaten raw. Viruses have been shown to persist for days on vegetation [16]. Spray irrigation is practised in Israel where viruses have been detected in the associated aerosols [17].

GROUND WATER

Deep underground aquifer water is usually of high microbiological quality. However, over a period of many months, poliovirus was detected in the absence of bacteria from a chalk borehole [18]. Shallow aquifers surrounded by permeable or fissured rock are likely to be more highly contaminated.

DRINKING WATER

Water destined for drinking originates from sources dependent on local availability. A ground water source such as an aquifer is ideal, as the water should contain few infectious agents. Other sources include reservoirs or rivers which contain a proportion of sewage effluent. Treatment of these waters includes storage, to allow heavy particles to be deposited, flocculation and deposition by use of a coagulant such as aluminium sulphate, and rapid or slow sand filtration. These processes will remove the majority of organic materials and produce clean water. Finally, sufficient chlorine is added to oxidise the ammonia in the water to chloramines and to produce a residue of free chlorine [11]. Recommended chlorination procedures should produce virus-free water.

Detection of virus from water in the distribution system, ie post-chlorination, has been sporadic and usually associated with operational failures, especially at small treatment works [19,20]. The reports highlight the potential for contamination if operational procedures fail.

The arid area of Arizona, which relies almost totally on effluent as source water, has now included virus monitoring as a standard for drinking water quality [21].

SEAWATER

Seawater receives large quantities of untreated sewage all around the UK coastline. In the past, discharge outfalls ended just beyond low water mark, but now these are being replaced by long sea outfalls. The microbial content of the water surrounding the outfall reflects the sewage content. Its dispersal will depend on geography, currents and wind. Beach areas will become contaminated according to local conditions. Heavy particles will settle to the sediment and are likely to contain viruses [19,22–26]. Water disturbance will resuspend the particles and may periodically release the enclosed viruses.

Virus denaturation rates will depend on microbial activity, temperature, chemical composition of the water and release from protective material. Seawaters and sediments monitored from Welsh coastal areas produced 39% of samples positive for enterovirus (at 40pfu per litre) and 19% positive for rotavirus. The installation of a long sea outfall improved the quality and reduced the microbial content of the water, though virus numbers did not decline as much as bacterial counts. The limits set in the EC *Directive on Bathing Waters 1975* could be met for bacteria but not viruses. Virus numbers in water decreased more than those in sediment [15].

SHELLFISH

Bivalve shellfish are filter feeders of particulate organic matter. Viruses in seawater may therefore be ingested by mussels, cockles and oysters and may accumulate in the flesh. The quality of the seawater determines the contamination of the mollusc. Depuration of the shellfish prior to sale will eliminate the majority of bacterial pathogens but viruses may remain. Well-documented outbreaks of gastroenteritis caused by the so-called small round structured viruses (SRSV) and infectious hepatitis caused by hepatitis A virus (HAV) have occurred after shellfish consumption [27,28].

Indicators of the virus content of water

Due to the relatively specialised techniques involved in virus isolation and the time needed for a test result, alternative test systems have been actively sought. Total coliforms and faecal coliforms, faecal streptococci, enterococci, coliphages and bifidobacteria have all been compared with enteroviruses for presence in sewage and waters. Survival characteristics in the environment and resistance to chlorine have also been studied [29,30].

It is now recognised that no single bacterial indicator can predict the presence of enteroviruses, and that enteroviruses may be present in any type of water when coliforms are absent. A combination of two or three indicators such as faecal coliforms, faecal streptococci and coliphage may be needed to predict the presence of enteroviruses [31].

Poliovirus may serve as an indicator for other viruses in communities where Sabin vaccine use is continuous [32]. However, detection of poliovirus has not been consistent even from the sewage of these populations [33]. Enteroviruses in the plaque assay are in fact indicators for the viruses difficult or impossible to detect by current methods (eg SRSV, HAV and rotavirus).

PROCESSING SAMPLES FOR VIRUS CONTENT

The techniques used to detect virus from the environmental waters depend on the proportion of organic material contained in the water.

Clean, potable water: Tests to detect very low numbers of viruses in large volumes of water (100 litres or more) involve the use of specialised cartridge filters with on-site flow-through apparatus. This type of sampling is rarely undertaken in the UK. More commonly, volumes of 20 litres are sampled using the method described below [34].

Environmental water: river, reservoir, effluent, seawater, groundwater. These waters contain some organic matter, usually less than 2% of the total volume. Samples of two to 20 litres, usually 10 litres, are processed. The most widely used concentration technique is that of membrane filtration. The water is acidified to pH3.5. At this pH the viruses behave as positively-charged colloidal particles. As the water is passed, under pressure, through the negatively-charged cellulose nitrate membrane, the viruses will be adsorbed to the surface. The addition of a cation may increase the efficiency of the adsorption process. To elute the virus from the membrane, a volume of proteinaceous material such as beef extract or skimmed milk is passed through at high pH. The next stage is a concentration step involving the flocculation of the protein (including viruses). The eluant is acidified to the point where flocculation occurs. The protein is then deposited during centrifugation after which it is resuspended in 10ml of buffer. Thus a thousand-fold concentration has been achieved. The efficiency of recovery is not high however and may be 30%–70% of the initial virus content (Figure 1).

Other possible concentration methods for viruses in water include dialysis using PEG, 2-phase separation using PEG and sodium dextran sulphate, ultrafiltration under pressure or adsorption to a solid material such as PE60 or iron oxide.

Polluted water: raw sewage, sludge, sediment, soil. The aim for waters containing more than 4% solid material is to adsorb virus onto the organic material, elute and then concentrate. 50ml of sample is acidified

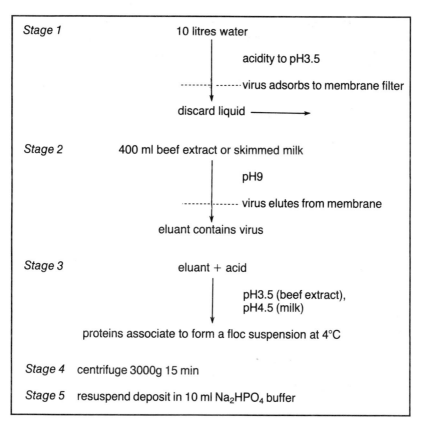

Figure 1 Clean water processing

in the presence of a cation and stirred to facilitate virus adsorption on to the suspended solids. These solids are deposited by centrifugation and resuspended in a proteinaceous liquid at high pH (beef extract or skimmed milk). Virus should be released into the liquid during mixing. The remaining deposit is discarded after centrifugation. Protein in the supernatant flocculates on acidification and is then deposited during further centrifugation. This is resuspended in 10ml of buffer. The efficiency of the method is not proven, though it is likely to be low (Figure 2).

DETOXIFICATION AND DECONTAMINATION
Contaminants such as heavy metals, fungi and bacteria can be removed by mixing equal portions of dithizone in chloroform with the sample concentrate. Storage of the buffer after chloroform removal should be at $-20°C$ until assayed for virus content.

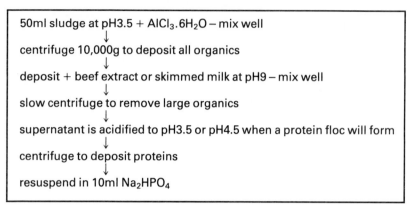

50ml sludge at pH3.5 + AlCl₃.6H₂O – mix well
↓
centrifuge 10,000g to deposit all organics
↓
deposit + beef extract or skimmed milk at pH9 – mix well
↓
slow centrifuge to remove large organics
↓
supernatant is acidified to pH3.5 or pH4.5 when a protein floc will form
↓
centrifuge to deposit proteins
↓
resuspend in 10ml Na₂HPO₄

Figure 2   Solids processing

Detection of virus

ROTAVIRUS

Rotavirus is detected by culture of the sample on LLCMK$_2$ or MA104 cells overnight. The fixed cells are then incubated with anti-rotavirus antibody and subsequently labelled with fluorescein or peroxidase markers. As there is little extra-cellular spread of the virus, the number of labelled foci gives a quantitative measurement of the number of virus infectious units in the original sample. Direct electron microscopy is not practical on these samples and immunochemical (ELISA) assays do not have sufficient sensitivity to detect the low numbers of virus in water samples.

ENTEROVIRUS

Poliovirus and coxsackievirus B can be detected when a sample is mixed with BGM cells and agar containing neutral red [35]. After 3–5 days' incubation, pale circular plaques occur in the agar where a virus infectious unit has infected and spread locally in cells. The number of 'plaque-forming units' (pfu) represents the virus content of the sample [36].

Liquid cell culture: The sample may be incubated with a range of cell types, eg BGM, Vero, BK, FL, HEL under liquid medium [37]. Any virus present will produce a cytopathic effect (CPE). Poliovirus, coxsackieviruses B, A7, A9, echovirus, adenovirus and reovirus will be detected by this method. Quantitation may be achieved by the MPN method.

THE FUTURE

Future developments may allow the use of gene probes, PCR or more refined immunochemical techniques. Methods for the detection of HAV and SRSVs are especially sought after.

The results produced by the above methods are dependent on sound technical practices throughout. Initial sampling must be from identified sites, with the relationship to important influences recognised. These could include time of sampling, proximity of sewage discharges, tide and current status, aquifer geology, river characteristics and treatment type for potable water. The equipment used for sampling must be sterile and the sample representative, as far as possible, of the whole body of water. Cross-contamination within the laboratory must be prevented when processing many samples through multiple stages.

Quality assurance must include internal laboratory controls for the efficiency of the processes used and of laboratories providing similar services. However tissue culture is not easily standardised and while detection methods depend largely on this highly variable basis it will be difficult to achieve comparable results.

Monitoring water for viruses

There is no general agreement on the extent to which potable water supplies should be monitored for the presence of viruses. Currently available techniques limit the possibilities to the detection of entero-viruses and perhaps rotaviruses. The necessity to maintain tissue culture systems and the time required to observe a cytopathic effect, or to detect antigen by immunofluorescence (or similar technique), will mean that useful information cannot be acquired rapidly. Moreover, because the amount of virus present is likely to be small, relatively large volumes of water have to be sampled and concentrated before detection methods can be employed. The questions of cost and the availability of technical expertise mean that relatively few laboratories can offer such a service.

In view of this, it is evident that routine monitoring of potable water for the presence of viruses is not practical in the UK. However, there are circumstances when detection of viruses in potable water may be of value. These would be, for example, when failures in the disinfection or filtration process have occurred, when bacteriological examination has revealed deficiencies in the quality of the supply, or as part of the investigation of an outbreak of suspected viral infection associated with a water supply. There may also be value in determining fluctuation in the viral load of surface waters entering a treatment works to enable appropriate adjustments to treatment to be made.

The development of new, and less time-consuming, techniques for detecting viruses could make a significant difference in attitudes to the routine virological monitoring of potable waters. However, these tech-niques may depend upon the detection of viral antigens or nucleic acid and may therefore not give an indication of infectivity.

The implementation of the EC bathing water directive has resulted in an increasing level of sampling of designated UK bathing waters for the presence of (usually) enteroviruses. Because of their greater resistance to environmental decay, a better indication of the potential infectivity of such waters may be obtained from virological than from bacteriological examination [38]. Nevertheless, there are many variables relating to any sampling site such as weather, tides etc, so that the presence or absence of enteroviruses in any one 10 litre sample (as required by the directive) is of questionable significance. Regular sampling from the same site over a period of time, and relating the findings to those obtained by bacteriological examination (and to environmental conditions), is more likely to yield information of value.

LEGISLATION AND DIRECTIVES

For many years, potable water supplied in the UK has been required to be 'wholesome' but this term has never been defined and no specific microbiological requirements have been included in the various water and public health acts. However, since 1985, potable water within the European community has been subject to the provisions of the appropriate directive [39] in which certain microbiological parameters are defined. These refer primarily to bacterial indicators of water quality, but enteroviruses are included in the list of 'pathogens' that may need to be looked for when it is 'necessary to supplement the microbiological analysis . . .'. In contrast, the earlier bathing water directive [40] includes a mandatory requirement that enteroviruses should not be detectable (as pfus) in 10 litres of water 'when an inspection in the bathing area shows that the substance (sic) may be present or that the quality of the water has deteriorated'.

No virological requirements are included in the directives relating to surface water intended for abstraction as drinking water [41,42] or to shellfish waters [43].

The World Health Organization recommends that drinking water should be free of any virus infectious for man and considers it 'desirable to examine raw water sources and finished drinking water for virus presence where suitable facilities can be provided' [44]. In the USA, Arizona has taken the lead by proposing that viruses should be absent in 1,000 litres of drinking water and in 40 litres of water used for other purposes [21].

Health effects

As enteric viruses are primarily gut-associated, the illnesses they can cause are predominantly gastrointestinal in nature, with diarrhoea and vomiting as the prominent features of infections with the SRSVs and rotaviruses. The hepatitis viruses, culturable adenoviruses and enteroviruses are capable of systemic invasion, with the potential to cause a wider spectrum of illness involving the liver, respiratory tract and central nervous system respectively. By contrast, other viruses such as reoviruses and parvo-like viruses, although detectable in faeces, do not appear to cause clinical illness; asymptomatic enterovirus infections are well recognised. The corona-like viruses appear to occupy an intermediate position – their presence in faeces has in some instances been associated with diarrhoeal illness but not in others [45].

All of the aforementioned viruses can be spread by the faecal–oral route, mostly by direct spread from person to person or from the immediately-contaminated environment. Sometimes food may be the vehicle of infection as a result of direct contamination by an infected food handler [46]. As indicated, there is also the potential for spread via the water route, but the full extent to which this may occur is not clear.

HEALTH RISKS FROM SEWAGE

Sewage, effluent and sludge all contain enteric viruses to a greater or lesser extent. Direct ingestion of these materials will therefore create a risk of infection. There will also be a risk from the inhalation of infected aerosols produced during processing or disposal, particularly from spray irrigation. There is, however, little epidemiological evidence of infections being acquired in this way. There is one UK report [47] of hepatitis A occurring in five workers involved in spreading sewage sludge. However, no increase in illness rates has been observed in farming families where spray sludge has been used, or in kibbutz members where effluent irrigation is widely practised [48,49].

There is a further potential for the spread of viral infection as a result of eating raw vegetable crops irrigated with sewage effluent. If the appropriate guidelines for treating such crops are followed, the risk appears to be small [50].

Hepatitis B and human immunodeficiency virus are viruses which cause public concern. It is possible that they might be shed in excreta in small amounts, but their further survival in sewage for any length of time is unlikely. There is no evidence to suggest that infection caused by these viruses has been acquired by this route.

HEALTH RISKS FROM SURFACE WATERS

The risk of acquiring a viral infection from recreational activities involving natural waters is largely unknown; it depends on the dose swallowed or possibly received via the conjunctiva, and the susceptibility to any viruses present. Children will be the most susceptible to the widest range of viruses. Some epidemiological studies have suggested an increased risk of acquiring a presumed viral infection (gastroenteritis) as a result of seawater bathing in certain locations [2,51]. Similar findings were obtained from a survey of snorkel swimmers in dock water [52]. However, such studies have the disadvantage of being difficult to control and are based largely on the self-reporting of symptoms. The possible effect of other factors such as dietary indiscretions or subsequent infections occurring within the community cannot readily be excluded.

At present it seems reasonable to suppose that there is a risk of acquiring a viral (or other) infection as a result of immersion in a natural water where there is evidence of faecal contamination, but that this is small when compared with other means of transmission. A more definite assessment will have to await the results of carefully planned, prospective, epidemiological studies.

HEALTH RISKS FROM POTABLE WATER

In the UK, some recorded outbreaks of gastroenteritis, presumed to be of viral origin, have resulted from the contamination of potable water supplies, both public and private. The largest of these was caused by sewage contamination, via the aquifer, of a borehole source at a time when the chlorination was defective [53]. The other outbreaks were also associated either with unchlorinated or defectively chlorinated water. Although in these instances there was no direct evidence of viral infection, Norwalk-like viruses were detected in the stools of some of those affected in a more recent large outbreak [54].

In other countries, outbreaks associated with potable water supplies have been recorded where there was either serological evidence of infection with Norwalk virus, or rotaviruses were detected in the stools of those affected. All these outbreaks were associated with defects in the supply system [55]. Outbreaks of viral hepatitis associated with potable water supplies have also been recorded [56].

It has been shown that enteroviruses (in small numbers) can be detected in water in the absence of the traditional bacterial indicators of faecal pollution [38] and it has been postulated that low-level viral contamination of drinking water could be occurring and so resulting in sporadic viral infections in the consumers, possibly causing further dissemination by person-to-person contact. Such infections would be difficult to detect against the background of viral infections generally prevalent within the community. Although this possibility is not afforded great significance in Europe [57], some virologists in the USA consider it to be worthy of considerable attention [58].

The risk of acquiring a viral infection from potable water that has been adequately treated and meets conventional bacteriological standards appears to be small. However, it must be appreciated that the infectious dose of many enteric viruses is low and that they are generally more resistant to disinfection processes than enteric bacteria. The margin for error is thus considerably diminished, and this is reflected in the number of reported outbreaks of probable and confirmed viral gastroenteritis associated with drinking water supplies in recent years.

Conclusion

Human enteric viruses can be detected at all stages of the water cycle and there is epidemiological evidence to indicate their potential to cause illness. In developed countries, with treated potable water supplies, the risk of infection appears to depend upon the adequacy of treatment and the degree of integrity of the distribution system. Continuing epidemiological surveillance will be required to enable waterborne infections to be recognised and other possible risk factors identified. The possibility of low-level transmission of virus via potable water remains to be disproved, although there is little evidence that this constitutes a significant risk of spreading viral infections within a community.

Current methods for detecting viruses in water and water-related materials are cumbersome, restricted in scope, and cannot be used routinely as a means for monitoring the adequacy of treatment of potable water supplies. Nor, as yet, do surrogate indicators of virus presence provide a practical alternative.

The detection of enteric viruses in natural bathing waters may provide a better indication of faecal pollution and therefore of potential infectivity than do traditional bacterial indicators. Their usefulness for this purpose is nevertheless restricted, as for potable water, by technical considerations. The precise significance of the presence of viruses in such waters in terms of health risks remains to be determined.

More information is required on the extent to which enteric viruses are found in surface and groundwaters that are to be abstracted for drinking water purposes, and the extent to which current and future treatment processes will remove them. Similar considerations apply to sewage effluent and sludge used for agricultural purposes.

The application of new virological techniques to 'water virology' should enable viruses to be detected more rapidly than at present and may challenge the use of conventional bacteriological monitoring in evaluating the efficacy of treatment processes. They may also allow the detection of a wider range of viruses, including those perhaps more relevant to the viral infections likely to be associated with both potable and recreational waters.

References

1   Galbraith NS, Barrett NJ, Stanwell-Smith R. Water and disease after Croydon: a review of water-borne and water-associated disease in the UK 1937–86. *J Inst Water Environ Manage* 1987, **1**, 7–21.

2   Brown JM, Campbell EA, Richards AD, Wheeler D. Sewage pollution of bathing water. *Lancet* 1987, **ii**, 1208–9.

3   Sockett PN, West PA, Jacob M. Shellfish and public health. *PHLS Microbiology Digest* 1985, **2**, 29–35.

4   Slade JS, Ford BJ. Discharge to the environment of viruses in wastewater, sludges, and aerosols. In: Berg G (ed). Viral Pollution of the Environment. Boca Raton: CRC Press, 1983, 3–15.

5   Sellwood J, Dadswell JV, Slade JS. Viruses in sewage as an indicator of their presence in the community. *J Hyg Camb* 1981, **86**, 217–25.

6   Cooper RC, Potter JL, Leong C. Virus survival in solid waste leachates. *Water Res* 1975, **9**, 733–9.

7   Morris R. Viral surveillance in the UK drinking, ground, surface and waste waters. In: Viral surveillance – needs and feasibility. Water Research Centre Seminar 1986, 50–60.

8   Butler M. Virus removal by disinfection of effluents. In: Goddard M, Butler M (eds). Viruses and Wastewater Treatment. London: Pergamon Press, 1981, 145–64.

9   Bitton G. Virus inactivation by disinfectants in water and wastewater. In: Introduction to Environmental Virology. New York: Wiley, 1980, 175–99.

10 Wellings FM, Lewis AL, Mountain CW, Pierce LV. Demonstration of virus in groundwater after effluent discharge onto soil. *Appl Environ Microbiol* 1975, **29**, 751–7.

11 Davies FG, Bassett WH. Water. In: Clay's Handbook of Environmental Health, 14th edn. London: HK Lewis, 1977, 593–639.

12 Ward RL, Ashley CS. Identification of the virucidal agent in wastewater sludge. *Appl Environ Microbiol* 1977, **33**, 860–4.

13 Williams FP, Hurst CJ. Detection of environmental viruses in sludge: enhancement of enterovirus plaque assay titers with 5-iodo-2'-deoxyuridine and comparison to adenovirus and coliphage titers. *Water Res* 1988, **22**, 847–51.

14 Jorgensen PH, Lund E. Detection and stability of enteric viruses in sludge, soil and groundwater. *Water Sci Technol* 1985, **17**, 185–95.

15 Tyler J. Occurrence in water of viruses of public health significance. *J Appl Bact Symp Suppl* 1985, no 14, 37S–46S.

16 Ward BK, Irving LG. Virus survival on vegetables spray-irrigated with wastewater. *Water Res* 1987, **21**, 57–63.

17 Shuval HI, Guttman-Bass N, Applebaum J, Fattal B. Aerosolized enteric bacteria and viruses generated by spray irrigation of wastewater. *Water Sci Technol* 1989, **21**, 131–35.

18 Slade JS. Viruses and bacteria in a chalk well. *Water Sci Technol* 1985, **17**, 111–25.

19 Tyler JM. Viruses in fresh and saline waters. In: Butler M, Medlem AR, Morris R (eds). Viruses and Disinfection of Water and Wastewater. Guildford: University of Surrey, 1982, 42–63.

20 Payment P, Trudel M, Plante R. Elimination of viruses and indicator bacteria at each step of treatment during preparation of drink-

ing water at seven water treatment plants. *Appl Environ Microbiol* 1985, **49**, 1418–28.

21 Rose JB, De Leon R, Gerba CP. Giardia and virus monitoring of sewage effluent in the state of Arizona. *Water Sci Technol* 1989, **21**, 43–7.

22 Rao VC, Metcalf TG, Melnick JL. Development of a method for concentration of rotavirus and its application to recovery of rotaviruses from estuarine waters. *Appl Environ Microbiol* 1986, **52**, 484–8.

23 Rao VC, Seidel KM, Goyal SM, Metcalf TG, Melnick JL. Isolation of enteroviruses from water, suspended solids, and sediments from Galveston Bay: survival of poliovirus and rotavirus adsorbed to sediments. *Appl Environ Microbiol* 1984, **48**, 404–9.

24 Watson PG, Inglis JM, Anderson KJ, Read PA. The effect of the introduction of a sewage scheme on the virological quality of an intertidal zone. *J Infect* 1982, **5**, 39–45.

25 Goyal SM, Adams WN, O'Malley ML, Lear DW. Human pathogenic viruses at sewage sludge disposal sites in the middle Atlantic region. *Appl Environ Microbiol* 1984, **48**, 758–63.

26 Lucena F, Bosch A, Jofre J, Schwartzbrod L. Identification of viruses isolated from sewage, riverwater and coastal seawater in Barcelona. *Water Res*, 1985, **19**, 1237–9.

27 O'Hara H, Naruto H, Watanabe W, Ebisawa I. An outbreak of hepatitis A caused by consumption of raw oysters. *J Hyg Camb* 1983, **91**, 163–5.

28 O'Mahony MC, Gooch CD, Smyth DA, Thrussell AJ, Bartlett CLR, Noah ND. Epidemic hepatitis A from cockles. *Lancet* 1983, **i**, 518–20.

29 Berg G, Dahling DR, Brown GA, Berman D. Validity of fecal coliforms, total coliforms, and fecal streptococci as indicators of viruses in chlorinated primary sewage effluents. *Appl Environ Microbiol* 1978, **36**, 880–4.

30 Fattal B, Katzenelson E, Guttman-Bass N, Sadovski A. Relative survival rates of enteric viruses and bacterial indicators in water, soil and air. In: Melnick JL, ed. Enteric Viruses in Water. Basel: Karger, 1984, 184–92. (Monographs in Virology, vol 15.)

31 Grabow WOK, Idema GK, Coubrough P, Bateman BW. Selection of indicator systems for human viruses in polluted seawater and shellfish. *Water Sci Technol* 1989, **21**, 111–7.

32 Payment P, Laose Y, Trudel M. Polioviruses as indicator of virological quality of water. *Can J Microbiol* 1979, **25**, 1212–4.

33 Katzenelson E, Kedmi S. Unsuitability of polioviruses as indicators of virological quality of water. *Appl Environ Microbiol* 1979, **37**, 343–4.

34   Berg G, Safferman RS, Dahling DR, Berman D, Hurst CJ. United States: Environmental Protection Agency. Manual of Methods for Virology. *EPA*, 1984, **600**, 13.

35   Dahling DR, Berg G, Berman D. BGM, a continuous cell line more sensitive than primary rhesus and African green kidney cells for the recovery of viruses from water. *Health Lab Sci* 1974, **11**, 275–82.

36   Schmidt NJ. Cell culture techniques for diagnostic virology. In: Lennette EH, Schmidt NJ, eds. Diagnostic Procedures for Viral, Rickettsial and Chlamydial Infections. Washington: American Public Health Association, 1979, 5th edn, 65–139.

37   Chonmaitree T, Ford C, Sanders C, Lucia HL. Comparison of cell cultures for rapid isolation of enteroviruses. *J Clin Microbiol* 1988, **26**, 2576–80.

38   Berg G, Metcalf TG. Indicators of viruses in waters. In: Berg G (ed.) Indicators of Viruses in Water and Food. Ann Arbor: Ann Arbor Sci Publ Inc, 1978, 267–96.

39   Council Directive of 15 July 1980 relating to the quality of water intended for human consumption (80/778/EEC). (Official Journal of the European Communities no L229, 30 August 1980, 11–29.)

40   Council Directive of 8 December 1975 concerning the quality of bathing water (76/160/EEC). (Official Journal of the European Communities no L31, 5 February 1976, 1–7.)

41   Council Directive of 16 June 1975 concerning the quality required of surface water intended for the abstraction of drinking water in the Member States (75/440/EEC). (Official Journal of the European Communities no L194, 25 July 1975, 26–31.)

42   Council Directive of 9 October 1979 concerning the methods of measurement and frequencies of sampling and analysis of surface water intended for the abstraction of drinking water in the Member States (79/869/EEC). (Official Journal of the European Communities no L271, 29 October 1979, 44–53.)

43   Council Directive of 30 October 1979 on the quality required of shellfish waters (79/923/EEC). (Official Journal of the European Communities no L281, 10 November 1979, 45–52.)

44   Guidelines for Drinking-water Quality. Volume 1: Recommendations. Geneva: World Health Organization, 1984.

45   Madeley CR. Viruses associated with acute diarrhoeal disease. In: Zuckerman AJ, Banatvala JE, Pattison JR (eds.). Principles and Practice of Clinical Virology. Chichester: John Wiley, 1987, 159–96.

46   PHLS Communicable Disease Report Weekly Edition 85/40, 4 October 1985.

47   PHLS Communicable Disease Report Weekly Edition 84/03, 20 January 1984.

48 Dorn CR, Reddy CS, Lamphere DN, Gaeuman JV, Lanese R. Municipal sewage sludge application on Ohio farms: health effects. *Environ Res* 1985, **38**, 332–59.
49 Fattal B, Margalith M, Shuval HI, Wax Y, Morag A. Viral antibodies in agricultural populations exposed to aerosols from wastewater irrigation during a viral disease outbreak. *Am J Epidemiol* 1987, **125**, 899–906.
50 Standing Committee on the Disposal of Sewage Sludge. Report of the Subcommittee on the Disposal of Sewage Sludge to land. Standing Technical Committee Report No 20. Department of the Environment and National Water Council 1981.
51 Cabelli VJ. Health effects quality criteria for marine recreational waters. United States: Environmental Protection Agency. 1983, EPA report 600/1-80-031.
52 Philipp R, Evans EJ, Hughes AO, Grisdale SK, Enticott RG, Jephcott AE. Health risks of snorkel swimming in untreated water. *Int J Epidem* 1985, **14**, 624–27.
53 Short CS. The Bramham incident, 1980: an outbreak of waterborne infection. *J Inst Water Envir Manage* 1988, **2**, 383–90.
54 Forbes GI, 1989, personal communication.
55 Williams FP, Akin EW. Waterborne viral gastroenteritis. *J Am Waterworks Assoc* 1986, **78**, 34–9.
56 Gerba CP, Rose JB, Singh SN. Waterborne gastroenteritis and viral hepatitis. *CRC Crit Rev Environ Control* 1985, **15**, 213–36.
57 Gamble DR. Viruses in drinking-water: reconsideration of evidence for postulated health hazard and proposals for virological standards of purity. *Lancet* 1979, **i**, 425–8.
58 Melnick JL. Etiologic agents and their potential for causing waterborne virus diseases. In: Melnick JL, ed. Enteric Viruses in Water. Basel: Karger, 1984, 1–16. (Monographs in Virology vol 15.)

Chapter 3

Mumps – worthy of elimination?

ND Noah

Mumps, like chickenpox, measles and rubella, is one of the 'acute crowd infections' of Cockburn [1], all of them relatively recently acquired infections of man. Mumps may have adapted to man from a virus of poultry [2]. Hippocrates described a disease resembling mumps in the fifth century BC, but historically mumps became prominent only in the last 200–300 years when outbreaks affecting armies were described, including the two world wars in this century. In modern times mumps is a common infection, mainly of childhood. An effective live vaccine affords the means to control, and possibly even to eradicate*, it. This review examines the overall burden of mumps, the cost effectiveness of attempting to control it with vaccine, and the appropriate vaccine strategy.

Epidemiology

PERIODICITY

Mumps is an endemic infection with a marked and unusual cyclic pattern of an epidemic period which lasts two years followed by a trough period of one year [3] (Figure 1). Both the laboratory reports to CDSC and the RCGP clinical reporting scheme [4] show this pattern. Superimposed on this pattern is a less marked seasonal variation showing a peak in spring and summer. The periodicity is fairly predictable, so that it is possible to say that without vaccine 1989 and 1990 would have been 'high' years and 1991 a 'low' year. It can be seen from Figure 1 that the introduction of the vaccine in 1988 has already altered both the incidence and the periodicity of mumps. A similar periodicity occurs in Spain [5].

---

*Eradication: disease and agent permanently removed.
 Elimination: agent remains, disease rare or non-existent.
 Containment: disease no longer constitutes a public health problem.

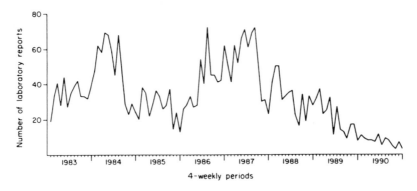

Figure 1   Mumps: laboratory reports, England and Wales, 1983–90

AGE AND SEX DISTRIBUTION AND INFECTIVITY

The age distribution of mumps in a stable unvaccinated population is important because of any shifts that may result from vaccination. In England and Wales, the overall incidence of clinical mumps as reported to GPs is 6.6–6.8 per 1,000 population per year, with the highest incidence (26.3–27.6/1,000/year) in 5–14-year-olds [4]. In a serological survey of 8,716 people aged 1–65 years, 11% were found to have antibodies by two years of age, 38% by four years, 63% by five years and 87% by ten years [6]. This suggests that the average age of infection was 4–5 years, compared with an estimate using mathematical modelling of 6–7 years [7]. An epidemic in a 'virgin' population of 561 eskimos on St Lawrence Island in the Bering Sea was studied in 1957 [8]. Three hundred and sixty-three of the 561 on the island (65%) developed clinical mumps, and a further 151 (27%) had asymptomatic infection, giving an overall infection rate of 92%. Ninety-five per cent of those with mumps had salivary gland swelling and 9% of the males had scrotal swelling alone. The clinical attack rates were highest in those aged 5–14 years (86%); 12% of those aged 60–69 years and none of those over 70 were clinically affected. In the survey in the Netherlands [9], a population similar to our own, the age group most affected was also found to be the 5–14-year-olds with the peak acquisition rate at 4–6 years. Ninety per cent of the population was affected by 14 years of age and 95% of those aged 18–65 years. The age distribution of mumps in Denmark was similar [10]. The sex distribution of clinical mumps shows only a slight preponderance in males, which is of interest when the sex ratio of complications is compared.

48

Complications

CENTRAL NERVOUS SYSTEM

The overall complication rate in mumps has been estimated at 24 per 1,000 cases [4] with the likelihood of a complication greatest in those over 15 years. In this study, unusually, no cases of mumps meningitis were reported. Meningitis is the most common complication of mumps, although it is almost always benign. In the USA up to 15% of mumps causes clinical meningitis [11]. As mumps is common it, in turn, accounts for a considerable proportion of all virus meningitis. In England and Wales, of 6,056 cases of virus meningitis reported to the Communicable Disease Surveillance Centre between 1978 and 1982, 1,269 (21.0%) were attributed to mumps, the percentage varying from 13.0% to 28.8% a year during this period [12]. Only the echoviruses were a more common cause of virus meningitis (40.5%). In France [13] the proportion of lymphocytic meningitis attributed to mumps was 10–15%. The age distribution of mumps meningitis is similar to that of clinical mumps, with perhaps a slight increase in those over 15 years of age, but the sex distribution shows a considerable male predominance of 7:3 [14].

The much rarer encephalitis of mumps has a more serious prognosis, and the risk was estimated at 1 in 6,000 cases [15].

DEAFNESS

Deafness is one of the more serious complications of mumps – transient perceptive deafness is common during an attack of clinical mumps but permanent sensorineural deafness is fortunately rare. It occurs in fewer than 5 cases per 100,000 [16,17] and accounts for about one case of deafness in 20,000 [13]. It is thought to result from endolymphatic labyrinthitis, and the virus has been isolated from the perilymph of the inner ear [17].

GONADAL INFECTION

In the eskimo 'virgin' outbreak [8], 52 of 205 males (25%) developed orchitis, and this complaint was especially common (38%) in puberty and young adults, and rare after 50 years. In westernised countries, the rates are similar [11,18]. Testicular atrophy undoubtedly occurs following mumps, and is common: some atrophy can be detected in 35% of cases [11] but sterility is rare. Indeed, although fertility is known to be maintained even after severe bilateral testicular atrophy, sterility occurring after an attack of mumps orchitis in a man known previously to have been fertile has not to the author's knowledge been recorded. Testicular neoplasms have developed in patients with atrophic testes [19], but the association was not necessarily causal.

Oophoritis is more rare, and estimated to be about 5% [20]. This may be related to the anatomical position of the ovaries, making clinical diagnosis less certain, but complications of mumps seem to be less common in women generally. Mumps virus has been isolated from the cervico-vaginal secretions of a woman with mild oophoritis [21]. There is, however, some evidence that mumps oophoritis may have more lasting functional sequelae than orchitis. Morrison *et al* [20] reviewed the evidence and also described three females who developed oligo- or amenorrhoea after mumps, one of whom had acquired the infection perinatally from her mother. Mumps has also been postulated (not very convincingly) to increase the risk for ovarian cancer [22].

Mumps infection may lead to abortion [23], as high as 27% in the first trimester, but there is little evidence that it can cause congenital malformations [11]. The virus has, however, been isolated from a fetus aborted spontaneously at ten weeks [24].

PANCREATITIS

Like oophoritis, mumps pancreatitis may be difficult to diagnose and is more common than expected. A syndrome of nausea, vomiting with epigastric and left upper quadrant pain, was diagnosed in 50% of the eskimos affected in the eskimo 'virgin' outbreak [8], but either the population had a propensity for pancreatitis or the diagnosis was incorrect as the complication rate for pancreatitis is probably nearer 5%. Diarrhoea is a prominent symptom of mumps pancreatitis [25].

RARE COMPLICATIONS

Myocarditis is usually asymptomatic. Transient renal involvement is probably quite common, but sequelae such as nephrotic syndrome, if related, are very rare [26]. If mumps does cause juvenile onset diabetes, it would account for less than 1% of this type of diabetes [27]. Neonatal respiratory distress has been described in a neonate with virologically proven mumps born to a mother who had asymptomatic mumps infection in late pregnancy [28]. Mumps may cause subacute thyroiditis, and the virus has been isolated from the affected gland [29].

DEATHS

Death from mumps is very rare. Between 1962 and 1981 in England and Wales, five deaths from mumps were certified, on average, each year, and a review of these cases showed that less than half could probably be directly attributable to mumps, giving an annual death rate of about 1–2 per 20 million population [30]. Encephalitis was the most common cause of death in mumps.

BURDEN OF MUMPS: SUMMARY

Although there seems to be a fairly lengthy list of indictments against mumps, of what serious crimes does it actually stand accused? Meningitis, though a common complication, is almost invariably benign. Death from it is almost unknown, and even sequelae are rare. Nevertheless, almost half of all patients with mumps meningitis in Sweden [31] (and possibly a considerably higher proportion in this country), are admitted to hospital for 2–4 weeks and remain 'disabled' in some way for 1–2 months. Ten per cent take a longer time to recover.

Encephalitis appears to be very rare, but clearly more serious.

Orchitis, although common and unpleasant, also does not appear to have serious consequences. Oophoritis is less common and may have some sequelae. The evidence for testicular or ovarian cancer as a late development is not convincing. Mumps in pregnancy can certainly predispose to abortion, especially in the first trimester, but there is little evidence that it causes congenital abnormalities.

The case for deafness of the rarer permanent sensorineural type as a serious complication of mumps is more convincing, but deaths from mumps are uncommon, and certainly fewer than those attributed to chickenpox and measles. On this evidence, the prosecution has a weak case.

For the defence, mumps remains an infection which produces a solid life-long immunity after only one attack, tends to attack at a time of life when illness is not important economically, does not usually require expensive treatment, or even an antibiotic, and nearly everyone recovers fully from it. Moreover, the rate of asymptomatic infection is high, so that solid immunity is provided 'free' in about 25% of the population.

Undoubtedly, on this evidence, the grounds for introducing mass vaccination against mumps were weak, but availability of the vaccine combined with two other vaccines of undoubted value has changed the balance of the case against it.

Cost benefits of mumps vaccination

Judgement on the value of a vaccine should take into account cost benefits or cost-effectiveness studies as an important part of the decision-making process, but should not make such studies the sole criterion on which a decision is made. Moreover, savings made in terms of time lost from school or work (indirect costs) may be of little interest to a health administration. Mumps vaccine is an example of a vaccine for which the benefit–cost ratio diminishes considerably from 4:1 to 1:1 when indirect costs are omitted [32]. In a detailed assessment of benefit-risk of mumps vaccine [33], Koplan and Preblud estimated that, if 90% of a population were infected by age 30 years and 60% of the infections were symptomatic, 540,000 cases of mumps and 23 deaths would be averted for every million persons followed up for 30 years. For reported cases these numbers would be somewhat smaller. Using MMR vaccine would increase the benefit–cost ratio for mumps from 6.7:1 to 14:1 [34] in one estimate, or from 3.3:1 to 39:1 in another [33]. Others

[35] were less convinced that the costs of mumps eradication were justified by the potential benefits, and in the few studies in Europe, estimates based on mumps vaccine alone also show a small benefit. In Austria [36] and Switzerland [37] the benefit–cost ratios were estimated at 3.6:1 and 2.1:1 respectively; however, in the Austrian study high vaccine costs were used, and in the Swiss study indirect costs were not considered. That these studies were based on estimates probably accounted for the variation in the degree of benefit, but it seems clear that the benefit–cost of mumps vaccine is marginal on its own, but increases greatly when combined with rubella and measles.

Mumps vaccination
Only some of the important epidemiological aspects of mumps vaccination will be considered here.

EFFICACY
The vaccine is considered to be about 95% effective in preventing disease [11]. Nevertheless, some disquieting features about mumps vaccination may prove to be important: even in the early trials [38], the vaccine was shown to produce a GMT of antibody considerably lower than that produced after natural infection, and the seroconversion rates were also lower for mumps (84%) than for measles (94%) and rubella (97%) [39]. However, the antibody response to mumps vaccine virus develops slowly, and the lower response may reflect too early testing for mumps antibody levels after MMR vaccine. Four weeks may be too early and 5–6 weeks more appropriate. Although protection after one dose of monovalent mumps vaccine appears to last at least 20 years [11], the low titre of antibody produced by mumps vaccine may mean that protection ceases in late adulthood or the elderly.

CONTRAINDICATIONS
The usual contraindications to any live virus vaccine apply to mumps vaccine, but need to be qualified. Mumps vaccine should not be given less than two weeks before immunoglobulin and for three months after [11]. Close contacts of the immunosuppressed can and should be vaccinated, although mumps is not especially severe in the immunosuppressed. Asymptomatic children who are HIV positive (HIV+) can be vaccinated at 15m; symptomatic HIV+ children should be given MMR if measles vaccination is indicated. It is safe to vaccinate children with leukaemia provided it is in remission and at least three months after chemotherapy. Local steroid therapy, including that applied to the respiratory tract or intra-articular steroid, and a short-term course of steroid therapy (less than two weeks) are not contraindications to mumps vaccination. Mumps vaccine virus is known to infect the

placenta [40], so it would be wise to avoid pregnancy for about three months after vaccination.

SIDE EFFECTS

In a double-blind placebo controlled trial of 581 twin pairs [41], the true frequency of side-effects to MMR vaccine was 0.5–4.0%, and the mumps component was thought to be harmless, as most reactions occurred within 7–11 days and were compatible more with the measles and rubella components. Nevertheless, there has been an increasing number of reports of meningitis characteristic of mumps occurring about four weeks after administration of MMR. In Canada, eight such cases (all meningitis) have been reported [42] following the use of a Smith Kline and French MMR vaccine containing the Urabe strain*. Other cases have been reported since [43,44]. In all these cases mumps virus was isolated from the CSF of the patient. Although we do not know whether the vaccine virus can be isolated from the CSF of those who do not develop CNS complications after MMR, the evidence that the meningitis in these cases was caused by the mumps component of MMR was fairly convincing. It is now possible to distinguish between wild and vaccine viruses, and the CSF isolate from one of these cases has been shown to be a vaccine strain [45]. In one report from England of encephalitis occurring 27 days after MMR vaccine [46], there was a rise in titre to mumps S antibody but the titres to measles and rubella antibodies were also raised, as expected after MMR, and the association could be attributed to chance.

MUMPS VACCINE IN OUTBREAKS

Unlike measles vaccine, mumps vaccine may not prevent an outbreak if given within 2–3 days of onset in a primary case. Nevertheless it may still be worth giving, because it may prevent the second and subsequent generations of cases. In an interesting study of 332 cases of mumps in a high school [47], the vaccine was given to 53 of 178 susceptible students after the peak of the outbreak had occurred. Within the next 21 days, 15 cases (28%) of mumps occurred in the 53 vaccinated students and 51 cases (41%) in the 125 unvaccinated students. No further cases of mumps occurred in the vaccinated group after this time, whereas eight more cases occurred in the unvaccinated group. This suggested that the vaccine did have some impact on the outbreak after 21 days, but the lower attack rate in the immunised group within the 21-day period is

*It is important, however, to note that although all these cases were associated with one brand of vaccine, it does not necessarily imply that that particular vaccine is at fault. If one brand has a monopoly in one country, and a complication is recognised, other cases will looked for actively, and all will be associated with the same vaccine. More soundly based epidemiological studies are needed before such an association is proven.

difficult to explain. The relative inability of mumps vaccine to protect after exposure is compatible with the late rise in antibody after vaccine, and also with the late appearance of the main side effect of mumps vaccine-induced meningitis.

EPIDEMIOLOGICAL SIDE EFFECTS OF MUMPS VACCINE

Mass vaccination is ecological interference [48] and evidence is accumulating that mumps vaccine, like measles vaccine, may alter the normal age distribution of mumps, shifting it towards the older age groups. In the USA, where more than 85 million doses of mumps vaccine were used between 1967 and 1988, attack rates of clinical mumps are now very low [11]. In 1985 only 2,982 cases were reported [11]. In 1986–87 there was a relative resurgence of mumps to 7,790 and 12,848 cases respectively, mostly in high school and middle school populations in states without comprehensive school mumps vaccinations laws [49]. Mumps increased in all age groups in these two years but especially in the 10–14 years (a seven-fold increase) and 15–19 years (an eight-fold increase); more than one third of the cases were over 15 years compared with 8% before mass vaccination was introduced. This was shown to be due not to the waning immunity from mumps vaccine but to those unimmunized persons who had hitherto been 'protected' by a low mumps incidence, ie by herd immunity. Between licensure of the vaccine in 1967 and the recommendation for mass immunisation in 1977, mumps vaccine was in limited use [50]. The increase in reported cases in local outbreaks [50,51] was in adolescent and young adults. The shift in age is said to be 'more dramatic for mumps than for the other two diseases (measles and rubella)' [52]. This has led to recommendations to support school laws for mumps vaccine [49] and even to vaccinate susceptible employees of institutions such as banks, which employ large numbers of young adults [50]. These changes in the age distribution of mumps are of course disturbing, but it is important to note that the actual incidence of mumps has not risen in the older age groups after mass vaccination; the shifts have only been proportional. Also, the 'relative resurgence' of mumps in 1986 and 1987 was probably only a phase in a natural cycle of mumps, a cycle whose periodicity may have been affected by the high vaccine uptake rates. The resurgence will almost certainly be followed by one or more 'low' years from 1988.

Can mumps be eliminated?

'Mumps is a disease subject to eradication. It is not highly contagious and has no animal reservoir' [53]. If the characteristics, as described by Evans [54], of smallpox and smallpox vaccine which favoured eradication are applied to mumps and mumps vaccine (Table 1), there are many that favour mumps, but some important ones do not. With smallpox, it was possible to use a strategy that did not require the

Table 1 Characteristics of smallpox/smallpox vaccine favouring eradication, adapted from Evans [54]: comparison with mumps and mumps vaccine

Smallpox	Mumps
Low infectivity	Probably not low enough
High average age at infection	Yes
Disease transmitted person to person (no animal/insect reservoirs)	Yes
Characteristic easily diagnosed clinical disease	Yes
Few or no subclinical cases	No
No long-term carriers	Yes
Only one agent/serotype	Yes
Short period of infectivity pre- and post-disease	Yes
Immunity after disease:	
of long duration	Yes
not subject to reinfection or reactivation	Yes
carriage and transmission of organism not possible	Yes
Evidence of immunity visible (scars)	No
Disease is seasonal (permitting vaccine strategies)	Marginal
Disease has significant impact on economy	Probably yes
Immunity persists	Yes, probably
Immunity after one injection	Yes
Evidence of previous vaccination detectable (scars)	No
Vaccine is:	
safe, few side effects	Yes
stable, resists physical and genetic change	Yes? (freeze dried)
inexpensive to produce or purchase	Yes ⎱ as
simple to administer	Yes ⎰ MMR

wholesale vaccination of large populations; with mumps, this is probably the only strategy available, which makes eradication possible but administratively too difficult, except possibly in the few countries where vaccination is compulsory and enforced. Thus eradication of mumps is at present probably a fairly optimistic objective and elimination or containment may be more realistic aims in most countries with finite health resources.

What is the best strategy to use and what level of uptake should we aim for? In an attempt to avoid building up a large population of susceptible older persons, Sweden implemented a two-dose schedule for MMR in 1982 [55,56] giving the first dose at 18 months and the second at 12 years. This has proved popular and other Scandinavian countries have also implemented similar schedules (with the second dose at six years in Finland). The USA has relied more on school laws to ensure high coverage [49]. The critical proportion of the population (Pc) that needs to be immunised depends on the efficacy of the vaccine, the typical course of the infection, including the length of the latent and infectious periods, the infectiousness of the agent and the demographic and behavioural features of the community [57]. The median age of infection does not appear to have been taken into account in making these calculations. The level of uptake needed was found to be 93% for mumps vaccination given at 12–15 months. The authors commented that the whole country would be required to reach this uptake rate otherwise transmission of mumps would still occur in districts with lower rates. Other estimates have been made of 85% [7] (with a 75% uptake predicted to be adequate to reduce the incidence of complications of mumps) and 75% [58].

Conclusion: should mumps be eliminated?

'Both the individual and society should be protected against measles, mumps and rubella' [55].

Such statements tacitly assume that all three diseases should be eliminated, almost because both the diseases and vaccines are there. Cost and other studies show beyond reasonable doubt that measles vaccination is extremely worthwhile, and few would argue against rubella vaccination after considering the devastating effects of congenital rubella. For mumps, however, the case for prevention with a vaccine has not been quite so clear cut.

Further detailed discussion of the pros and cons of mumps vaccine has been virtually pre-empted by the opportunity to give it as MMR, and this has swung the balance much more heavily towards it. Having made the decision to use it, however, a high uptake of probably over 90% is almost certainly going to be necessary if outbreaks in adults, who have a greater risk of complications, are to be prevented in the future.

Acknowledgements
I am grateful to Dr Susan Hall and Dr Elizabeth Miller for helpful comments on this paper, and to Mrs CE Owers for secretarial help.

References
1   Cockburn A. Infectious diseases: their evolution and eradication. Springfield, Ill: Thomas, 1967.

2   Black FL. Modern isolated pre-agricultural populations as a source of information on prehistoric epidemic patterns. In: Stanley NF, Joske RA, eds. Changing disease patterns and human behaviour. London: Academic Press, 1980, 37–54.

3   Noah ND. Cyclical patterns and predictability in infection. *Epidemiol Infect* 1989, **102**, 175–90.

4   Research Unit, Royal College of General Practitioners. The incidence and complication of mumps. *J R Coll Gen Pract* 1974, **24**, 545–51.

5   *Bol Epidemiol Semanal* 1985, **33**, no 1704.

6   Morgan-Capner P, Wright J, Miller CL, Miller E. Surveillance of antibody to measles, mumps, and rubella by age. *BMJ* 1988, **297**, 770–2.

7   Anderson RM, Crombie JA, Grenfell BT. The epidemiology of mumps in the UK: a preliminary study of virus transmission, herd immunity and the potential impact of immunization. *Epidemiol Infect* 1987, **99**, 65–84.

8   Philip RN, Reinhard KR, Lackman DB. Observations on a mumps epidemic in a 'virgin' population. *Amer J Hyg* 1959, **69**, 91–111.

9   Wagenvoort JHT, Harmsen M, Boutahar-Trouw BJK, Kraaijeveld CA, Winkler KC. Epidemiology of mumps in the Netherlands. *J Hyg* 1980, **85** 313–26.

10   Glikmann G, Petersen I, Mordhorst CH. Prevalence of IgG-antibodies to mumps and measles virus in non-vaccinated children. *Dan Med Bull* 1988, **35**, 185–7.

11   Immunization Practices Advisory Committee (ACIP). Mumps prevention. *MMWR* 1989, **38**, 388–92, 397–400.

12   Communicable Disease Surveillance Centre. Virus meningitis and encephalitis 1978–82. *BMJ* 1985, **290**, 921–2.

13   Rey M. Mumps immunization. *Rev Prat (Paris)* 1985, **35**, 410–2.

14   Mumps surveillance. *Weekly Epidemiological Record* 1973, **48**, 65–67.

15   Russell RR, Donald JC. The neurological complications of mumps. *Brit Med J* 1958, **2**, 27–30.

16   Vuori M, Lahikainen EA, Peltonen T. Perceptive deafness in connection with mumps. A study of 298 servicemen suffering from mumps. *Acta Otolaryngol* (Stockholm) 1962, **55**, 231–6.

17 Westmore GA, Pickard BH, Stern H. Isolation of mumps virus from the inner ear after sudden deafness. *Brit Med J* 1979, **1**, 14–5.

18 Association for the Study of Infectious Disease. A retrospective survey of the complications of mumps. *J Roy Coll GPs* 1974, **24**, 552–6.

19 Beard CM, Benson RC, Kelalis PP, Elveback LR, Kurland LT. The incidence and outcome of mumps orchitis in Rochester, Minnesota, 1935–1974. *Mayo Clin Proc* 1977, **52**, 3–7.

20 Morrison JC, Givens JR, Wiser WL, Fish SA. Mumps oophoritis: a cause of premature menopause. *Fertil Steril* 1975, **26**, 655–9.

21 Taparelli F, Squadrini F, De Rienzo B, Lami G, Fornaciari A. Isolation of mumps virus from vaginal secretions in association with oophoritis. *J Infect* 1988, **17**, 255–8.

22 Cramer DW, Welch WR, Cassells S, Scully RE. Mumps, menarche, menopause and ovarian cancer. *Am J Obstet Gynecol* 1983, **147**, 1–6.

23 Siegel M, Fuerst HT, Peress NS. Comparative foetal mortality in maternal virus disease. *New Engl J Med* 1966, **274**, 768–71.

24 Kurtz JB, Tomlinson AH, Pearson J. Mumps virus isolated from a foetus. *BMJ* 1982, **284**, 471.

25 Imrie CW, Ferguson JC, Sommerville RG. Coxsackie and mumps virus infection in a prospective study of acute pancreatitis. *Gut* 1977, **18**, 53–6.

26 Helin I, Carstensen H. Nephrotic syndrome after mumps virus infection. *Am J Dis Child* 1983, **137**, 1126.

27 Gamble DR. Relationship of antecedent illness to development of diabetes in children. *BMJ* 1980, **281**, 99–101.

28 Reman O, Freymuth F, Laloum D, Bonte JF. Neonatal respiratory distress due to mumps. *Arch Dis Child* 1986, **61**, 80–1.

29 Eylan E, Zmucky R, Sheba C. Mumps virus and subacute thyroiditis. Evidence of a causal association. *Lancet* 1957, **i**, 1062–3.

30 Galbraith NS, Pusey JJ, Young SEJ, Crombie DL, Sparks JP. Mumps surveillance in England and Wales 1962–1981. *Lancet* 1984, **i**, 91–4.

31 Bjorvatn B, Skoldenberg B. Parotitmeninget och-orkiti Stockholm – en epidemiologisk bakgrund till vaccinations policy. *Lakartidningen* 1978, **75**, 2295–8.

32 Koplan JP. Benefits, risks and costs of immunization programmes. In: The value of preventive medicine. London: Pitman, 1985. (*Ciba Foundation Symposium* **110**, 55–68.)

33 Koplan JP, Preblud SR. A benefit-cost analysis of mumps vaccine. *Am J Dis Child* 1982, **136**, 362–4.

34 White CC, Koplan JP, Orenstein WA. Benefits, risks and costs of immunization for measles, mumps and rubella. *Amer J Publ Hlth* 1985, **75**, 739–44.

35 Yorke JA, Nathanson N, Pianigiani G, Martin J. Seasonality and the requirements for perpetuation and eradication of viruses in population. *Amer J Epidemiol* 1979, **109**, 103–23.

36 Wiedermann G, Ambrosch F. Costs and benefits of measles and mumps immunization in Austria. *Bull WHO* 1979, **57**, 625–9.

37 Just M. Rentiert die Masem-und/oder Mumps: Impfung für schwerzerische Verhältnisse? *Schweiz Med Wochenschr* 1978, **108**, 1763–8.

38 Brunell PA, Brickman A, Steinberg S. Evaluation of a live attenuated mumps vaccine (Jeryl Lynn) with observations on the optimal time for testing serologic response. *Amer J Dis Child* 1969, **118**, 1863–8.

39 Eguiluz GC, Trallero EP. Mumps meningitis following measles, mumps, and rubella immunisation. *Lancet* 1989, **ii**, 394.

40 Yamauchi T, Wilson C, St Geme Jr JW. Transmission of live attenuated mumps virus to the human placenta. *New Engl J Med* 1974, **290**, 710–2.

41 Peltola H, Heinonen OP. Frequency of true adverse reactions to measles-mumps-rubella vaccine. *Lancet* 1986, **i**, 939–42.

42 Mumps meningitis, possibly vaccine-related-Ontario. *Can Dis Wkly Rep* 1988, **14**, 209–11.

43 von Mühlendahl KE. Mumps meningitis following measles, mumps, and rubella immunisation. *Lancet* 1989, **ii**, 394.

44 Gray JA, Burns SM. Mumps meningitis following measles, mumps, and rubella immunisation. *Lancet* 1989, **ii**, 98.

45 Gray JA, Burns SM. Mumps vaccine meningitis. *Lancet* 1989, **ii**, 927.

46 Crowley S, al-Jawad ST, Kovar IZ. Mumps, measles and rubella vaccination and encephalitis. *BMJ* 1989, **299**, 660.

47 Wharton M, Cochi SL, Hutcheson RH, Bistowish JM, Schaffner W. A large outbreak of mumps in the postvaccine era. *J Infect Dis* 1988, **158**, 1253–60.

48 Noah ND. Measles. In: Silman AJ, Alwright SPA (eds). Elimination or reduction of diseases? Oxford: Oxford University Press 1988, 46–59.

49 Cochi SL, Preblud SR, Orenstein WA. Perspectives on the relative resurgence of mumps in the United States. *Amer J Dis Child* 1988, **142**, 499–507.

50 Kaplan KM, Marder DC, Cochi SL, Preblud SR. Mumps in the workplace: further evidence of the changing epidemiology of a childhood vaccine-preventable disease. *J Amer Med Assoc* 1988, **260**, 1434–8.

51 Pachman DJ. Mumps occurring in previously vaccinated adolescents. *Amer J Dis Child* 1988, **142**, 478–9.

52 Cochi SL, Preblud SR, Orenstein WA. Perspectives on the relative resurgence of mumps in the United States. *Amer J Dis Child* 1988, **142**, 1021–2.

53   Bader M. Perspectives on the relative resurgence of mumps in the
     United States. *Amer J Dis Child* 1988, **142**, 1021.
54   Evans AS. The eradication of communicable diseases: myth or
     reality? *Amer J Epidemiol* 1985, **122**, 199–207.
55   Rabo E, Taranger J. Scandinavian model for eliminating measles,
     mumps, and rubella. *BMJ* 1984, **289**, 1402–4.
56   Christenson B, Bottiger M, Heller L. Mass vaccination programme
     aimed at eradicating measles, mumps, and rubella in Sweden: first
     experience. *BMJ* 1983, **287**, 389–91.
57   Nokes DJ, ANderson RM. Measles, mumps, and rubella vaccine:
     what coverage to block transmission? *Lancet* 1988, **ii**, 1374.
58   Cox MJ, Anderson RM, Bundy DAP, *et al*. Seroepidemiological
     study of the mumps virus in St Lucia, West Indies. *Epidemiol Infect*
     1989, **102**, 147–60.

Chapter 4

Norwalk-like viruses and winter vomiting disease

T Riordan

The last 20 years have seen an explosion in our knowledge of enteric pathogens, with a long list of new agents being identified. Although the Norwalk agent was one of the first of these new enteric pathogens to be described it has been overshadowed by organisms causing a higher morbidity and mortality, such as rotaviruses and campylobacters. This review aims to describe current knowledge of the Norwalk group of viruses and in particular to highlight their growing epidemiological, if not clinical, significance.

Historical
Acute infective gastrointestinal illnesses are extremely common world-wide. Thus, for example, a study of prospective illness in a group of Cleveland families estimated an annual incidence of 1.5 per person per year [1]. Fortunately, the vast majority of cases of gastroenteritis are mild, short-lived and do not result in hospitalisation. Until the 1970s the microbiological investigation of gastroenteritis was very unsatisfactory: the known bacterial pathogens such as salmonellas, shigellas and enteropathogenic *Escherichia coli* were detectable in only a minority of patients.

Zahorsky, an American paediatrician, gave a superbly perceptive description of a syndrome which could be identified as a distinct entity among the many cases of non-bacterial gastroenteritis he saw [2]. His epidemic winter vomiting or winter vomiting disease (WVD) remains a valid entity today and the features are listed in Table 1. Although subsequent authors did not refer to Zahorsky's paper, it is interesting that over the next three decades several papers appeared describing institutional outbreaks of typical winter vomiting disease, in which infection was believed to be transmitted from person to person [3–7].

The failure to isolate a bacterial pathogen from these outbreaks and from most sporadic cases of gastroenteritis led to the suspicion that

61

Table 1   Clinical features of winter vomiting disease

Occurs in winter epidemics
Vomiting the main symptom
Mild diarrhoea with pale stool and no cells
Pyrexia absent or low grade
Spontaneous and rapid recovery with no mortality

---

viruses might be involved. However, despite considerable effort and increasingly sophisticated cell cultures, and even organ cultures, no virus could be isolated.

Supportive evidence for a viral aetiology did, however, come in one or two studies in which illness was successfully transmitted to volunteers by means of bacteria-free faecal filtrates [8].

The important advance occurred following an incident in an elementary school in the town of Norwalk, Ohio. This was first reported in 1969 as an explosive outbreak of WVD affecting 50% of the students and teachers at the school, with a secondary attack rate among family contacts of 32.3% and an average incubation period of 48 hours [9]. As usual in such incidents, no bacterial or viral pathogen could be isolated from victims. However, in a subsequent paper, Dolin *et al* induced WVD in two of three volunteers by oral administration of a filtrate prepared from a rectal swab from an adult secondary case in the Norwalk outbreak [10]. It was then successfully passaged serially through volunteers, again inducing typical illness. The 'agent' was shown to be less than 66nm in diameter and was not inactivated by ether, acid or heating at 60°C for 30 min. All attempts to isolate a virus failed. However, Kapikian *et al* later examined material from the Norwalk outbreak by immune electron microscopy (IEM) [11]. This involved reacting a stool filtrate, shown to be capable of transmitting illness to volunteers, with a volunteer's convalescent serum. The rationale for this approach was that the agent was presumed to be in low titre and it was hoped that virus particles in the filtrate would be aggregated by antibody, enabling them to be visualised by electron microscopy.

The technique did indeed result in the appearance of clusters of virus-like particles. The antibody-coated particles were said morphologically to resemble picorna or parvoviruses. The diameter of the particles was recorded at 27nm. Using IEM these workers also demonstrated rising antibody titres in paired sera from three of five of the original victims of the Norwalk outbreak. Thus for the first time a virus had been detected in association with non-bacterial gastroenteritis.

These observations set the ball rolling for an explosion of interest in viral gastroenteritis. Sadly, however, the progress made with the Norwalk agent in the last 18 years has been painfully slow, with many blind alleys.

Virology

Following the initial reports of identification of the Norwalk agent, many other workers reported visualization of virus particles in stool samples. Two problems rapidly emerged. Firstly, it became clear that some viruses could be detected in stools of healthy individuals, particularly children. This meant that before a virus could be regarded as established as an enteric pathogen, certain criteria must be fulfilled (Table 2). Some viruses detected in stools, such as enteric corona viruses, do not fulfil these criteria.

The second problem concerned the characterisation and nomenclature of viruses detected in material from winter vomiting disease outbreaks. Besides the Norwalk agent, several other small round viruses were described. Because none of these viruses could be cultured, comparisons have been made largely on the basis of size and morphology. There has been considerable confusion as to the relationships between these agents. This has been fostered by some authors who have tended to lump them all together as small round viruses, or parvoviruses, although there was almost no virological basis for the latter approach.

The confusion was considerably reduced by a simple and logical classification scheme proposed by Caul and Appleton for small round viruses detected in faeces [12]. As Figure 1 shows, the fundamental sub-division is into those viruses which, in the absence of antibody, are morphologically featureless and those which have a definite surface structure. The small round viruses implicated in diarrhoeal disease will be considered in detail.

FEATURELESS VIRUSES

Among the featureless viruses are several agents identified in stools from outbreaks in the 1970s, including Wollan [13], Ditchling [14], Cockle [15] and Paramatta [16]. These agents are now regarded as candidate human parvoviruses, their morphology closely resembling that of known animal parvoviruses (Figure 2). Their size and buoyant densities would also be compatible with this, although definite classifi-

Table 2 Criteria for assigning a pathogenic role to viruses detected in stools

1	Present in stools with a temporal association with illness
2	Low incidence in symptomless individuals
3	Correlation between illness and detection of virus in material from outbreaks
4	Transmission of illness to volunteers by virus containing material
5	Evidence of seroconversion in association with illness

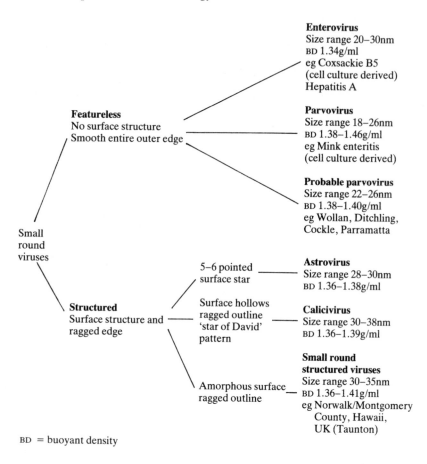

Figure 1 Scheme for grouping of small round viruses. Reprinted from *J Med Virol* 1982, **9**, 260.

cation is not possible yet. When first reported, these agents were considered to be causally related to the disease. However, they do not fulfil all the criteria for establishing them as pathogens and in the view of many virologists their role as a cause of diarrhoea has not been established [17]. They should not be lumped with Norwalk and other structured viruses, although authors have continued to do this, thereby fostering unnecessary confusion.

STRUCTURED VIRUSES

In Caul and Appleton's scheme, structured viruses are subdivided into three groups on the basis of morphology, size and buoyant density.

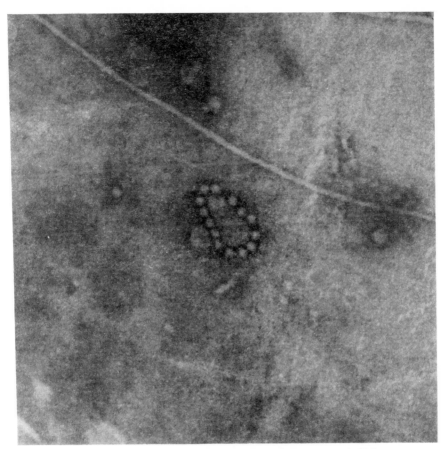

Figure 2   Group of small round featureless viruses, ×120,000

Astroviruses:
Astroviruses have a classical structure with a five-pointed star (Figure 3) and are serologically quite distinct from the other two groups. Although nosocomial outbreaks associated with astroviruses have been described, epidemiological studies have shown them to be more related to sporadic gastroenteritis in infants than epidemic winter vomiting disease [18].

Caliciviruses:
The second group of structured viruses, caliciviruses, had been described in a number of animals including cats and pigs but they were first reported in human faeces by Madeley and Cosgrove in 1976 [19]. Caliciviruses classically show fivefold symmetry with surface hollows or cups, giving a 'star of David' appearance (Figure 4). They also have an

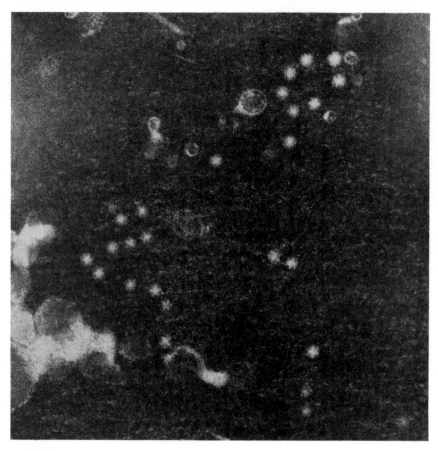

Figure 3 Astrovirus particles showing classical five-pointed stars, ×145,000

amorphous or ragged edge. Caliciviruses contain single stranded RNA as their genome and were originally classified as a genus of the family Picornaviridae. However, with the recognition of their distinct morphology and the presence of a dominant 60,000 MD polypeptide [20] they have been reclassified as a separate family, Caliciviridae.

Although the first reports of caliciviruses in humans came from children with diarrhoea, their role as pathogens was initially in doubt since they were also identified in stools of symptomless children. However, evidence supporting their pathogenic role began to accumulate when institutional outbreaks of winter vomiting disease, in which virus detection correlated closely with clinical illness, were reported from the UK and Japan [21,22]. In addition, seroconversions and rises

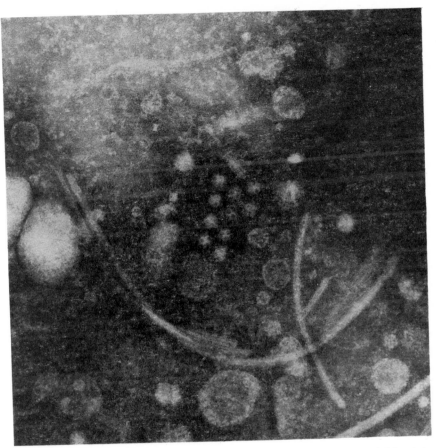

Figure 4   Calicivirus particles showing surface hollows and twofold axis of symmetry, ×145,000

in calicivirus specific IgM antibody have been demonstrated in ill subjects by IEM [23].

Small round structured viruses:
The final group of structured viruses, which includes the original Norwalk agent, are referred to as Norwalk-like viruses in American literature. In the UK they have come to be known by the accurate but rather unhelpful term small round structured viruses (SRSV). The two terms tend to be used interchangeably although it should be borne in mind that use of the term SRSV refers purely to morphology. By contrast, in many American papers no virus has been visualised and diagnosis has been serological, using reagents from the original Norwalk outbreak.

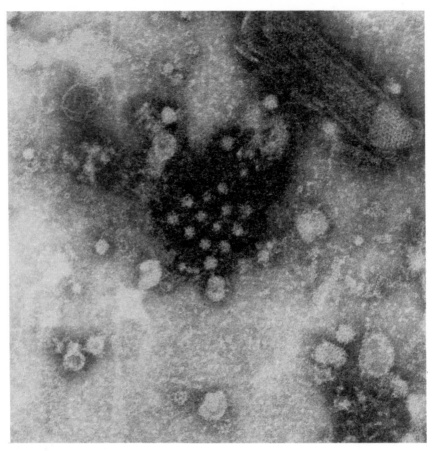

Figure 5 Small round structured viruses with typical ragged edges, ×145,000

The characteristic features of the SRSV particles are their amorphous surface and ragged edge (Figure 5). Considerable confusion has arisen over the size of these agents. In the American literature, the surprisingly precise figure of 27nm is quoted repeatedly. By contrast, measurement of these particles in the UK, including material containing the original Norwalk agent, has given diameters in the region of 30–35nm [12]. These discrepancies probably relate, at least in part, to the difficulties in accurate calibration of instruments and measurement of such ragged particles.

A number of studies have attempted to examine the serological relationship between SRSVs. This has been done principally by IEM. It is clear from Table 3 that there are serologically distinct strains, but

Table 3 Serological relationships between some of the small round structured viruses detected in published outbreaks

Agent	Relationship	Community affected	Country	Reference
Norwalk ⎫ Related	School	USA	[11]
Montgomery County ⎭	Family	USA	[24]
Hawaii	Distinct	Family	USA	[24]
Otofuke ⎫	Related to each other. Distinct	Institution	Japan	[25]
Sapporo ⎭	from Norwalk and Hawaii	Children's home	Japan	[26]
*Snow Mountain	Distinct from Norwalk and Hawaii	Resort camp	USA	[27]

*There are no published electron micrographs of this agent in the absence of convalescent serum and its morphology is therefore speculative

studies so far have been severely constrained by the limitations of IEM.

The taxonomic status of Norwalk and other SRSVs remains indeterminate. It was originally suggested that Norwalk was a parvovirus and this has tended to perpetuate the confusion between SRSVs and the featureless viruses such as cockle virus. It is now abundantly clear, on the basis of size, buoyant density, morphology and lack of any serological cross reaction, that SRSVs are not parvoviruses. There is substantial evidence to suggest that SRSVs may be members of the caliciviridae. Thus, they have remarkably similar sizes and buoyant density [12], and have a major protein of very similar molecular weight to that of caliciviruses [28]. Although the classical structure of calicivirus differs significantly from that of SRSVs, feline caliciviruses are morphologically indistinguishable from SRSVs and in some faecal extracts both classical caliciviruses and SRSVs can be seen in the same preparation. Finally, there is increasing serological evidence of cross-reaction between strains of calicivirus and SRSV [29].

Despite these data it is probably premature to regard caliciviruses and SRSVs as entirely interchangeable; for example, there are seroepidemiological data which suggest that there is a much higher prevalence of antibody to calicivirus in young children [29,30].

Pathogenesis and immunity

Our understanding of the pathogenesis of infection by SRSVs is based largely on volunteer studies. It is assumed that the virus replicates in mucosal cells. No lesions have been detected in gastric or colonic biopsies. Small intestinal biopsies from affected volunteers have shown an intact mucosa but other changes have been noted, including blunting of villi, shortening of microvilli, dilation of endoplasmic reticulum and an increase in intracellular multivesiculate bodies [31]. Brush-border enzyme studies revealed a reduction in activity of alkaline phosphatase, sucrase and trehalase compared to that in baseline and convalescent studies. Interestingly, these changes have also been detected in a few asymptomatic volunteers who responded serologically to the virus.

Biochemical studies performed on volunteers have shown evidence of malabsorption of D xylose, fat and lactose [32]. Although clinical illness persisted for only 1–2 days, malabsorption persisted for at least one week.

Information on immunity also comes largely from human volunteer studies [33]. Only half of a group of unselected volunteers developed illness in response to an oral challenge. Individuals who became ill produced an antibody response, but this clearly lacked a protective role. Indeed, it seems that illness commonly occurs in the presence of antibody, whereas clinically immune volunteers usually have low level or absent antibody both before and after challenge. Susceptible individuals remain susceptible on repeated long-term challenge. These unusual observations have led to speculation as to the mechanism of this striking clinical resistance: one possibility would be that resistance is genetically controlled on the basis of intestinal receptor sites. Unfortunately, there are insufficient data available from volunteer studies to evaluate this; the second hypothesis is that repeated exposure to the virus is required to generate an immune response as well as illness. Thus, clinical immunity would relate to the fact that the individual had not been fully primed. This hypothesis is consistent with the low prevalence of antibody in the young and with its gradual appearance during early adulthood in developed countries.

In contrast to the response to long-term (ie 24–30 month elapsed) rechallenge, short-term clinical immunity can be demonstrated when previously ill volunteers are rechallenged 6–14 weeks later.

Diagnosis

For the individual patient, the laboratory diagnosis of infection by SRSVs poses considerable difficulties. There are three principal approaches: visualisation of virus particles by electron microscopy, detection of rising antibody titres in paired blood samples and detection of specific antigen in faecal samples by immunoassay.

ELECTRON MICROSCOPY

This was the technique by which these viruses were first discovered and in the UK it remains the only method routinely available. It suffers from a number of well-recognised limitations: firstly, the equipment is expensive and the technique highly labour intensive; secondly, detection of SRSVs is technically demanding and requires considerably more observer skill than for other agents, such as rotaviruses. To compound this, the period of virus excretion is short, usually less than four days, and the number of particles present in faeces is comparatively low (of the order of 10^5–10^7 virus particles/g faeces compared to rotavirus which may be excreted in numbers between 10^{11}–10^{12}/g faeces [34]. Techniques are available which can be used to concentrate virus from faeces, such as ultracentrifugation and ammonium sulphate precipitation which can increase sensitivity ten- or 100-fold [35]. However, even with a concentration step, EM remains a disappointingly insensitive technique and not infrequently in a confirmed outbreak only 10–30% of specimens are found to be positive. To improve the chance of detecting SRSVs, it is important to obtain specimens as soon as possible after the onset of illness. Certainly there is little point in examining specimens collected more than three days after onset. Another practical point is that in the laboratory the virus particles appear rather fragile and freezing and thawing may have a deleterious effect. Specimens should, therefore, be stored at +4°C rather than frozen.

Kapikian *et al* attempted to enhance the sensitivity of EM by an immunological method when they detected virus particles in faecal material from the original Norwalk outbreak [11]. Their concept was that antibody in convalescent serum would aggregate virus particles, rendering them more readily visible. While clumping does indeed occur, there is no evidence that for an experienced observer this actually increases sensitivity. Solid phase immune electron microscopy (SPIEM) is a more promising immunological approach. It involves coating EM grids with specific antibody which can then capture virus particles present in the specimen. Rubenstein and Miller showed that for detection of rotavirus, the antibody coated grid technique was nine times more sensitive than conventional EM [36]. Wilkinson *et al* have recently achieved a significant increase in sensitivity and ease of detection of SRSVs using this technique [37]. The main disadvantage of SPIEM over conventional EM is that it is virus specific and is, therefore, likely to be applicable only to outbreaks where SRSVs are strongly suspected on clinical and/or epidemiological grounds.

ANTIBODY DETECTION

Kapikian *et al* [11], in their early papers on the Norwalk agent, described detection of rising antibody titres in patients using IEM. The technique is very simple in concept: acute or convalescent serum is reacted with faecal material known to contain SRSVs; the preparation is then examined by EM. Virus aggregation is taken to indicate the presence of specific antibody. IEM has been used extensively in subsequent research but suffers from many disadvantages: it is extremely time consuming and there are severe limits on numbers of sera that can be examined; also the technique is technically demanding and somewhat subjective. Development of other immunoassays has been severely inhibited by the failure to grow the virus *in vitro* and the difficulty of purifying virus from faeces. Despite these difficulties, a radio-immunoassay was described by Greenberg *et al* [38]. The assay was a blocking test and employed virus partially purified from faeces. The antibody trap and conjugate were either convalescent human sera or sera from experimentally-infected chimpanzees. Despite the rather crude antigen preparation and the unavailability of hyperimmune or monoclonal antibodies, the test appeared to perform with reasonable specificity and produced results closely comparable with those of IEM. Gary *et al* have subsequently described an enzyme linked immuno-sorbent assay (ELISA) for Norwalk virus antibody [39].

While attractive for detailed study of major outbreaks and for sero-epidemiological studies, these techniques suffer from the disadvantage of any test involving antibody detection, in that paired blood samples must be collected and laboratory diagnosis must, of necessity, be retrospective. They also have the problem of being strain specific. In the USA there seemed to be a rapid change to the use of RIA and ELISA for serological diagnosis of Norwalk in published outbreaks. This did not occur in the UK, where EM remains the only method routinely available.

ANTIGEN DETECTION

Immunological techniques such as ELISA have been successfully developed for detection of rotavirus in faeces and have an acceptable sensitivity and specificity. A similar technique for SRSVs would be of considerable benefit, since it would enable a rapid diagnosis without the expense and technical difficulty of EM.

RIA and ELISA techniques have been described [38–40] but suffer from the problems of crude antisera, which tend to trap non-viral antigens from faeces causing high background readings with a resulting loss in sensitivity. Thus, the RIA described by Greenberg *et al* [38] was only comparable in sensitivity to EM, even when examining material from experimentally-infected volunteers from whom specimens could be collected at the optimal time. In the more demanding situation of an outbreak investigation, the results from one incident, shown serologically to be due to Norwalk-like virus, were very disappointing in that Norwalk antigen was detected in only three of 51 stools examined [41].

Herrmann *et al* recently described an ELISA for detection of Norwalk antigen in stool [40]. Evaluation on stool samples from infected volunteers showed it to have a similar sensitivity to the RIA antigen test. When this test was applied in an outbreak [42] shown serologically to be due to a Norwalk-like virus, stools from four of seven patients tested were positive. Further refinement in ELISA techniques should lead to sensitive specific assays which can be widely applied to the diagnosis of infection in both sporadic cases and outbreaks.

PRACTICAL APPROACH TO DIAGNOSIS

As mentioned previously, EM is the only method of detecting SRSVs routinely available in the UK. The cost and low sensitivity make it unsuitable for diagnosis of sporadic cases, for which laboratory confirmation will have no management implications anyway. The position is somewhat different with outbreaks, in that the management of the outbreak will be materially altered if the agent is known. Kaplan *et al* [43] have highlighted the fact that SRSV outbreaks have a very characteristic pattern, so that if an outbreak fulfils certain clinical and epidemiological criteria (Table 4) there is a high chance of it being caused by SRSVs. Given that these criteria have been fulfilled, then detection of SRSVs in even a small proportion of faeces from an incident should be regarded as highly significant.

Epidemiology

Despite the diagnostic difficulties described above, our knowledge about the epidemiology of SRSV infection has gradually increased. The subject divides naturally into two areas, epidemic and sporadic infection.

SRSV OUTBREAKS

To date no study has given a completely satisfactory insight into the true incidence of outbreaks due to SRSV compared to other agents. However, several studies have provided evidence that SRSVs are a very significant cause of outbreaks [43–45]. Detailed information on outbreaks will be discussed according to the source of infection.

OUTBREAKS INVOLVING PERSON TO PERSON SPREAD

Person to person spread is probably a component of almost all outbreaks involving SRSVs, but a distinction must be drawn between those in which it is the primary mode of infection and those in which it is the route of infection for secondary cases only. Most of the descriptions of outbreaks of WVD published before the identification of the Norwalk agent seem likely to have involved person-to-person spread, in that they occurred in institutions and had an epidemic curve typical of person-to-person spread. There are now several well-documented institutional

Table 4   Criteria for suspecting outbreak is due to Norwalk

1	Stool cultures negative for bacterial pathogens
2	Mean (or median) duration of illness 12–60 hours
3	Vomiting in >50% of cases
4	Incubation period (if known) of 24–48 hours

From Kaplan et al [43]

Table 5 Institutional outbreaks involving person-to-person spread

Institution	Patients/residents		Staff			Duration	Methods of Diagnosis	Reference
	no affected	attack rate	no affected	attack rate				
Nursing home	46	46%	32	46%		14 days	Rising antibody titre	[46]
Hospital	22	61%	31	55%		11 days	Rising antibody titre	[47]
Psychogeriatic hospital	67	59%	30	50%		28 days	EM	[48]
Hospital	57	28%	69	27%		26 days	EM	[49]
Hospital	NA	NA	635	27%		22 days	EM + rising antibody titre	[50]

outbreaks in which SRSVs have been implicated (Table 5). One particularly well investigated outbreak [46] occurred in a nursing home. As is typical in such outbreaks, the attack rate was high (46%) for both staff and residents. Person-to-person transmission was demonstrated by the following: firstly, the epidemic curve was typical; secondly, there was temporal clustering of cases. Thus there was a higher rate of illness among residents exposed to an ill room-mate one or two days previously than among those not similarly exposed (relative risk = 3.74). Thirdly, employees having daily contact with residents had a higher rate of illness than those without such contact (57% vs 17%, p < 0.01). Finally, there was evidence of secondary transmission to household contacts of ill employees, the secondary attack rate being 33%.

Published reports of hospital-based outbreaks of SRSV infection involving person-to-person spread demonstrate the facility with which SRSVs can spread in this milieu. Outbreaks can rapidly involve several different wards. Practical difficulties are often experienced in providing adequate facilities and staff for isolating cases during an outbreak. It is fortunate that such incidents usually terminate spontaneously in 7–14 days and rarely lead to severe illness, even in those who are elderly or immunocompromised.

Following the Stanley Royd Hospital outbreak of salmonellosis, an enormous amount of attention has been focused on the problems of nosocomially-acquired gastrointestinal infection. There are countless papers describing institutional outbreaks of salmonellosis, whereas, in contrast, there are relatively few published SRSV outbreaks. It might be concluded from this that salmonella outbreaks are numerically more important. It is my belief that, in reality, the position is quite different. We have studied all properly investigated institutional outbreaks occurring in the Greater Manchester area over a five-year period (October 1982–September 1987) [51]. If one bears in mind the insensitivity of

Table 6 Institutional outbreaks involving person to person spread, Greater Manchester October 1982–September 1987 [49]

Agent	No of outbreaks
Shigella sonnei	42
SRSV	29
Cryptosporidium sp	12
Salmonella sp	6
Giardia lamblia	5
Rotavirus	4
Unknown	34
Total	132

electron microscopy and the fact that, because of the milder illness, SRSV outbreaks are much less likely to be investigated than salmonella incidents, the results in Table 6 are really quite striking.

MODE OF TRANSMISSION FROM PERSON TO PERSON

As with most enteric infections, it has been assumed that SRSV have been transmitted from person to person by the faecal oral route. Although this is undoubtedly a major means of transmission, there is increasing evidence for other routes. Firstly, virus has been demonstrated in vomit [52] and it seems likely that heavy environmental contamination from projectile vomiting could be important. Such environmental contamination may account in part for repeated outbreaks on successive voyages of cruise liners that have been reported [41]. Even more intriguing is the possibility that airborne transmission can occur. A recent paper described an outbreak in a hospital affecting 65 of 2379 staff over a three-week period [50]. The attack rate was highest for staff in the emergency room. It was, therefore, of considerable interest that of 100 patients and their companions, who visited the emergency room on 11–12 November for unrelated problems, 33 developed gastroenteritis 24–28 hours after their visit versus 1 of 18 who attended on 8 November ($p < 0.001$). Similarly, among housekeepers working in the hospital between 9–13 November, the risk of becoming ill was four times greater for those who visited or walked through the emergency room than for those who did not ($p = 0.028$). The housekeepers had no routine interaction with staff or patients as part of their job, and wore gloves while working. Most of the affected housekeepers had merely walked through the emergency room when either entering or leaving the hospital and had probably touched little except door handles as they passed through. Similarly, among the cases occurring in visitors to the emergency room, several had made only fleeting visits, such as a person who stepped into the emergency room to use the money-changing machine. These data would seem to suggest very strongly that airborne transmission can occur.

OUTBREAKS ASSOCIATED WITH POTABLE WATER

Several outbreaks associated with contaminated potable water supplies have been reported in the USA. Three of the published outbreaks involved small private water supplies, two to camps [53,54] and the other to a school [55], and in both of these there was evidence of cross-contamination from septic tanks. The remaining outbreaks involved municipal water supplies and resulted in cases running into thousands [56,57]. In one instance surface runoff from heavy rain (4.5 inches in two days) preceding the outbreak may have contaminated the system [56]. In another major outbreak the source of contamination was believed to be a cross-connection between the water supply and an industrial water system [57].

In the UK, no outbreaks related to drinking water have been unequivocally linked to SRSV. However, in a review of waterborne outbreaks in the UK spanning the period 1937–83, Galbraith *et al* [58] considered that 8/33 outbreaks were likely to have been viral, on the basis of the clinical and epidemiological features and the failure to find a bacterial pathogen. This represented the single most important cause numerically. The most dramatic of the outbreaks resulted in over 3,000 cases in North East Leeds in July 1980. A public borehole supply had been contaminated by a sewage-polluted stream, and to compound the problem there was a failure of chlorination.

It is likely that waterborne viral gastroenteritis is considerably under-reported. Indeed, one of the outbreaks included in Galbraith *et al*'s review came to light only as a result of a questionnaire administered as part of the investigation of an unrelated outbreak of hepatitis A. Although outbreaks associated with potable water from mains supplies are likely to be rare, the possibility should always be considered when an explosive outbreak occurs, particularly if it affects a whole community. Virological methods for examining water supplies are not sensitive enough to detect SRSVs even if large volumes are concentrated. Thus, the demonstration of an association must at present be largely epidemiological. Considerable debate has centred around the relevance of bacteriological indicator organisms, particularly *E coli* and coliforms, in assessing the possibility of viral contamination. On the one hand, unsatisfactory bacteriological results suggest faecal contamination of the water and, therefore, indirectly support the possibility of SRSV in the water. On the other hand, there is increasing evidence that viral contamination can occur in water supplies that are bacteriologically satisfactory [59]. Equally, viruses appear to survive in water with significant levels of residual chlorine [60]. This does, of course, raise many crucial questions about both routine monitoring and investigation of outbreaks.

OUTBREAKS ASSOCIATED WITH SWIMMING IN CONTAMINATED LAKES ETC

Outbreaks of SRSV have also been associated with exposure to recreational water [61–64). One outbreak involved visitors to a leisure park. [61], where there was a strong association between illness and swimming in the lake. The outbreak was explosive in nature suggesting that contamination of water was only transient. Although segments of the park's sewer line passed under the lake, there was no evidence of leakage and no other obvious source of faecal pollution of the lake could be identified. The bacteriological quality of the lake water was monitored regularly and samples before and after the outbreak showed levels of faecal coliforms which were acceptable (<200/100ml).

Although there are no published reports of outbreaks related to bathing beaches, epidemiological studies have shown a correlation between rates of acute gastrointestinal illness resembling WVD and the bacteriological quality of bathing beach waters [65]. With the advent of EEC directives on bathing beaches, this issue is becoming particularly relevant.

NON-SHELLFISH FOOD-BORNE INFECTION

It had long been known that a significant proportion of apparently clear-cut food poisoning incidents had no identifiable cause. In 1981 Appleton and Palmer pointed out that a third of unexplained outbreaks reported by MOSEH and EHOs had incubation periods in the range 24–48 hours and were associated with symptoms consistent with WVD [66]. On this basis, they suggested that these incidents could have been caused by viruses. The following year their hypothesis was vindicated: Griffin *et al* reported an outbreak involving three luncheon banquets [67]. The illness had a median incubation period of 31 hours and was typical of WVD. Serological studies showed rising antibody titres to Norwalk agent in seven of 12 patients tested. Epidemiological investigations showed a strong association between illness and eating green salad. Several of the restaurant's employees had displayed symptoms of WVD, including one of the two staff who prepared the lettuce for all salads. It was suggested that she had contaminated the lettuce and thereby transmitted infection.

Since this report, a substantial number of others have appeared. As regards the types of food implicated, it can be seen from Table 7 that they represent a quite different spectrum from those typically associated with bacterial food poisoning. The items tend to be extensively handled in the kitchen, for example sandwiches and salads; also, in all but one outbreak, they have been items eaten cold. All these features clearly favour successful contamination by a food handler. In some outbreaks, several items have been implicated, and in others it has proved impossible to identify any association, possibly because multiple items were contaminated. Salads clearly have a particularly prominent place in the list of foods. Whether this has any particular significance beyond the way they are prepared is uncertain.

Since infected food handlers seem to play a central role in outbreaks of this type, it is obviously important to have a clear understanding as to how, and under what circumstances, infection can be transmitted in this manner. In almost all the published accounts, there has been evidence of typical illness in kitchen staff. In some, the individual who prepared the suspected item fell ill. In those instances where illness was denied by employees, it is difficult to be sure whether fear of possible recriminations in the aftermath of an outbreak may have discouraged individuals from admitting to mild illness which they had not previously reported,

Table 7 Non-shellfish food-borne outbreaks associated with SRSV/NLV

Location	No affected/ no at risk	Implicated food	Method of diagnosis	Food handlers	Reference
Restaurant	38/41 25/31 0/12	Green salads	Rising antibody titre	ill	[67]
Hospital	129/248	Chicken sandwiches	EM	ill	[68]
Bakery	130/220	Cake icing	Rising antibody titre	ill	[69]
Hotel banquets	275/790	Cooked ham	EM	virus in stool	[70]
College cafeteria	220/383	Tossed salad	Rising antibody titre	no evidence of illness	[71]
Hotel banquet	129/375	Salads	Rising antibody titre	ill	[72]
School cafeteria		French fries/ hamburgers	Rising antibody titre	ill	[73]
Hotel banquets	282/353	Melon/vermicelli	EM	serological evidence of illness	[74]
University football game		Ice	EM	no evidence of illness	[75]
Hotel	125/248	Cold smoked trout/ soup	EM	ill	[76]
School banquet	106/211	Salad/salad dressing	Rising antibody titre	ill	[77]
Restaurant	350/700	Unknown	Rising antibody titre + antigen detection by ELISA	ill	[78]
Party	26/87	Fruit salad	Rising antibody titre + antigen detection by ELISA	serological evidence of illness	[78]

or whether genuinely asymptomatic individuals can contaminate food.

The period during which an individual is able to transmit infection via food is clearly likely to be related to the period of virus excretion. Volunteer studies suggest that this begins a few hours before onset of illness and continues for only 2–3 days [79]. Although one study provided convincing epidemiological evidence that transmission could occur up to 48 hours after recovery [72], transmission is much more likely when an individual is symptomatic. A very prolonged period of infectivity (one month) was suggested by Iversen *et al* [74] but the interpretation of their data has been questioned by others, and at present it seems reasonable to regard 48 hours after recovery as a reasonable limit to the period when an individual may infect food.

It has been assumed that the main means by which transmission occurs is by the faecal oral route involving poor hand hygiene. Kuritsky *et al* gave a very graphic account of an outbreak related to a bakery [69]. The employee who prepared the implicated butter cream frosting had been ill just before, during and after the particular shift in question. Observations after the outbreak revealed that in preparing the 76-litre batches of frosting, he often submerged his arm up to the elbow in the frosting to break up lumps. This would obviously have been an ideal way to transfer any virus particles on the hands to the frosting. Subsequent mixing would have ensured even distribution of virus throughout the product. However, the faecal oral route is not necessarily the only means of contaminating food. Virus has been detected in vomitus and in some incidents staff have vomited in the kitchen [76]. Reid *et al* suggest that vomiting could liberate over 20 million virus particles in the kitchen [76]. Aerosols could contaminate exposed food directly, or alternatively lead to environmental contamination with the potential for subsequent transfer to foods.

Even more worrying is the possibility that food could be contaminated by an airborne route. Having established that direct person-to-person transmission can occur by this means, it seems possible that the same process could lead to food contamination. This would have major implications, in that food need not have been handled by staff at all. At the moment this is still speculative, but would certainly account for observations made in some outbreaks; for example, in one large outbreak that we investigated, we believe that food was contaminated not by kitchen workers but by waitresses. It could also explain the fact that, in some outbreaks, multiple items seem to have been contaminated.

An obvious question regarding food-borne SRSV infection is how common such incidents are. On the basis of individual published reports, they would appear to be rare. However, the only way to assess this adequately is in prospective studies in which all outbreaks of food poisoning in a defined community are investigated. Unfortunately, very few such studies have been performed. Kuritsky *et al* reported their experience of 34 food-borne outbreaks investigated by the Minnesota Department of Health in a three-year period [80]. The results of their studies are shown in Table 8. Unfortunately, only two of the 12 SRSV outbreaks were virologically confirmed, the remainder being defined purely on the clinical and epidemiological criteria defined by Kaplan [44]. Despite these limitations, the findings are really most striking, in that the presumptive NLV outbreaks considerably overshadowed any of the bacterial pathogens in numerical importance. The seven outbreaks with no known cause had a different pattern to that of NLVs.

OUTBREAKS ASSOCIATED WITH MOLLUSCAN SHELLFISH

The hazards to the gastrointestinal tract of consuming molluscan shellfish are almost legendary. That molluscs should be a source of enteric pathogens is hardly surprising, in that nearly all the shellfish consumed in this country are harvested from estuaries where exposure to sewage contamination is likely. To compound this, the bivalves such as oysters, clams, cockles and mussels are filter feeders and draw large volumes of water across gill structures which act as a sieve. The process is, of course, completely undiscriminating so that enteric pathogens inevitably present in sewage are trapped and indeed concentrated.

Although gastroenteritis was commonly associated with consumption of shellfish, the investigation of outbreaks was generally unproductive bacteriologically. Thus, between 1941 and 1970 81% of outbreaks were of unknown cause [81]. The symptoms in such incidents were certainly compatible with viral gastroenteritis and in 1977 Appleton and Pereira reported finding small round viruses in a high proportion of stools from outbreaks associated with cockles [15]. The virus particles were featureless and 25–26nm diameter. Subsequent investigation suggests that the cockle agent is probably a parvovirus. Although, at the time, this

Table 8 Food poisoning outbreaks, Minnesota, January 1981– December 1983

SRSV/NLV	12
Salmonella sp	4
Clostridium perfringens	3
Campylobacter jejuni	2
Staphyloccus aureus	1
Bacillus cereus	1
Enteropathogenic *E coli*	1
Shigella sp	1
Hepatitis A virus	1
Unknown	7

From: Kuritsky *et al* [80]

seemed to be a highly significant finding, as discussed previously, the pathogenic role of the cockle agent and similar featureless viruses has not been adequately established.

Two years after the cockle agent was described, Murphy *et al* reported their findings in a huge nationwide oyster-associated outbreak in Australia. Featureless viruses similar to cockle agent were detected by EM, but in addition 27–30nm SRSV particles were also detected [82]. These were shown to be antigenically related to Norwalk agent by IEM and RIA [83]. In addition, paired sera showed a serological response to Norwalk by both IEM and RIA [83]. Thus the involvement of SRSVs in shellfish food poisoning was established.

Although there have been comparatively few published outbreaks of gastroenteritis related to molluscs in which SRSVs have been detected (Table 9), there is a widespread impression that this is an extremely common problem. Thus, Morse *et al* described 103 well-documented outbreaks in New York state in an eight-month period [86]. Sockett *et al* reviewed molluscan shellfish outbreaks reported to the CDSC between 1965 and 1983 [81]. Unfortunately, they did not subdivide small round viruses according to the currently-accepted scheme, but it is clear that viruses are the major agents found. Appleton reported experience of investigating six outbreaks of shellfish-associated gastroenteritis over a period of four months. SRSVs were detected in stool samples in three of these incidents [88].

The absence of bacterial pathogens from the list of Sockett *et al* [81] is quite striking and represents a considerable public health triumph. Salmonella infections, including typhoid fever, were a significant hazard from shellfish in the early part of this century; this led to the introduction of depuration processes to cleanse the shellfish. After harvesting, the live shellfish are put into tanks of clean water under controlled conditions such that they continue filter feeding and thereby self-purify. This process, if correctly carried out, is remarkably successful for removal of bacterial pathogens but is much less so for viruses, as witnessed by the numbers of outbreaks that occur in spite of depuration. This may be because virus particles are sequestered from the tissues of

Table 9 Norwalk-like virus/SRSV outbreaks associated with molluscs

Number affected	Mollusc involved	Diagnosis	Reference
2000+	Oysters	EM, RIA, antibody detection	[82] [83]
6/13	Oysters	Antibody detection	[84]
181/451	Oysters	EM	[85]
1017	Oysters + clams	Antibody detection	[86]
28/64	Clams		[87]

the digestive tract [89]. This also means that bacteriological monitoring of shellfish, eg by *E coli* counts, correlates poorly with the likelihood of viral contamination. One is thus in the uneasy position of having no way of assessing the infectivity of batches of shellfish.

The problem is most acute with those molluscs traditionally eaten uncooked, such as oysters. Provided the cooking process is correctly applied, cockles, mussels and winkles should be rendered virologically safe. If the safety of eating uncooked molluscs is to be improved, considerable research and expenditure will have to be put into improving the conditions in which shellfish are cultured and into developing methods of cleansing that are effective for viruses. There is also an urgent need for techniques by which the virological safety of batches of shellfish can be assessed.

SPORADIC DIARRHOEA/GASTROENTERITIS
The Cleveland studies of illnesses occurring in families, referred to above [1], demonstrated that acute episodes of self-limited diarrhoeal illness are second only to respiratory infections as causes of illness and episodes typical of winter vomiting disease feature prominently among these.

It might, therefore, be predicted that SRSVs feature prominently as a cause of sporadic gastroenteritis. Published studies have produced widely differing overall results, depending particularly on whether the study involved infants and pre-school children or adults, and also whether the cases were severe enough to be hospitalised or were seen in general practice (Table 10). However, in none of the studies do SRSVs feature as a significant cause. The highest positive rate (3.6%) was reported by Ellis *et al* in children under the age of two years admitted to hospital [97]. However, very interestingly, this study included a comparison group of children admitted to hospital with predominantly respiratory symptoms and 3.7% of these controls were also excreting SRSVs. If the evidence were not so strong that SRSVs cause WVD, one might become suspicious in the face of these studies.

It would also be tempting to believe from the published studies that SRSVs are an uncommon cause of sporadic illness. This would certainly be backed up by the paucity of laboratory reports of SRSVs compared with the huge numbers of reports of salmonellas, campylobacters and rotaviruses. However, only a tiny proportion of the total cases of SRSV are diagnosed.

Two sets of data give a clue to the true prevalence of SRSVs in the community: the first consists of results from family outbreaks. Thus, in our own studies we have looked at 30 incidents of food poisoning affecting families. In many of these incidents it was clear from the wide range of incubation periods after eating the food that illness was most unlikely to have been caused by food. However, the association in the

Table 10 Prevalence of viruses in published series of sporadic gastroenteritis

Population studied	Age range	No studied	Laboratory results						Reference
			Rota	SRSV	Adeno	Calici	Astro	B*act	
1 Hospital admissions	0–2 years	183	55	14†	11	–	26	34	[90]
2 Hospital admissions	Children	669	385	75†	86	–	19	?	[91]
3 Hospital admissions	0–13 years	150	74	–	3	2	–	23	[92]
4 Hospital admissions	Adults	71	3	2	–	–	–	34	[93]
5 Paediatric clinic	Children	361	58	–	11	–	–	82	[94]
6 Hospital admissions	Children	1537	438	23	67	–	–	?	[95]
7 General practice	All ages	73	13	–	1	–	–	14	[96]
8 Hospital admissions	0–2 years	447	153	16	77	11	16	82	[97]
9 Hospital admissions/ outpatients	0–13 years	416	168	3?	32	?	?	58	[98]
10 General practice	0–14 years	143	28	–	10	2	10	15	[99]

*Bact, all bacterial pathogens isolated
†Small round viruses recorded without subdivision

victims' minds had ensured that the episodes were reported and investigated. No bacterial pathogens were identified in any of the stool samples received, but SRSVs were detected in at least one patient in 10/30 incidents. Similarly Pickering *et al*, in a prospective study of diarrhoea in families containing infants, provided serological evidence suggesting that four (25%) of 16 family outbreaks were associated with Norwalk [100]. The second piece of evidence comes from seroepidemiological studies, which show that more than 50% of adults have serological evidence of past exposure [101–103].

The most important reason for the underreporting of sporadic cases is the mild nature of the illness: most cases recover without consulting a doctor. Even among those who do present, GPs are unlikely to investigate the patient since the clinical syndrome is so characteristic. Finally, even in those cases which are investigated, electron microscopy is so insensitive that only a small proportion (10–20%) of the cases investigated will be confirmed. Thus one arrives at the paradoxical situation that SRSVs, despite being one of the commonest causes of gastroenteritis, are among the least commonly diagnosed.

Prevention and control of outbreaks

The first aspect of prevention must be to prevent person-to-person transmission. In view of the information now emerging about airborne transmission, this may be very difficult. However, one should certainly emphasize the importance of personal hygiene, particularly hand washing. Removal of the environmental contamination generated by vomiting or profuse diarrhoea is also important.

Prevention of food contamination falls into two distinct areas, molluscan shellfish and other foods that have been secondarily contaminated. Contamination of shellfish almost certainly occurs as a result of culturing in sewage-polluted waters. Resiting of shellfish beds and/or reduction of sewage pollution are clearly the most fundamental steps. Another alternative is to move shellfish to cleaner water for a period of several weeks to reduce the level of contamination. Consideration has also been given to removal of viruses from sewage effluent by treatments such as chlorination, ozone or ultraviolet light.

The second line of defence against infection from shellfish is heat treatment. Experiments performed on heat inactivation of hepatitis A in shellfish have led to the introduction of a continuous flow process for cockles which ensures that the centre of the meat reaches 90°C and is maintained for 90 seconds [104]. Although there is no direct evidence, it is likely that this procedure would also be effective for SRSVs. The problem remains, however, that oysters which are the commonest vehicle are eaten without any heating at all. In the kitchen, care should be taken to avoid cross-contamination to other food.

Since other implicated foods have usually been contaminated by food handlers, attention must be focused on the food handler. The most fundamental aspect is to have a system which ensures that food handlers who develop acute gastrointestinal infection do not continue to work. They should be able to report such illness and take time off work without financial penalty. Despite the evidence suggesting that, in outbreaks, food handlers have contaminated foods 48 hours after recovery, it would be unrealistic, for sporadic cases, to expect exclusion to be enforced beyond the point when full symptomatic recovery has occurred.

Allied to this must be the enforcement of a scrupulously high standard of personal hygiene for food handlers. This does of course necessitate adequate toilet and hand washing facilities which are regularly cleaned and serviced with toilet paper, soap and paper towels/hand driers. Finally, attempts should be made to rethink food handling practices to try to reduce the amount of handling of cooked foods or foods such as salads or fruit which are eaten uncooked. Consideration has been given to the use of rubber gloves, but it is debatable whether this would be workable or indeed useful.

If preventive measures fail and an outbreak occurs, then control measures must be instituted. In the case of institutional outbreaks involving person-to-person spread this is often very difficult, particularly as outbreaks are often well advanced before they are reported. However, measures should be implemented to try to reduce further spread. These may include isolation or cohorting of affected individuals. If isolation is not possible in a hospital outbreak, then the movements of symptomatic patients should be restricted. Similarly, transfer of staff from ward to ward should be avoided since this may spread infection to other wards. The importance of hand washing, particularly for staff, and environmental decontamination after vomiting or uncontrolled diarrhoea should be stressed.

In food-borne outbreaks, the measures described above may also apply but in addition, if the outbreak is ongoing, potentially contaminated items still present in the kitchen such as salads or cakes should be removed and steps taken to prevent further contamination. This will involve thorough cleaning of work surfaces to remove any possible environmental contamination. In addition, it may be prudent to buy in 'high risk' foods such as salads and sandwiches until the outbreak is clearly at an end. In the face of an established outbreak it is essential to ensure affected food handlers are excluded until 48 hours after recovery.

The problems posed in trying to control SRSV outbreaks have been most clearly demonstrated by incidents on cruise liners [41,105]. The relatively crowded conditions on board, with sharing of toilets, favour person-to-person transmission. The rapid turnround between cruises, with regular influxes of new susceptibles, has resulted in repeated outbreaks on successive cruises.

Conclusions

Eighteen years after the first reports of the original Norwalk agent, we still know very little about the group of viruses known in the UK as SRSVs. However, it is now clear that they are not parvoviruses and appear to fit best among caliciviridae.

Laboratory diagnosis of SRSVs remains unsatisfactory. The only method routinely available in the UK is electron microscopy which is expensive, labour intensive and insensitive. Sensitive immunoassays are urgently required.

SRSVs are remarkable for the mildness of the illness they induce throughout the age spectrum. As a result, despite evidence that they are widely prevalent in the community, they are not a significant cause of gastroenteritis severe enough to be investigated virologically.

The importance of SRSVs lies in their tendency to cause epidemics of gastroenteritis, and indeed they are probably the most common pathogen in this situation. Outbreaks may result from person-to-person spread in an institution or from a common source. A range of common sources has been identified, including food, drinking water and swimming in recreational waters. Food-borne outbreaks have been linked to either contamination of food by infected food handlers or contamination of shellfish harvested from polluted waters. These outbreaks have significant public health and economic implications and further study is required to try to reduce the risk of infection from these sources.

Acknowledgements

I would like to thank Dr A Curry for supplying the electron micrographs and Mrs S Montgomery for typing the manuscript.

References

1   Hodges RG, McCorkle LP, Badger GF, *et al*. A study of illness in a group of Cleveland families, XI. The occurrence of gastro-intestinal symptoms. *Am J Hyg* 1956, **64**, 349–56.

2   Zahorsky J. Hyperemesis hiemis or winter vomiting disease. *Arch Pediat* 1929, **46**, 391–5.

3   Miller R, Raven M. Epidemic nausea and vomiting. *BMJ*, 1936i, 1242–4.

4   Gray JD. Epidemic nausea and vomiting. *BMJ*, 1939i, 209–11.

5   Bradley WH. Epidemic nausea and vomiting. *BMJ*, 1943i, 309–13.

6   Reimann HA, Hodges JH, Price AH. Epidemic diarrhea, nausea and vomiting of unknown cause. *JAMA* 1945, **127**, 1–6.

7   Ingalls TH, Britten SA. Epidemic diarrhea in a school for boys. *JAMA* 1951, **146**, 710–2.

8   Jordan WS, Gordon I, Dorrance WR. A study of illness in a group of Cleveland families, VII. Transmission of acute non-bacterial gastroenteritis to volunteers: evidence for two different etiologic agents. *J Exp Med* 1953, **98**, 461–75.

9   Adler JL, Zickl R. Winter vomiting disease. *J Inf Dis* 1969, **119**, 668–73.

10 Dolin R, Blacklow NR, DuPont H, *et al*. Transmission of acute infectious nonbacterial gastroenteritis to volunteers by oral administration of stool filtrates. *J Infect Dis* 1971, **123**, 307–12.

11 Kapikian AZ, Wyatt RG, Dolin R, Thornhill TS, Kalica AR, Chanock RM. Visualization by immune electron microscopy of a 27-nm particle associated with acute infectious nonbacterial gastroenteritis. *J Virol* 1972, **10**, 1075–81.

12 Caul EO, Appleton H. The electron microscopical and physical characteristics of small round human fecal viruses: an interim scheme for classification. *J Med Virol* 1982, **9**, 257–65.

13 Paver WK, Caul EO, Ashley CR, Clarke SKR. A small virus in human faeces. *Lancet* 1973, **i**, 237–40.

14 Appleton H, Buckley M, Thom BT, Cotton JL, Henderson S. Virus-like particles in winter vomiting disease. *Lancet* 1977, **i**, 409–11.

15 Appleton H, Pereira MS. A possible viral aetiology in outbreaks of food-poisoning from cockles. *Lancet* 1977, **i**, 780–1.

16 Christopher PJ, Grohmann GS, Millsom RH, Murphy AM. Parvovirus gastroenteritis. A new entity for Australia. *Med J Aust* 1978, **1**, 121–4.

17 Caul EO. Small round human fecal viruses. In: Pattison JR, ed. Parvoviruses and human disease. Boca Raton: CRC Press, 1988, 139–63.

18 Kurtz JB, Lee TW. Astroviruses: human and animal. In: Bock G, Whelan J, eds. Novel diarrhoea viruses. Chichester: Wiley, 1987, 92–107. (CIBA Foundation Symposium No 128)

19 Madeley CR, Cosgrove BP. Caliciviruses in man. *Lancet* 1976, **i**, 199–200.

20 Terashima H, Chiba S, Sakuma Y, *et al*. The polypeptide of a human calicivirus. *Arch Virol* 1983, **78**, 1–7.

21 Cubitt WD, McSwiggan DA, Moore W. Winter vomiting disease caused by calicivirus. *J Clin Path* 1979, **32**, 786–93.

22 Chiba S, Yasuhiko Y, Kogasaka R, *et al*. An outbreak of gastro-enteritis associated with calicivirus in an infant home. *J Med Virol* 1979, **4**, 249–54.

23 Nakata S, Chiba S, Terashima H, Yokoyama T, Nakao T.

Humoral immunity in infants with gastroenteritis caused by human calicivirus. *J Infect Dis* 1985, **152**, 274–9.

24 Thornhill TS, Wyatt RG, Kalica AR, Dolin R, Chanock RM, Kapikian AZ. Detection by immune electron microscopy of 26 to 27-nm viruslike particles associated with two family outbreaks of gastroenteritis. *J Infect Dis* 1977, **135**, 20–7.

25 Taniguchi K, Urasawa S, Urasawa T. Virus-like particle, 35 to 40 nm, associated with an institutional outbreak of acute gastroenteritis in adults. *J Clin Micro* 1979, **10**, 730–6.

26 Kogasaka R, Nakamura S, Chiba S, *et al*. The 33- to 39-nm virus-like particles, tentatively designated as Sapporo agent, associated with an outbreak of acute gastroenteritis. *J Med Virol* 1981, **8**, 187–93.

27 Dolin R, Reichman RC, Roessner KD, *et al*. Detection by immune electron microscopy of the snow mountain agent of acute viral gastroenteritis. *J Infect Dis* 1982, **146**, 184–9.

28 Greenberg HB, Valdesuso JR, Kalica AR, *et al*. Proteins of Norwalk virus. *J Virol* 1981, **37**, 994–9.

29 Cubitt WD, Blacklow NR, Herrmann JE, Nowak NA, Nakata S, Chiba S. Antigenic relationships between human caliciviruses and Norwalk virus. *J Infect Dis* 1987, **156**, 806–14.

30 Sakuma Y, Chiba S, Kogasaka R *et al*. Prevalence of antibody to human calicivirus in general population of Northern Japan. *J Med Virol* 1981, **7**, 221–5.

31 Agus SG, Dolin R, Wyatt RG, Tousimis AJ, Northrup RS. Acute infectious nonbacterial gastroenteritis: Intestinal histopathology. *Ann Int Med* 1973, **79**, 18–25.

32 Blacklow NR, Dolin R, Fedson DS, *et al*. Acute infectious non-bacterial gastroenteritis: Etiology and pathogenesis. *Ann Int Med* 1972, **76**, 993–1008.

33 Parrino TA, Shreiber DS, Trier JS, Kapikian AZ, Blacklow NR. Clinical immunity in acute gastroenteritis caused by Norwalk agent. *NEJM* 1977, **297**, 86–9.

34 Curry A. Personal communication.

35 Roberts JL. Personal communication.

36 Rubenstein AS, Miller MF. Comparison of an enzyme immuno-assay with electron microscopic procedures for detecting rotavirus. *J Clin Micro* 1982, **15**, 938–44.

37 Wilkinson N, Ashley C, Caul EO. In preparation.

38 Greenberg HB, Wyatt RG, Valdesuso J et al. Solid-phase microtiter radioimmunoassay for detection of the Norwalk strain of acute nonbacterial epidemic gastroenteritis virus and its antibodies. *J Med Virol* 1978, **2**, 97–108.

39 Gary GW, Kaplan JE, Stone SE, Anderson LJ. Detection of

Norwalk virus antibodies and antigen with a biotin-avidin immunoassay. *J Clin Micro* 1985, **22**, 274–8.

40 Herrmann JE, Nowak NA, Blacklow NR. Detection of Norwalk virus in stools by enzyme immunoassay. *J Med Virol* 1985, **17**, 127–33.

41 Gunn RA, Terranova WA, Greenberg HB, *et al.* Norwalk virus gastroenteritis aboard a cruise ship: an outbreak on five consecutive cruises. *Am J Epidemiol* 1980, **112**, 820–7.

42 Herrmann JE, Kent GP, Nowak NA, Brondum J, Blacklow NR. Antigen detection in the diagnosis of Norwalk virus gastroenteritis. *J Infect Dis* 1986, **154**, 547–8.

43 Kaplan JE, Feldman R, Campbell DS, Lookabaugh C, Gary GW. The frequency of a Norwalk-like pattern of illness in outbreaks of acute gastroenteritis. *Am J Public Health* 1982, **72**, 1329–32.

44 Kaplan JE, Gary GW, Baron RC, *et al.* Epidemiology of Norwalk gastroenteritis and the role of Norwalk virus in outbreaks of acute nonbacterial gastroenteritis. *Ann Int Med* 1982, **96**, 756–61.

45 Greenberg HB, Valdesuso J, Yolken RH, *et al.* Role of Norwalk virus in outbreaks of nonbacterial gastroenteritis. *J Infect Dis* 1979, **139**, 564–8.

46 Kaplan JE, Schonberger LB, Varano G, Jackman N, Bied J, Gary GW. An outbreak of acute nonbacterial gastroenteritis in a nursing home. Demonstration of person-to-person transmission by temporal clustering of cases. *Am J Epidemiol* 1982, **116**, 940–8.

47 Gustafson TL, Kobylik B, Hutcheson RH, Schaffner W. Protective effect of anticholinergic drugs and psyllium in a nosocomial outbreak of Norwalk gastroenteritis. *J Hosp. Infect* 1983, **4**, 367–74.

48 Riordan T, Wills A. An outbreak of gastroenteritis in a psychogeriatric hospital associated with a small round structured virus. *J Hosp Infect* 1986, **8**, 296–9.

49 Leers W-D, Kasupski G, Fralick R, Wartman S, Garcia J, Gary W. Norwalk-like gastroenteritis epidemic in a Toronto hospital. *Am J Publ Health* 1987, **77**, 291–5.

50 Sawyer LA, Murphy JJ, Kaplan JE, *et al.* 25 to 30-nm Virus particle associated with a hospital outbreak of acute gastroenteritis with evidence for airborne transmission. *Am J Epidemiol* 1988, **127**, 1261–71.

51 Riordan T. Unpublished observations.

52 Greenberg HB, Wyatt RG, Kapikian AZ. Norwalk virus in vomitus. *Lancet* 1979, **i**, 55.

53 Morens DM, Zweighaft RM, Vernon TM, *et al.* A waterborne outbreak of gastroenteritis with secondary person-to-person spread. Association with a viral agent. *Lancet* 1979, **i**, 964–6.

54 Wilson R, Anderson LJ, Holman RC, Gary GW, Greenberg HB.

Waterborne gastroenteritis due to the Norwalk agent: clinical and epidemiologic investigation. *Am J Publ Health* 1982, **72**, 72–4.

55   Taylor JW, Gary GW, Greenberg HB. Norwalk-related viral gastroenteritis due to contaminated drinking water. *Am J Epidemiol* 1981, **114**, 584–92.

56   Goodman RA, Buchler JW, Greenberg HB, McKinley TW, Smith JD. Norwalk gastroenteritis associated with a water system in a rural Georgia community. *Arch Env Health* 1982, **37**, 358–60.

57   Kaplan JE, Goodman RA, Schonberger LB, Lippy EC, Gary GW. Gastroenteritis due to Norwalk virus: an outbreak associated with a municipal water system. *J Infect Dis* 1982, **146**, 190–7.

58   Galbraith NS, Barrett NJ, Stanwell-Smith R. Water and disease after Croydon: a review of water-borne and water-associated disease in the United Kingdom 1937–1986. *J Inst Water Environ Manage* 1987, **1**, 7–20.

50   Slade JS. Viruses and bacteria in a chalk well. *Water Sci Technol* 1985, **17**, 111–25.

60   Keswick BH, Satterwhite TK, Johnson PC, *et al.* Inactivation of Norwalk virus in drinking water by chlorine. *Appl Env Micro* 1985, **50**, 261–4.

61   Baron RC, Murphy FD, Greenberg HB, *et al.* Norwalk gastro-intestinal illness. An outbreak associated with swimming in a recreational lake and secondary person-to-person transmission. *Am J Epidemiol* 1982, **115**, 163–72.

62   Koopman JS, Eckert EA, Greenberg HB, Strohm BC, Isaacson RE, Monto AS. Norwalk virus enteric illness acquired by swimming exposure. *Am J Epidemiol* 1982, **115**, 173–7.

63   Cabelli VJ, Dufour AP, McCabe LJ, Levin MA. Swimming-associated gastroenteritis and water quality. *Am J Epidemiol* 1982, **115**, 606–16.

64   Kappas KD, Marks JS, Holman RC, *et al.* An outbreak of Norwalk gastroenteritis associated with swimming in a pool and secondary person-to-person transmission. *Am J Epidemiol* 1982, **116**, 834–9.

65   Dufour AP. Bacterial indicators of recreational water quality. *Can J Public Health* 1984, **75**, 49–56.

66   Appleton H, Palmer SR, Gilbert RJ. Foodborne gastroenteritis of unknown aetiology: a virus infection? *BMJ* 1981, **282**, 1801–2.

67   Griffin MR, Surowiec JJ, McCloskey DI, *et al.* Foodborne Norwalk virus. *Am J Epidemiol* 1982, **115**, 178–84.

68   Pether JVS, Caul EO. An outbreak of food-borne gastroenteritis in two hospitals associated with a Norwalk-like virus. *J Hyg (Camb)* 1983, **91**, 343–50.

69   Kuritsky JN, Osterholm MT, Greenberg HB, *et al.* Norwalk gastroenteritis: A community outbreak associated with bakery

product consumption. *Ann Int Med* 1984, **100**, 519–21.

70   Riordan T, Craske J, Roberts JL, Curry A. Foodborne infection by a Norwalk like virus (small round structured virus). *J Clin Path* 1984, **37**, 817–20.

71   Lieb S, Gunn RA, Medina R, *et al*. Norwalk virus gastroenteritis. An outbreak associated with a cafeteria at a college. *Am J Epidemiol* 1985, **121**, 259–68.

72   White KE, Osterholm MT, Mariotti JA, *et al*. A foodborne outbreak of Norwalk virus gastroenteritis. Evidence for post-recovery transmission. *Am J Epidemiol* 1986, **124**, 120–6.

73   Guest C, Spitalny KC, Madore HP, *et al*. Foodborne snow mountain agent gastroenteritis in a school cafeteria. *Paediatrics* 1987, **79**, 559–63.

74   Iversen AM, Gill M, Bartlett CLR, Cubitt WD, McSwiggan DA. Two outbreaks of foodborne gastroenteritis caused by a small round structured virus: Evidence of prolonged infectivity in a food handler. *Lancet* 1987, **ii**, 556–8.

75   Talbot GH, Broom EA, Collins M, *et al*. Outbreak of viral gastroenteritis – Pennsylvania and Delaware. *MMWR* 1987, **36**, 709–11.

76   Reid JA, Caul EO, White DG, Palmer SR. Role of infected food handler in hotel outbreak of Norwalk-like viral gastroenteritis: implications for control. *Lancet* 1988, **ii**, 321–3.

77   Heun EM, Vogt RL, Hudson PJ, Parren S, Gary GW. Risk factors for secondary transmission in households after a common-source outbreak of Norwalk gastroenteritis. *Am J Epidemiol* 1987, **126**, 1181–6.

78   Fleissner ML, Herrmann JE, Booth JW, Blacklow NR, Nowak NA. Role of Norwalk virus in two foodborne outbreaks of gastro-enteritis: definitive virus association. *Am J Epidemiol* 1989, **129**, 165–72.

79   Thornhill TS, Kalica AR, Wyatt RG, Kapikian AZ, Chanock RM. Pattern of shedding of the Norwalk particle in stools during experimentally induced gastroenteritis in volunteers as deter-mined by immune electron microscopy. *J Infect Dis* 1975, **132**, 28–34.

80   Kuritsky JN, Osterholm MT, Korlath JA, White KE, Kaplan JE. A statewide assessment of the role of Norwalk virus in outbreaks of food-borne gastroenteritis. *J Infect Dis* 1985, **151**, 568.

81   Sockett PN, West PA, Jacob M. Shellfish and public health. *PHLS Microbiology Digest* 1985, **2**, 29–35.

82   Murphy AM, Grohmann GS, Christopher PJ *et al*. An Australia-wide outbreak of gastroenteritis from oysters caused by Norwalk virus. *Med J Aust* 1979, **2**, 329–33.

83   Grohmann GS, Greenberg HB, Welch BM, Murphy AM. Oyster-

associated gastroenteritis in Australia: The detection of Norwalk virus and its antibody by immune electron microscopy and radio-immunoassay. *J Med Virol* 1980, **6**, 11–19.

84   Gunn RA, Janowski HT, Lieb S, Prather EC, Greenberg HB. Norwalk virus gastroenteritis following raw oyster consumption. *Am J Epidemiol* 1982, **115**, 348–51.

85   Gill ON, Cubitt WD, McSwiggan DA, Watney BM, Bartlett CLR. Epidemic of gastroenteritis caused by oysters contaminated with small round structured viruses. *BMJ* 1983, **287**, 1532–4.

86   Morse DL, Guzewich JJ, Hanrahan JP, *et al*. Widespread outbreaks of clam- and oyster-associated gastroenteritis. Role of Norwalk virus. *NEJM* 1986, **314**, 678–81.

87   Anon. Raw clam associated gastroenteritis – Suffolk county. *State Dept of Health Food Protection Bull* 1986, **2**, no 6.

88   Appleton H. Small round viruses: classification and role in food-borne infections. In: Bock G, Whelan J, eds. Novel diarrhoea viruses. Chichester: Wiley, 1987, 108–25. (Ciba Foundation Symposium No. 128)

89   Metcalf TG. Indicators of viruses in shellfish. In: Berg G, ed. Indicators of viruses in water and food. Michigan: Ann Arbor Science, 1978, 383–415.

90   Madeley CR, Cosgrove BP, Bell EJ, Fallon RJ. Stool viruses in babies in Glasgow. 1. Hospital admissions with diarrhoea. *J Hyg Camb* 1977, **78**, 261–73.

91   Middleton PJ, Szymanski MT, Petric M. Viruses associated with acute gastroenteritis in young children. *Am J Dis Child* 1977, **131**, 733–7.

92   Lewis HM, Parry JV, Davies HA, *et al*. A year's experience of the rotavirus syndrome and its association with respiratory illness. *Arch Dis Child* 1979, **54**, 339–46.

93   Jewkes J, Larson HE, Price AB, Sanderson PJ, Davies HA. Aetiology of acute diarrhoea in adults. *Gut* 1981, **22**, 388–92.

94   Sekine S, Yamada S, Matsushita S, *et al*. Virological and bacteriologic studies on the cause of acute diarrheal diseases in a paediatric clinic in Tokyo. *Ann Rep Tokyo Metr Res Lab Public Health* 1982, **33**, 59–65.

95   Brandt CD, Kim HW, Rodriguez WJ, *et al*. Pediatric viral gastroenteritis during eight years of study. *J Clin Micro* 1983, **18**, 71–8.

96   Rousseau SA. Investigation of acute gastroenteritis in general practice – relevance of newer laboratory methods. *J Royal Coll Gen Pract* 1983, **33**, 514–6.

97   Ellis ME, Watson B, Mandal BK, *et al*. Micro-organisms in gastroenteritis. *Arch Dis Child* 1984, **59**, 848–55.

98   Uhnoo I, Olding-Stenkvist E, Kreuger A. Clinical features of acute gastroenteritis associated with rotavirus, enteric adeno-

viruses and bacteria. *Arch Dis Child* 1986, **61**, 732–8.

99 Issacs D, Day D, Crook S. Childhood gastroenteritis: a population study. *BMJ* 1986, **293**, 545–6.

100 Pickering LK, DuPont HL, Blacklow NR, Cukor G. Diarrhea due to Norwalk virus in families. *J Infect Dis* 1982, **146**, 116–7.

101 Blacklow NR, Cukor G, Bedigian MK, *et al*. Immune response and prevalence of antibody to Norwalk enteritis virus as determined by radioimmunoassay. *J Clin Micro* 1979, **10**, 903–9.

102 Echeverria P, Burke DS, Blacklow NR, Cukor G, Charoenkul C, Yanggratoke S. Age-specific prevalence of antibody to rotavirus, *Escherichia coli* heat-labile enterotoxin, Norwalk virus and hepatitis A virus in a rural community in Thailand. *J Clin Micro* 1983, **17**, 923–5.

103 Ryder RW, Singh N, Reeves WC, Kapikian AZ, Greenberg HB, Sack RB. Evidence of immunity induced by naturally acquired rotavirus and Norwalk virus infection on two remote Panamanian Islands. *J Infect Dis* 1985, **151**, 99–105.

104 Early JC, Nicholson FJ. Specification for a model cockle processing plant. MAFF Torry Research Station, 1987. (Torry document 2106)

105 Caul EO. Viral gastroenteritis. In: Reeves DS, Geddes AM, eds. Recent advances in infection **3**. Edinburgh: Churchill Livingstone, 1989: 105–17.

Chapter 5

HTLV I

P White

The Retroviridae family is characterised by the presence of reverse transcriptase, an RNA-dependent DNA polymerase first described in 1970 by Baltimore [1] and Temin [2]. Human T cell leukaemia virus 1 (HTLV I) was the first human retrovirus to be isolated. It is a member of the Oncovirinae subfamily with the morphological features of a type C oncovirus and therefore HTLV I is only distantly related to human immunodeficiency virus (HIV).

History

Retroviruses may be regarded by some as a new family of viruses that have only come to the fore in the 1980s. In fact, members of the family were among the first viruses to be recognised. In 1908 Ellerman and Bang transferred a leukosis of chickens from diseased birds to healthy ones in cell-free filtrates of tissue and in 1911 Rous transmitted a sarcoma of chickens to susceptible fowl, also via cell-free filtrates [3]. Rous postulated that this was due to a minute parasitic organism, and this led to the recognition that viruses play a major role in the aetiology of many neoplastic processes in numerous animal species including primates.

The discovery of T-cell growth factor in the late 1970s [4] allowed the establishment of continuously-growing T-cell lines from normal individuals and from patients with T-cell neoplasms. In 1980 workers in Gallo's laboratory isolated the first human retrovirus from a 28-year-old Negro who had been diagnosed initially as having cutaneous T-cell lymphoma or mycosis fungoides [5]. The diagnosis was revised later when it become apparent that he had an aggressive T-cell leukaemia [6]. Two cell lines were established, one from a lymph node biopsy and the other from peripheral blood cells, and a retrovirus with type C morphology was isolated. It was called human T-cell leukaemia virus, later modified to human T-cell leukaemia virus type I after the discovery of other human retroviruses.

In 1977, Takatsuki and colleagues described a new clinical syndrome, adult T-cell leukaemia (ATL), in a group of patients from a localised area of the islands of south west Japan [7]. In 1981, Hinuma [8] found that sera from all patients with this new syndrome reacted in an indirect immunofluorescence test with a cell line derived from a patient with ATL. Electron microscopy showed that these cells contained a type C retrovirus which Hinuma called adult T-cell leukaemia virus (ATLV). In areas where ATL was common Hinuma reported that 26% of healthy individuals had anti-ATLV compared to only 2% in non-endemic areas. Gotoh [9] and Yoshida [10] have shown that all anti-ATLV positive individuals have T-cells infected with ATLV in their peripheral blood. Hybridisation studies have shown that HTLV I and ATLV are the same virus [11].

Clinical features

HAEMATOLOGICAL

Most individuals infected with HTLV I have no signs or symptoms of disease [8] but independent observations in Japan and the United States strongly suggest that HTLV I is aetiologically related to ATL. The features of ATL include onset in adult life and the presence in the peripheral blood of pleomorphic leukaemic cells with deformed nuclei [7,12]. ATL cells usually carry the surface markers CD2, CD3, CD4, CD25 and HLA-DR, but not CD8, which suggests that they originate from the CD4 (helper) subset of T-cells. Hepatosplenomegaly, lymphadenopathy and hypercalcaemia are commonly observed. A variety of skin lesions may be present such as systemic erythroderma or violaceous papules, nodules or plaques and it is these features that were responsible for the initial confusion with mycosis fungoides and the Sezary syndrome. The disease usually follows an aggressive course and is resistant to current chemotherapeutic regimes.

In a study of 95 consecutive patients with non-Hodgkin's lymphoma in Jamaica, Gibbs [13] found that 52 (55%) had anti-HTLV I compared to 5.4% in a study of persons without malignancy [14]. Phenotypic classification of 36 patients showed that 27 had adult T-cell leukaemia-lymphoma (ATLL), of which 24 had anti-HTLV I. The median survival time of the patients with ATLL was 17 weeks but a subgroup of nine patients followed a more chronic course, with a median survival of 81 weeks. Hypercalcaemia was identified as the single most important prognostic feature. Patients with hypercalcaemia had a median survival of only 12.5 weeks compared to 50 weeks for patients with a normal serum calcium. Hypercalcaemia was the primary cause of death in 14 patients.

Takatsuki now divides ATL into five types [12].

1 Acute – the most common type; most patients die within a few months of the onset of disease.
2 Chronic – previously called chronic T-cell lymphocytic leukaemia.
3 Smouldering – may be difficult to distinguish from the carrier state but HTLV I pro-viral DNA is often integrated.
4 Crisis – following progression from chronic or smouldering ATL.
5 Lymphoma – previously considered to be a T-cell malignant lymphoma.

In addition, there are intermediate types that cannot be allocated to any of these groups. It appears that there is a broad spectrum of T-cell malignancy associated with HTLV I. Support for this comes from the studies of Gibbs [14] and Blattner [15] in Jamaica. There is clearly a need to establish a comprehensive nomenclature for HTLV I-associated lymphoid malignancies.

NEUROLOGICAL

Myelopathies are common in tropical areas. Most are of unknown aetiology but are classified into two groups.

1 Tropical ataxic neuropathy, where there is a predominantly posterior column deficit with sensory ataxia.
2 Tropical spastic paraparesis (TSP), which is a slowly progressive myelopathy affecting mainly the pyramidal tracts with minimal sensory deficit.

The geographical distribution of these myelopathies suggests that environmental factors, such as nutrition, toxins or infectious agents, may play an aetiological role.

TSP was described in 1964 by Montgomery [16] in Jamaica. Symptoms usually began in the third or fourth decade and both sexes were equally affected. The onset was abrupt in 20% but in a third weakness developed first in one leg. The disease was slowly progressive, without remissions and relapses, but in many less severe cases progression ceased. Bladder dysfunction sometimes occurred early but upper limb involvement was rare. Posterior column abnormalities could be detected in half the cases and a minority had a unilateral laryngeal nerve palsy or other mononeuropathies, tremor of the tongue, bizarre asymmetrical pupillary reactions or bilateral ptosis. Tendon reflexes in the upper limb were frequently exaggerated and the jaw jerk often positive. Protein and gammaglobulin levels in the cerebrospinal fluid were frequently raised and this was accompanied by a mild lymphocytosis.

During a large retrospective survey of haematological, infectious and other diseases on the island of Martinique, anti-HTLV 1 was found in two patients with TSP. As a result, Gessain and his colleagues [17] conducted a systematic study of patients with unexplained myelopathies with and without neurological or systemic disease. The mean age of the 13 women and four men identified as cases of TSP without systemic symptoms was 55 years (range 12–66). A further five patients with TSP had systemic disease. Ten of the 17 patients in the first group (59%) and all five of the patients with systemic disease had anti-HTLV 1. Only 13 of 303 controls (4%) who included blood donors, medical staff and other neurological patients were seropositive. Gessain concluded that HTLV I may be aetiologically associated with TSP.

In 1986 Osame [18] reported six Japanese patients with a myelopathy of gradual onset and a slowly progressive course over a period of three months to 11 years before hospital admission. The patients had a spastic paraparesis with pyramidal signs associated with mild sensory and sphincter disturbances. There were no abnormalities on myelography, computerised tomography of the spine or magnetic resonance imaging. Evidence of infection with HTLV I was sought because one patient had atypical lymphocytes characteristic of ATL in both serum and cerebrospinal fluid (CSF). All six patients were found to have high titres of anti-HTLV I in both serum and CSF. Only 12 of 78 (15%) control patients with other neurological disorders were seropositive. Osame felt that HTLV I had a causal role in the myelopathy and proposed the term HTLV I-associated myelopathy (HAM). Four patients demonstrated a striking response to steroid therapy but this has not been the experience of other workers. Matsuo reported an improvement in gait, sensory and/or sphincter disturbances in 11 of 18 patients with HAM following plasmapheresis [19].

Azizuki [20] reported that the most striking post mortem findings in a case of HAM were proliferation of capillaries, perivascular cuffing with lymphocytes and loss of myelin and axons. These changes were most pronounced in the lateral and posterior columns but were also seen in the medulla, pons, cerebrum and cerebellum. Subsequent studies have reported a high prevalence of anti-HTLV I in patients with TSP/HAM (they are synonymous) from Colombia, Jamaica [21], Trinidad [22], the Ivory Coast [23] and the Seychelles [24].

In 1987 [25] Newton reported a group of 13 West Indian-born UK residents with TSP. In contrast to Montgomery's original description, all were female with a mean age of 52 years (range 41–65). They had been in the UK for 26 years (range 25–35) and had been ill for 11 years (range 1.5–20). Seven of the 11 patients tested had oligoclonal bands in the CSF and all 11 had anti-HTLV I compared to none in a group of 48 Caucasian patients with multiple sclerosis.

Gessain reported [26] evidence for intra blood-brain barrier synthesis of HTLV I-specific IgG in 14 of 19 (74%) patients with TSP. IgG oligoclonal bands were present in the CSF of 13 of these 14, supporting a causal link between HTLV I and TSP. Further evidence was provided by Bhagavati [27] in a study of patients with chronic progressive myelopathy. HTLV I DNA was detected in activated lymphocytes from three of seven anti-HTLV I-positive patients using southern blot and in all of 11 seropositive individuals by gene amplification techniques. Viral sequences were also detected in DNA extracted from T cells co-cultured with CSF from three anti-HTLV I positive patients and HTLV I antigen was present in the CSF of all of seven seropositive individuals.

The mechanism of neurotropism and neurovirulence of HTLV I is unknown. However the fact that TSP affects 10–100/100,000 of the population in tropical areas where HTLV I is endemic can only increase the public health interest in HTLV I-associated disease.

MISCELLANEOUS

It has been suggested that HTLV I has an aetiological role in a number of other diseases, the most notable being multiple sclerosis and systemic lupus erythematosis (SLE). Koprowski [28] reported that cells cultured from CSF from four of eight patients with multiple sclerosis reacted with an HTLV I RNA probe at low stringency. Despite extensive efforts by other workers this finding has not been repeated. It is already widely known that in the CSF of these patients, antibody levels to other viruses, such as measles and Epstein-Barr virus, are frequently raised, making it difficult to assess the significance of Koprowski's report.

SLE is another disease of unknown aetiology for which an infectious cause has been proposed. Reports have been published both in favour and against a causal role for HTLV I. Murphy [29] found no evidence of excess HTLV I infection in 63 Jamaican patients with SLE compared to a control group of healthy individuals.

Sugimoto [30] reported that five of six patients with HAM had a raised total cell count with an increased proportion of lymphocytes in bronchial lavage fluid. Five of the patients had normal chest X-rays and pulmonary function. The sixth patient had micronodular infiltrates on the chest X-ray and a decreased diffusion capacity. A transbronchial lung biopsy revealed a lymphocytic infiltration of predominantly T cells without the morphological features of leukaemic cells. Sugimoto concluded that HTLV I may be associated with a subclinical T-lymphocytic alveolitis.

Transmission

NATURAL TRANSMISSION

When Takatsuki described ATL he noted that the patients came from an area restricted to the islands of south west Japan, in particular Kyushu, southern Shikoku and the Kii peninsula. When Hinuma [8] tested sera from patients with ATL and from healthy adults he found that 26 of 100 healthy individuals from an ATL-endemic area had anti-HTLV I, compared to only two of 105 from a non-endemic area. It was clear from the start that the geographical distribution of this new virus was going to be interesting.

From 1980–84 an extensive serological survey was carried out by Kajiyama [31] in the main islands of Okinawa, the most south-western part of Japan. In total 3,978 serum samples were collected from healthy individuals and tested for anti-HTLV I. This represented 71% of the male population and 79% of the female. The overall prevalence rate of anti-HTLV I was 15.3%. However, the prevalence increased with age from 2.2% in those under 10 years of age to 30.1% in adults over the age of 70 years. A striking feature was that the seroprevalence was significantly higher in females (18.1%) than in males (12.2%). In both sexes the prevalence increased with age, but the increase was much greater in females than in males. In males aged 30–39 years, 8.8% were seropositive and this rose to 22.4% in those over 70 years. In females the prevalence rose steadily from 7.8% in those aged 20–29 years to 36.4% in those aged 70 years and over.

Kajiyama [32] went on to look at the position within families. Of 627 families studied, 129 of the heads of the household (index cases) had anti-HTLV I. This led to a study of 1,333 members of these families. The overall prevalence rate in families with a seropositive head was 38.5% compared to 7.7% when the head was seronegative. Again, significantly, more women were infected; the prevalence rate rose from 6.3% in the 10–19 year age group to 71.4% in the 50–59 year group. These observations confirmed that anti-HTLV I-positive individuals cluster within family units. More than half of all anti-HTLV I-positive individuals belonged to families with a high prevalence rate, although these families accounted for only 12% of the families surveyed.

The possible mode of transmission within families was then investigated [33]. Among 2,252 males and 2,397 females, 275 males and 444 females were found to be seropositive. The prevalence of anti-HTLV I was significantly higher in the children of seropositive mothers than of seropositive fathers. None of the children with a seropositive father and a seronegative mother had anti-HTLV I, compared to a seropositivity rate of 27.9% for children both of whose parents were seropositive and 19.9% where only the mother had anti-HTLV I. These results suggest that HTLV I is transmitted from mother to child but not from father to child. Data from a further study of married couples suggests that transmission occurs predominantly from husband to wife.

Further support comes from a study by Kusuhara [34] of 311 mother–child pairs over 15 years. Sixty-five (20.9%) of the mothers and 10 (3.2%) of the children had anti-HTLV I at the initial sampling; the first samples were taken from five at aged three years, one at eight years, one at 10 years and three at 11 years. None of the seronegative mothers had a seropositive child. Three of the seronegative mothers seroconverted during the study period but none of their children developed anti-HTLV I. There were no cases of HAM or ATL in the group.

These studies support two routes of natural transmission – vertically from mother to child and horizontally from husband to wife, probably by the sexual route. Presumably equal numbers of males and females are infected by the vertical route and this would explain the lack of significant difference in the prevalence between the sexes from birth to young adulthood. Vertical transmission may occur in the perinatal or postnatal period or possibly *in utero*, although it is known that HTLV I is not endogenous to the human species [10]. In later years males infected with HTLV I may transmit the virus to their sexual partners and this would account for the greater prevalence of anti-HTLV I in females belonging to the older age groups.

In order to identify the route of vertical transmission Hino [35] looked for evidence of infection with HTLV I in the cord blood of babies born to seropositive mothers and in the milk of these mothers. None of 115 cord bloods tested were positive for anti-HTLV I IgM, nor could the virus be isolated from cord lymphocytes, evidence against transmission *in utero*. In a follow-up study, two of 38 babies born to seropositive mothers seroconverted between the ages of 12 and 19 months suggesting infection in the postnatal period.

Hino found that milk from seropositive mothers contained HTLV I-infected lymphocytes [35,36]. The concentration of infected T lymphocytes in milk was estimated to be about 10^3/ml. It is known that as few as 2×10^3 infected cells are required to transmit bovine leukaemia virus. The number of HTLV I-infected cells received by different babies will vary considerably and this may be one of the reasons why not all babies born to seropositive mothers become infected.

This group provided experimental support for the role of breast feeding by establishing HTLV I infection in a marmoset following oral inoculation of HTLV I-infected cells [37] and milk from a seropositive mother [38]. Hino suggested that the titre of maternal anti-HTLV I may be a marker of the risk of transmitting HTLV I to babies. None of 11 seropositive mothers with a titre of less than 4,000 had seropositive children compared to 11 of 17 mothers with a titre of 256,000 or higher who had transmitted the virus to their child. HTLV I antigen was detected by culture in the blood in a higher proportion of mothers with a titre of greater than 4,000 than in those with a lower titre.

The evidence suggests that it is possible to prevent the endemic transmission of HTLV I from mother to child by carrier mothers refraining from breast feeding. Studies on the effectiveness of this approach are currently in progress and initial results are promising [39,40]. In addition Ando [41] reported that after freezing breast milk at $-20°C$ overnight no HTLV I-antigen positive cells were detected by cocultivation with cord lymphocytes. It is important to balance the benefits of breast feeding against the risk of transmitting HTLV I and appropriate advice may vary in different parts of the world. If the titre of maternal antibody proves to be a significant indicator of the risk of transmission it may be very useful in offering the correct advice to seropositive mothers.

IATROGENIC TRANSMISSION

Blood transfusion: The identification of HTLV I or HTLV I-associated antigen in peripheral blood lymphocytes of healthy anti-HTLV I-positive individuals raised the possibility that HTLV I could be transmitted by blood transfusion [9]. Okochi [42] reviewed retrospectively transfusion recipients on the island of Kyushu where HTLV I is endemic. From December 1981 to April 1983, 62 (1.9%) of 3,342 blood donors were anti-HTLV I positive and 33 (6%) of the 554 recipients of blood products were seropositive before transfusion. Twenty-six (63.4%) of 46 recipients who received at least one anti-HTLV I-positive unit containing cellular blood products (whole blood, packed cells, platelet concentrate) seroconverted. Blood units stored less than four days resulted in seroconversion in 13 of 15 patients, while those units stored longer did so in 12 of 25 patients, a significant difference. Seroconversion usually occurred within 20–50 days after transfusion.

None of 14 recipients who received at least one unit of anti-HTLV I-positive fresh frozen plasma or 252 recipients of anti-HTLV I-negative blood units seroconverted, which suggests that infectious HTLV I is strongly associated with a cellular component, presumably lymphocytes. This is the probable explanation for the general observation that haemophiliacs appear to have been spared HTLV I [43,44].

Okochi also commented that three young children seroconverted, since it is unusual to find anti-HTLV I in children. He postulated that a small infecting dose transmitted from mother to child produces a particular immunological state that results in a delayed antibody response, whereas the larger infecting dose from a blood transfusion results in the establishment of infection and antibody production. There may be age-related factors that alter the response to the virus.

As a result of Okochi's findings, the Red Cross Centres in Kyushu began to screen all blood donors for anti-HTLV I in February 1986. In a follow-up study of recipients of screened blood products [45], there was only one seroconversion. Investigations revealed that the patient had received one unit of platelet concentrate from a donor whose serological status had been incorrectly recorded as negative. Okochi had speculated that as many as 20,000 infections per year may have occurred in Kyushu alone as a result of blood transfusions. Natural transmission tended to occur within families but blood transfusion allowed HTLV I to be introduced into previously uninfected families.

To date no case of transfusion-associated ATL has been reported. However, insufficient time may have elapsed to exclude the possibility if there is a prolonged period of latency, particularly when experience with HIV is taken into account. In contrast, Osame [46] has reported that nine of 23 patients with HAM had received a blood transfusion six months to eight years prior to the onset of disease. Again, this response may be related to the size of the infecting dose.

Blood donors have been studied in a number of countries. In the United States, sera from almost 40,000 donors attending eight geo-graphically-distinct centres were tested [47]. Ten donors (0.025%) had anti-HTLV I and the prevalence rates ranged from 0 to 0.1% at the different centres. Most seropositive donors were in the south-eastern and south-western United States. Seven of the seropositive donors were interviewed. There were four black females, two black males and one white female. Four of the seven had either a history of intravenous drug abuse (IVDA), or had had sexual contact with a drug abuser, or both. One male had had sexual contact with an inhabitant of an HTLV I endemic area.

Tedder [43] found no seropositive individuals amongst 500 con-secutive donors in the UK nor in 440 donors selected because they had been born outside north-western Europe.

There is still a need to establish the prevalence rate of anti-HTLV I in the blood donor population in many countries. The recognition of yet another infection in which blood transfusion may play a significant role in transmission is another reminder that the benefits of blood trans-fusion are not without risk.

Drug abusers: It is not surprising that HTLV I was implicated in the search for the aetiological agent of AIDS. Since HTLV I, HTLV II and HIV are transmitted by similar pathways, it should not be unexpected that there is an increased prevalence of anti-HTLV I amongst IVDA. The first indication of this was in 1984 when Tedder [43] screened high-risk populations for evidence of infection with HTLV I and II and found that of 113 IVDA, one had a high titre of anti-HTLV I and three had high titres of anti-HTLV II. In a study of AIDS patients and others at risk, Robert-Guroff [48] identified 19 individuals with anti-HTLV I from 248 patients with AIDS or HIV-related disease. One was homo-sexual and three were heterosexual IVDA. In a later study of 56 IVDA in New York [49], the same workers found that 9% had anti-HTLV I and they noted that 11% of black IVDA were seropositive compared to only 5% of white IVDA.

Such reports have initiated studies elsewhere. Fuchs [50] did not detect anti-HTLV I in 100 IVDA in Innsbruck, Austria. However, Gradilone [51] reported that HTLV I was widespread in IVDA in Italy. He found that 32 (27%) of 120 IVDA in Rome and 19 (12%) of 164 IVDA in Naples had anti-HTLV I. Six of these individuals had only anti-HTLV I but 45 were seropositive for both HTLV I and HIV. The authors suggested that this may indicate reciprocal enhancement of infectivity by the two viruses. Robert-Guroff, however, had drawn a different conclusion from her study of IVDA in New York, where approximately half anti-HIV positive individuals had anti-HTLV I.

These studies suggest that there is a potential for HTLV I to spread among IVDA in a similar way to HIV. It appears that there may be a prolonged period of latency between infection and the onset of clinical disease, and it may be some time before the importance of infection with HTLV I among IVDA becomes clear.

Epidemiology

In a follow-up study to his original description of ATL, Takatsuki [52] paid particular attention to the birthplace of the 35 patients reported. Although most of them lived near the city of Kyoto, 22 had been born on the southern island of Kyushu and 11 of these in the Kagoshima Prefecture at the southern tip of the island. Other Japanese workers have found clusters in the southern island of Shikoku and the adjacent Kii peninsula of Honshu. Detailed tracing of birthplaces revealed that they were frequently seaside places or small islands.

The most extensive prevalence studies have been performed by Japanese workers, such as those already described by Kajiyama [31]. However, wherever HTLV I has been sought it has been found. The Caribbean has been identified as a second major endemic area with a seroprevalence rate of between 5–10% [14]. Murphy [29] found that 135 (3.2%) of 4,208 male foodhandlers and 577 (6.7%) of 8,617 female foodhandlers in Jamaica had anti-HTLV I. Blattner [15] reported that 19 (34%) of 56 Jamaicans with lymphoproliferative neoplasms had anti-HTLV I and 11 (69%) of 16 consecutive patients with non-Hodgkin's lymphoma (NHL) were seropositive. This finding was confirmed by Gibbs [13] who found that 52 (55%) of 95 consecutive patients with NHL had anti-HTLV I which suggests that in Jamaica HTLV I is a major contributor to NHL.

Information from other parts of the world is patchy, but infection with HTLV I has been detected worldwide. The prevalence in US blood donors has already been discussed.

The diagnosis of four cases of ATL in the south-eastern United States led to a seroepidemiological survey [53]. Two (2.1%) of 95 inhabitants of Georgia were seropositive; both were black. Similar pictures are emerging from elsewhere. Reeves [54] reported that 71 (4.9%) of 1,451 sera collected in Panama had anti-HTLV I.

There is little information on the position in Europe. The prevalence in Italian IVDA has been discussed. Those workers also identified an endemic area in Apulia in southern Italy, where 9% of the population was seropositive [51].

The situation in Africa is far from clear. Initial studies using an ELISA test suggested that the virus was widely dispersed in many African countries [55]. However these results have not been confirmed by other groups. Weiss, using a competitive test, failed to detect any seropositive individuals among 1,200 sera collected in African countries [56]. Similar difficulties have been experienced with sera collected from Falasha immigrants to Israel [57].

The United Kingdom

The first report of infection with HTLV I in the UK was made by Catovsky [58] in 1982. He described six patients with ATLL, all of whom were black. Four of the patients were women, five had been born in the West Indies and one in Guyana. Severe hypercalcaemia was present in five and it was felt to be an important prognostic feature. The recognition that all the patients were of Caribbean origin, as were all other cases of T-cell lymphoma-leukaemia reported outside Japan, led to sero-epidemiological studies in the West Indies and the recognition of the Caribbean basin as an endemic area.

The next report came from Tedder [43] during the search for the epidemiological agent of AIDS. He found a low level of infection with HTLV I among homosexuals and IVDA. In the same year, Greaves [59] reported a study of ATL. He found ten of 26 first-generation Caribbean immigrants with a haematopoietic malignancy had anti-HTLV I and a clinical picture compatible with ATL. Nine were female. None of a control group of 55 Caucasian patients with T-cell leukaemia had anti-HTLV I; nor did 22 patients with leukaemia or lymphoma who had emigrated from Asia, Africa or the Far East, or 55 hospital workers. However, six of 70 outpatients of Caribbean descent and one of 17 healthy controls of the same ethnic origin were seropositive. All the anti-HTLV I-positive individuals were first generation immigrants. Studies within the families of patients revealed further seropositive individuals, all born in the same place as the index case. Eight of the seropositive patients had left the West Indies 15–30 years previously, again suggesting a long latent period between infection and the manifestation of disease. These workers felt that anti-HTLV I is present in 5–10% of non-leukaemic individuals of Caribbean origin. In addition they noted that seropositive individuals were first generation immigrants, whereas those born in the UK were seronegative, suggesting that environmental factors may be important.

Subsequent reports in the UK have been spasmodic. Cunningham [60] reported a 68-year-old white woman with a T-cell lymphoma/

leukaemia and anti-HTLV I. This patient had never been outside the UK and there was no history of blood transfusions, parenteral therapy or sexual contact with persons from an endemic area: the source of her infection was a mystery. Goorney [61] reported a case of TSP in a 62-year-old Jamaican woman, resident in the UK for 35 years, who presented with a nine-year history of backache and difficulty in walking. The patient had anti-HTLV I but her husband, to whom she had been married for 35 years, was seronegative. One of the couple's five healthy children consented to be tested and was found to be seropositive. In this family vertical transmission appeared to have occurred.

The report by Newton [25] has already been discussed. The chronicity and long latent period are in marked contrast to the aggressive course of ATL reported elsewhere.

At present the prevalence of anti-HTLV I in the UK is unknown because there is no coordinated reporting system. Tables 1 and 2 show details of the anti-HTLV I positive sera received at the Virus Reference Laboratory, Colindale, from January 1987 to December 1988 (J Tosswill, personal communication). Approximately two thirds of the patients

Table 1 Anti-HTLV I positive sera, received at PHLS Virus Reference Laboratory, January 1987–December 1988

Clinical diagnosis	Total	Male	Female	Sex unknown
Tropical spastic paraparesis	23	6	16	1
T cell lymphoma/leukaemia	12	7	5	
Asymptomatic family contacts of above	3*	1	2	
No clinical information	8†	6	2	
Total	46			

*1 husband, 1 wife, 1 daughter
†includes 2 blood donors

Table 2 Country of origin of anti-HTLV I-positive individuals identified at PHLS Virus Reference Laboratory, January 1987–December 1988

West Indies	41
UK	2
Iran	1
Brazil	1
Not stated	1

with TSP are female compared to almost equal numbers of males and females with ATL. This picture is seen in other reports and may represent a difference in the pathogenesis or mode of transmission of the virus, resulting in the different clinical conditions.

The two seropositive individuals of UK origin were a married couple. The only history of travel abroad was from the husband, who had served in Africa during the second world war. His wife was healthy. Forty-one of the 46 seropositive individuals identified so far were of West Indian origin. Now that the Caribbean basin is well recognised as an endemic area, there is a danger that a false impression is created due to selective testing of those of Caribbean origin. However, it is clear that HTLV I is present in several sections of the UK population. There is a need for further studies to establish the true prevalence, and to provide firm knowledge about the prognosis associated with being seropositive.

Serological assays

In the initial serological studies of Hinuma [8], an indirect immuno-fluorescence test (IF) was used. Diluted test serum was applied to HTLV I-infected cells on a glass slide. After incubation, the slide was washed and bound anti-HTLV I detected with a fluorescein-conjugated rabbit anti-human IgG. In a limited number of tests on ten ATL patients and ten healthy adults, Hinuma found that the titres of anti-HTLV I ranged from 20 to 640 with no significant difference between patients and healthy carriers.

IF is time-consuming and subjective. Saxinger [55] developed an ELISA test for sero-epidemiological surveys. However there was a high rate of false positives, particularly when the test was used to examine sera collected on the African continent. In addition to IF, Tedder [43] described a competitive radioimmunoassay, a syncytium inhibition assay and a pseudoneutralisation assay. Ikeda [62] developed a gelatin particle agglutination test (PA) in which particles were coated in detergent-disrupted viral antigen. This test is now commercially available (Serodia-ATLA, Fujirebio Inc., Tokyo). Kobayash [63] compared two commercial anti-HTLV I kits, Serodia-ATLA and Eitest-ATL, an ELISA kit available in Japan. 2,316 serum samples from Japanese blood donors were tested by both kits and IF. It was concluded that non-specific reactions may occur in each assay and therefore positive results must be confirmed by a different technique. The results suggested that PA was the most sensitive test. White [64] compared two commercially-available kits, Serodia-ATLA and an ELISA test (du Pont Ltd), with IF and a competitive radioimmunoassay. Sera from 185 women of West African origin, 106 IVDA and 39 sera submitted for investigation of possible infection with HTLV I were examined by at least three of the assays, and most by all four. PA was found to be the most sensitive test but the least specific; IF was the most specific. Titres of anti-HTLV I measured by PA were 10–100 times that obtained in other assays. In the small group tested it appeared that patients with TSP generally had a higher titre than carriers or those with haematological disease.

There are few diagnostic facilities for the diagnosis of HTLV I infection in the UK at present. Laboratories serving a population with a high proportion of those of Caribbean or African origin may wish to do diagnostic work or undertake epidemiological studies. Both these studies concluded that the particle agglutination test was simple and sensitive but not very specific; positive results must be confirmed by another technique such as IF.

HTLV II

Kalyanaraman [65] isolated a virus closely related to HTLV I from a 37-year-old white male who had a benign T-cell leukaemia. The patient's lymphocytes had typical hairy cell morphology and expressed retroviral antigens, which were highly cross-reactive with HTLV I antigens. However the two viruses could be distinguished in a homologous competitive radioimmunoassay. It was proposed that the two viruses should be called HTLV I and II.

In 1984, Tedder [43] reported that three of the 113 IVDA had anti-HTLV II in the UK. The study of IVDA in New York [49] found that 10 of 56 (18%) had anti-HTLV II, an unexpectedly high result. Nine of these seropositive individuals were black. To date, HTLV II has not been associated specifically with any human disease or with a reservoir in a particular geographic area or ethnic group, and its significance is not clear.

Conclusion

HTLV I and II are the forgotten human retroviruses, eclipsed by the giant HIV. However, much is known of the former but little of the latter. The discovery of T-cell growth factor opened the door on the world of human retroviruses. The association of HTLV I and leukaemia has provided the opportunity to study the mechanisms of neoplastic disease. It is not known why so few individuals infected with HTLV I develop disease, nor the factors that determine whether it is an aggressive haematological condition or a chronic neurological disease.

The geographical distribution of the virus is fascinating and unexplained and has led to much historical conjecture. Knowledge of the prevalence of HTLV I in the UK is patchy. Studies are needed to establish the prevalence within different communities and to see whether the virus behaves in families in the UK as it does in Japanese families. It has been established that HTLV I is readily transmitted by blood transfusion and probably in breast milk. Future control of HTLV I may require screening of some blood products and avoidance of breast feeding by anti-HTLV I-positive mothers. Appropriate advice will have to be offered to individuals, preferably based on firmer knowledge about the prognosis than is currently available.

Unless the issue is dealt with sensitively, carriers are likely to be unjustifiably alarmed to learn that they are carrying 'a leukaemia virus'. A report in the *Wall Street Journal* [66], describing a programme to screen some American blood donors for anti-HTLV I, had the headline 'Concerns grow that cancer virus is spreading in same ways as AIDS', and may give a warning of the sort of attention that HTLV I may attract.

References

1   Baltimore D. RNA-dependent DNA polymerase in virions of RNA tumour viruses. *Nature* 1970, **226**, 1209–11.

2   Temin H, Mizutani S. RNA-dependent DNA polymerase in virions of Rous sarcoma virus. *Nature* 1970, **226**, 1211–3.

3   Rous P. A sarcoma of the fowl transmissable by an agent separable from the tumor cells. *J Exp Med* 1911, **13**, 397–411.

4   Morgan DA, Ruscetti FW, Gallo RC. Selective in vitro growth of T-lymphocytes from normal human bone marrow. *Science* 1976, **193**, 1007–8.

5   Poiesz BJ, Ruscetti FW, Gazadar AF, Bunn PA, Minna JD, Gallo RC. Detection and isolation of type C retrovirus particles from fresh and cultured lymphocytes of a patient with cutaneous T-cell lymphoma. *Proc Natl Acad Sci USA* 1980, **77**, 7415–9.

6   Robert-Guroff M, Gallo RC. HTLV and leukaemia in man. In: Gale RP, Golde DW eds. Leukemia: recent advances in biology and treatment. New York: Liss, 1985, 115–36.

7   Uchiyama T, Yodoi J, Sagawa K, *et al.* Adult T-cell leukemia: clinical and hematologic features of 16 cases. *Blood* 1977, **50**, 481–92.

8   Hinuma Y, Nagata K, Hanaoka M, *et al.* Adult T cell leukaemia: Antigen in an ATL cell line and detection of antibodies to the antigen in human sera. *Proc Natl Acad Sci USA* 1981, **78**, 6476–80.

9   Gotoh Y-I, Sugamura K, Hinuma Y. Healthy carriers of a human retrovirus, adult T cell leukaemia virus (ATLV): demonstration by clonal culture of ATLV-carrying T cells from peripheral blood. *Proc Natl Acad Sci USA* 1982, **79**, 4780–2.

10 Yoshida M, Miyoshi I, Hinuma Y. Isolation and characterisation of retrovirus from cell lines of human adult T cell leukaemia and its implication in the disease. *Proc Natl Acad Sci USA* 1982, **79**, 2031–5.

11 Reitz MS, Popovic M, Haynes BF, Clark SC, Gallo RC. Relatedness by nucleic acid hybridization of new isolates of human T-cell leukaemia-lymphoma virus (HTLV) and demonstration of provirus in uncultured leukaemic blood cells. *Virology* 1983, **126**, 688–92.

12 Takatsuki K. Adult T-cell leukaemia. *Asian Med J* 1988, **31**, 287–96.

13 Gibbs WN, Wycliffe S, Lofters MB, *et al*. Non-Hodgkin lymphoma in Jamaica and its relation to adult T-cell leukaemia-lymphoma. *Ann Int Med* 1987, **106**, 361–8.

14 Clark J, Saxinger C, Gibbs WN, *et al*. Seroepidemiologic studies of human T-cell leukemia/lymphoma virus type I in Jamaica. *Int J Cancer* 1985, **36**, 37–41.

15 Blattner WA, Gibbs WN, Saxinger C, *et al*. Human T cell leukaemia/lymphoma virus-associated lymphoreticular neoplasia in Jamaica. *Lancet* 1983, **ii**, 61–4.

16 Montgomery RD, Cruickshank EK, Robertson WB, *et al*. Clinical and pathological observations on Jamaican neuropathy; a report on 206 cases. *Brain* 1964, **87**, 425–62.

17 Gessain A, Barin F, Vernant JC, *et al*. Antibodies to human T-lymphotropic virus type-I in patients with tropical spastic paraparesis. *Lancet* 1985, **ii**, 407–10.

18 Osame M, Usuku K, Izumo S, *et al*. HTLV-I associated myelopathy, a new clinical entity. *Lancet* 1986, **i**, 1031–2.

19 Matsuo H, Nakamura T, Tsujihata M, *et al*. Plasmapheresis in treatment of human T-lymphotropic virus type-I associated myelopathy. *Lancet* 1988, **ii**, 1109–13.

20 Azizuki S, Nakazato O, Higuchi Y, *et al*. Necropsy findings in HTLV-I associated myelopathy. *Lancet* 1987, **i**, 156–7.

21 Rodgers-Johnson P, Gajdusek DC, Morgan OStC, Zaninovic V, Sarin PS, Graham DS. HTLV-I and HTLV-III antibodies and tropical spastic paraparesis. *Lancet* 1985, **ii**, 1247–8.

22 Bartholomew C, Cleghorn F, Charles W, *et al*. HTLV-I and tropical spastic paraparesis. *Lancet* 1986, **ii**, 99–100.

23 Gessain A, Francis H, Sonan T, *et al*. HTLV-I and tropical spastic paraparesis in Africa. *Lancet* 1986, **ii**, 698.

24 Roman GC, Schoenberg BS, Madden DL, *et al*. Human T-lymphotropic virus type I antibodies in the serum of patients with tropical spastic paraparesis in the Seychelles. *Arch Neurol* 1987, **44**, 605–7.

25 Newton M, Cruickshank K, Miller D, *et al*. Antibody to human T-lymphotropic virus type I in West-Indian-born UK residents with spastic paraparesis. *Lancet* 1987, **i**, 415–16.

26 Gessain A, Caudie C, Gout O, *et al*. Intrathecal synthesis of antibodies to human T lymphotropic virus type I and the presence of IgG oligoclonal bands in the cerebrospinal fluid of patients with endemic tropical spastic paraparesis. *J Inf Dis* 1988, **157**, 1226–34.

27 Bhagavati S, Ehrlich G, Kula RW, *et al*. Detection of human T-cell lymphoma/leukaemia virus type I DNA and antigen in spinal fluid

and blood of patients with chronic progressive myelopathy. *N Eng J Med* 1988, **318**, 1141–7.

28 Koprowski H, DeFreitas EC, Harper ME, *et al*. Multiple sclerosis and human T-cell lymphotropic retroviruses. *Nature* 1985, **318**, 154–60.

29 Murphy EL, De Ceulaer K, Williams W, *et al*. Lack of relation between human T-lymphotropic virus type I infection and systemic lupus erythematosus in Jamaica, West Indies. *J Acquir Immune Defic Syndr* 1988, **1**, 18–22.

30 Sugimoto M, Nakashima H, Watanabe S. T-lymphocyte alveolitis in HTLV-I-associated myelopathy. *Lancet* 1987, **ii**, 1220.

31 Kajiyama W, Kashiwagi S, Nomura H, Ikematsu H, Hayashi J, Ikematsu W. Seroepidemiologic study of antibody to adult T-cell leukaemia virus in Okinawa, Japan. *Am J Epidemiol* 1986, **123**, 41–7.

32 Kajiyama W, Kashiwagi S, Hayashi J, Nomura H, Ikematsu H, Okochi K. Intrafamilial clustering of anti-ATLA-positive persons. *Am J Epidemiol* 1986, **124**, 800–6.

33 Kajiyama W, Kashiwagi S, Ikematsu H, Hayashi J, Nomura H, Okochi K. Intrafamilial transmission of adult T cell leukaemia virus. *J Inf Dis* 1986, **154**, 851–7.

34 Kusuhara K, Sonoda S, Takahashi K, Tokugawa K, Fukushige J, Ueda K. Mother-to-child transmission of human T-cell leukemia virus type I (HTLV-I): a fifteen-year follow up study in Okinawa, Japan. *Int J Cancer* 1987, **40**, 755–7.

35 Hino S, Yamaguchi K, Katamine S, *et al*. Mother-to-child transmission of human T-cell leukemia virus type-I. *Gann* 1985, **76**, 474–80.

36 Kinoshita K, Hino S, Amagasaki T, *et al*. Demonstration of adult T-cell leukemia virus antigen in milk from three seropositive mothers. *Gann* 1984, **75**, 103–5.

37 Yamanouchi K, Kinoshita K, Moriuchi R, *et al*. Oral transmission of human T-cell leukemia virus type-I into a common marmoset (*Callithrix jacchus*) as an experimental model for milk-borne transmission. *Gann* 1985, **76**, 481–7.

38 Kinoshita K, Yamanouchi K, Ikeda S, *et al*. Oral infection of a common marmoset with human T-cell leukemia virus type-1 (HTLV-I) by inoculating fresh human milk of HTLV-I carrier mothers. *Gann* 1985, **76**, 1147–53.

39 Hino S, Sugiyama H, Doi H, *et al*. Breaking the cycle of HTLV-I transmission via carrier mothers' milk. *Lancet* 1987, **ii**, 158–9.

40 Ando Y, Nakano S, Saito K, *et al*. Transmission of adult T-cell leukemia retrovirus (HTLV-I) from mother to child: comparison of bottle- with breast-fed babies. *Gann* 1987, **78**, 322–4.

41 Ando Y, Nakano S, Saito K, *et al.* Prevention of HTLV-I transmission through the breast milk by a freeze-thawing process. *Gann* 1986, **77**, 974–7.

42 Okochi K, Sato H, Hinuma Y. A retrospective study on transmission of adult T cell leukemia virus by blood transfusion: seroconversion in recipients. *Vox Sang* 1984, **46**, 245–53.

43 Tedder RS, Shanson DC, Jeffries DJ, *et al.* Low prevalence in the UK of HTLV-I and HTLV-II infection in subjects with AIDS, with extended lymphadenopathy, and at risk of AIDS. *Lancet* 1984, **ii**, 125–8.

44 White PMB, unpublished.

45 Okochi K, Sato H. Transmission of adult T-cell leukaemia virus (HTLV-I) through blood transfusion and its prevention. *AIDS Res* 1986, **2 suppl i**, S157–61.

46 Osame M, Izumo S, Igata A, *et al.* Blood transfusion and HTLV-I associated myelopathy. *Lancet* 1986, **ii**, 104–5.

47 Williams AE, Fang CT, Slamon DJ, *et al.* Seroprevalence and epidemiological correlates of HTLV-I infection in US blood donors. *Science* 1988, **240**, 643–6.

48 Robert-Guroff M, Blayney DW, Safai B, *et al.* HTLV-I specific antibody in AIDS patients and others at risk. *Lancet* 1984, **ii**, 128–31.

49 Robert-Guroff M, Weiss SH, Giron JA, *et al.* Prevalence of antibodies to HTLV-I, -II and -III in intravenous drug abusers from an AIDS endemic region. *JAMA* 1986, **255**, 3133–7.

50 Fuchs D, Unterweger B, Hausen A, *et al.* Anti-HIV-1 antibodies, anti-HTLV-I antibodies and neopterin levels in parenteral drug addicts in the Austrian Tyrol. *J Acquir Immune Defic Syndr* 1988, **1**, 65–6.

51 Gradilone A, Zani M, Barillari G, *et al.* HTLV-I and HIV infection in drug addicts in Italy. *Lancet* 1986, **ii**, 753–4.

52 Takatsuki K, Uchiyama T, Ueshima Y, *et al.* Adult T cell leukaemia: further clinical observations and cytogenetic and functional studies of leukaemic cells. *Jpn J Clin Oncol* 1979, **9 (suppl)**, 317–24.

53 Blayney DW, Blattner WA, Robert-Guroff M, *et al.* The human T-cell leukemia-lymphoma virus in the southeastern United States. *JAMA* 1983, **250**, 1048–52.

54 Reeves WC, Saxinger C, Brenes MM, *et al.* Human T-cell lymphotropic virus type I (HTLV-I) seroepidemiology and risk factors in metropolitan Panama. *Am J Epidem* 1988, **127**, 532–9.

55 Saxinger W, Blattner WA, Levine PH, *et al.* Human T-cell leukemia virus (HTLV-I) antibodies in Africa. *Science* 1984, **225**, 1473–6.

56 Weiss RA, Cheingsong-Popov R, Clayden S, *et al*. Lack of HTLV-I antibodies in Africa. *Nature* 1986, **319**, 794–5.
57 Karpas A, Maayan S, Raz R. Lack of antibodies to adult T-cell leukaemia virus and to AIDS virus in Israeli Falashas. *Nature* 1986, **319**, 794.
58 Catovsky D, Greaves MF, Rose M, *et al*. Adult T-cell lymphoma-leukaemia in blacks from the West Indies. *Lancet* 1982, **i**, 639–43.
59 Greaves MF, Verbi W, Tilley R, *et al*. Human T-cell leukemia virus (HTLV) in the United Kingdom. *Int J Cancer* 1984, **33**, 795–806.
60 Cunningham D, Gilchrist NL, Jack A, Dalgleish AG, Soukop M, Lee FD. T-lymphoma associated with HTLV-I outside the Caribbean and Japan. *Lancet* 1985, **ii**, 337–8.
61 Goorney BP, Lacey CJN, White PMB. A case of tropical spastic paraparesis in the United Kingdom. *J Infect* 1988, **16**, 105–6.
62 Ikeda M, Fujino R, Matsui T, Yoshida T, Komoda H, Imai J. A new agglutination test for serum antibodies to adult T-cell leukemia virus. *Gann* 1984, **75**, 845–8.
63 Kobayashi S, Yoshida T, Hiroshige Y, Matsui T, Yamamoto N. Comparative studies of commercially available particle agglutination assay and enzyme-linked immunosorbent assay for screening of human T-cell leukemia virus type I antibodies in blood donors. *J Clin Micro* 1988, **26**, 308–12.
64 White PMB. Comparison of assays for antibody to HTLV I. *J Clin Pathol* 1988, **41**, 700–2.
65 Kalyanaraman VS, Sarngadharan MG, Robert-Guroff M, *et al*. A new subtype of human T-cell leukemia virus (HTLV-II) associated with a T-cell variant of hairy cell leukemia. *Science* 1982, **218**, 571–3.
66 Edit, *Wall Street Journal*, Jan 20 1987.

Chapter 6

The diagnosis of Epstein-Barr virus-associated disease

NA Hotchin and DH Crawford

Epstein-Barr virus (EBV) is a member of the human herpesvirus group. It has a worldwide distribution, with approximately 90% of the adult population showing evidence of past infection [1]. Primary infection is usually subclinical occurring during childhood, but when delayed until adolescence or early adult life it may manifest as acute infectious mononucleosis (IM) [2]. Infection occurs via the mouth, with a lytic infection of squamous epithelial cells of the pharynx [3]. Here a small number of B lymphocytes, the presence of which has been demonstrated in the circulation by their expression of the EB viral nuclear antigen, probably become infected [4].

In common with other herpesviruses, EBV persists in the body following primary infection. The exact site of persistence has yet to be established, although two postulated sites are the basal epithelial cells of the pharynx [5] and the small resting B-cells [6], since low-level virus shedding into the pharynx is detectable in the majority of seropositive individuals [7], and virus carrying B cells can be recovered from peripheral blood [8]. In normal individuals a delicate balance between the host and virus exists, controlled mainly by EBV-specific cytotoxic T cells, but also by natural killer cells and the humoral immune response [9]. Increased viral shedding is found in immunosuppressed individuals, where this balance is tipped in favour of virus replication [10]. On occasions this may lead to proliferation of virus-infected B cells and lymphoma development [11].

In addition to being the causative agent of IM, EBV is associated with a variety of other diseases. In particular, it is believed to be a cofactor in the development of two human malignancies, Burkitt's lymphoma (BL) and nasopharyngeal carcinoma (NPC).

The viral genome

The EB viral genome is present in the viral particle as a double-stranded, linear DNA molecule of 172kb. This DNA is large enough to code for over 100 proteins, and has been completely sequenced (for a review see Dambaugh *et al* [12]). Like other herpesviruses, the genome is organised into repetitive (I_R) and unique long and short (U_L and U_s) sequences flanked by terminal repeat (T_R) sequences:

$$T_R - U_S - I_R - U_L - T_R$$

Viral isolates from various disease processes as well as from normal seropositive individuals are remarkably similar, with no disease-associated variation. However, a variant viral type has recently been detected which differs from the sequenced prototype virus (B95-8) in the U_L region of the genome, giving rise to an alternative EB viral nuclear antigen (EBNA) 2 gene product [13].

Viral infection *in vitro*

On infection of B lymphocytes, the virus binds to the CR2 receptor (CD21 – the physiological receptor for the C3D component of complement) and enters the cell, leaving the viral envelope at the cell surface [14]. *In vitro* viral infection of some B lymphocytes leads to cell immortalisation and establishment of lymphoblastoid cell lines (LCL), whereas in others the infection is abortive [15]. No lytic cycle, with the production of progeny virus, occurs at this stage.

Little is yet known about infection of epithelial cells by EBV. Recently, however, *in vitro* lytic infection of epithelial cells via a CR2-like molecule has been reported [16].

Viral gene expression in B cells can be detected eight hours post infection, and is restricted to the latent gene products: that is, those which are consistent with continued cell proliferation [17]. These consist of the EB viral nuclear antigen complex (EBNA 1–6), the latent membrane protein (LMP), and the recently described terminal protein (TP) complex.

Sixteen to 20 hours after infection of B cells, a single viral genome circularises by covalent linkage of the terminal repeat sequences, to form an episome which resides in the nucleus. Amplification of this viral DNA gives rise to multiple copies of the nuclear episome, which replicate synchronously with the cell and are equally partitioned to the daughter cells, leaving the viral genome copy number within any one cloned LCL constant [18]. Within an established LCL a low, but constant, number of cells enter a lytic cycle, with the expression of lytic cycle genes, virus production and cell death [19].

EBV carrying cell lines can also be established direct from biopsy material obtained from endemic BL patients. Compared to LCLs these cell lines show a restricted viral gene expression in early passage, expressing only EBNA–1. However, in long-term culture they tend to undergo phenotypic drift, with expression of EBV genes similar to that seen in LCLs [20]. BL cell lines commonly used for the detection of antibodies to EBV antigens include P3HR-1 [21] and Raji [22], both of which were derived from African BLs. Another cell line commonly used as a control is Ramos, which was derived from an EBV-negative BL [23].

Viral antigens

LATENT CYCLE ANTIGENS

EB nuclear antigen complex: The nuclear antigen complex (EBNA), which is now known to comprise six components (EBNA 1–6), was first detected by immunofluorescence (IF) in the nuclei of EBV-immortalised B cells fixed in acetone and methanol [24]. EBNA-1 has a molecular weight of 65–85kDa, depending on the isolate. The size variation derives from a 20–45kDa glycine-alanine repeat sequence. Expression of EBNA-1 is required for the stable maintenance of the virus in episomal form, and as a consequence is expressed in all EBV latently-infected cells [25].

EBNA-2 is an 86kDa protein which is thought to be required for the immortalisation of B-cells [26]. There are two distinct EBNA-2 types – A and B – with less than 50% homology between the two [13]. There is no apparent disease correlation with EBNA-2 type, and the only biological difference found so far is that LCLs immortalised with type 2A virus are faster growing and proliferate at higher density than those immortalised by type 2B virus [27].

Little is known about the function of EBNA-3 (140–157kDa), EBNA-4 (148–180kDa), EBNA-5 (46kDa) and EBNA-6 (160kDa).

Latent membrane protein (LMP): LMP is a 63kDa protein found in the cytoplasm and on the membrane of infected cells, as well as in the virion particle. The precise function of this protein is not yet fully elucidated, but is thought to be involved in immortalisation [28]. Part of the molecule also acts as an epitope for EBV-specific cytotoxic T cells [29].

LYTIC CYCLE ANTIGENS

There are three lytic cycle antigen complexes: viral capsid antigen (VCA), early antigen (EA) and membrane antigen (MA). For a review of the properties of these antigens see Pearson and Luka [30].

Viral capsid antigen: This complex forms the structural component of the virus capsid and is produced late in the lytic cycle. VCA is detectable by indirect IF in the nucleus and cytoplasm of lytically-infected cells fixed in acetone. To date at least three major polypeptides have been identified, with molecular weights of 125, 152 and 160kDa, and monoclonal antibodies have been produced against the 125 and 160kDa proteins. All sera from EBV-infected individuals have been shown to react with the 125kDa polypeptide, but not all with the 160kDa component. Monoclonal antibody studies have also shown that the indirect IF method used to detect VCA will also detect components of the MA complex.

Membrane antigen complex: The MA complex comprises three glycoproteins – gp85/90, gp200/250 and gp300/350. These three glycoproteins have attracted a great deal of interest as all elicit production of neutralising antibodies, and thus have potential for use as subunit vaccines. In addition gp200/250 and gp300/350 appear to act as targets for antibody-dependent cell cytotoxicity (ADCC). Most research into potential vaccines has concentrated on gp300/350 [31].

Gp300/350 mediates virus binding to the cell surface molecule CR2 (CD21) and gp85/90 is involved in fusion to the cell membrane and internalisation of the virus. The function of gp200/250 is not fully understood, but the molecule is antigenically closely related to gp300/350.

Early antigen complex: EA is expressed only in cells which have entered the lytic cycle, and its expression is incompatible with continued cell proliferation. Two EA proteins are detectable by indirect IF on acetone-fixed cells. The diffuse component (EA-D) is present in the nucleus and cytoplasm, whereas the restricted form (EA-R) is confined to the cytoplasm. These two components can be distinguished by their differential sensitivity to solvents, the restricted component being destroyed by methanol and ethanol fixation [32].

EA-D is a 47–60kDa phosphorylated protein with DNA binding activity and EA-R is a 85kDa non-phosphorylated protein. In addition, a 140kDa protein has been identified, but not characterised. It is likely that other early antigens remain to be identified. Their biological function is not known, but they are thought to have enzymatic activity.

Expression of EA in an LCL can be induced or increased by use of chemical agents such as the phorbol ester, 12-O-tetradecanoyl-phorbol-1,3-acetate (TPA), or by superinfection with non-immortalising virus, recovered from the P3HR-1 cell line [33].

Methods of detection

CULTURE AND PREPARATION OF CELLS FOR USE IN SEROLOGICAL METHODS

Cell lines are grown in RPMI 1640 media, containing 10% heat inactivated, mycoplasma-free fetal calf serum, 2mM L-glutamine, 100IU/ml penicillin and 100μg/ml streptomycin. Cells are cultured at 37°C in 5% CO_2.

SEROLOGICAL METHODS
Indirect immunofluorescence:
1   VCA. Antibodies to VCA are determined by indirect IF using the virus-producing BL cell line P3HR-1. The method outlined below is an adaptation of the one originally used by Henle and Henle [1]:

   a)   Wash cells twice in phosphate buffered saline (PBS) and resuspend at approximately 2×10^6cells/ml in PBS.
   b)   Add 50μl aliquots to each well of a 12-well polytetrafluoroethylene (PTFE) coated glass slide and allow to air dry.
   c)   Fix in acetone at room temperature for 10 minutes (if not being used immediately the slides can be stored for up to six months in an airtight container at −20°C).
   d)   Make doubling dilutions (1:8 to 1:1024) of patients' sera in PBS.
   e)   Add 10μl aliquots of each dilution to the slides and incubate in a moist chamber for 60 minutes at 37°C.
   f)   Wash in PBS for 10 minutes and add 10μl of fluorescein isothiocyanate (FITC) conjugated goat anti-human IgG (or IgA or IgM), diluted in PBS to each well (dilution of conjugate to be used has to be determined empirically for each batch of conjugate).
   g)   Incubate for 60 minutes at room temperature and wash as before.
   h)   Mount in PBS/glycerol (1:1) and examine under an ultraviolet (UV) microscope at ×200 magnification (Figure 1).

2   EA. The method for determining antibody titres to EA [34] is similar to that used for VCA. Cells from the BL cell line, Raji, are treated with TPA (20ng/ml) 2–3 days before harvesting. Expression of EA is induced in 1–2% of cells. Raji is unusual that it does not express VCA and only expresses appreciable levels of EA when induced. Cells fixed in acetone will stain for both EA-D (Figure 2) and EA-R (Figure 3), whereas cells fixed in methanol or ethanol will show only EA-D staining. Because of problems associated with non-specific staining, it is advisable to use an EBV-negative BL cell line, such as Ramos, as a control.
3   MA. Antibodies to MA can also be detected by indirect IF on live cells in suspension [35]. Cells from an EBV-positive cell line, such as B95-8, are centrifuged, resuspended in patients' serum and incubated at 37°C for 30 minutes. Cells are washed in balanced salt solution (BSS) and incubated for 20 minutes at 37°C with FITC conjugated anti-human IgG. After a further wash cells are examined under a UV microscope for membrane-associated staining.

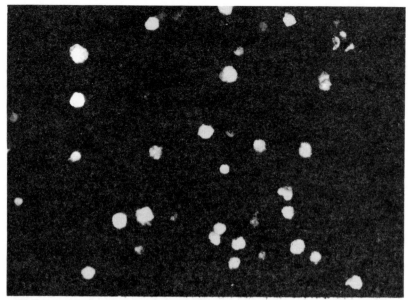

Figure 1 Indirect immunofluorescence staining of viral capsid antigen in P3HR-1 cells (×200 magnification)

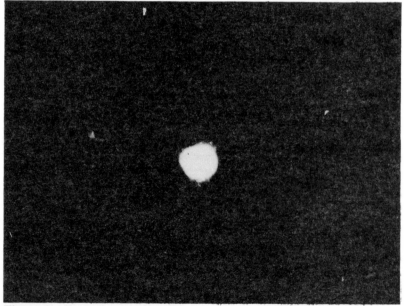

Figure 2 Detection of the diffuse component of early antigen in Raji cells by indirect immunofluorescence (×400 magnification)

Figure 3 Indirect immunofluorescence of Raji cells showing the restricted (cytoplasmic) component of early antigen (×400 magnification)

To avoid problems of interaction with non-specific membrane reactive antibodies in patients' sera, a modification of this method involving a blocking step has been devised [36]. In this modification the test cells are incubated with a mixture of test serum and FITC conjugated antibodies to MA, containing no non-specific binding activity, in a competitive binding assay. Hence, if the test serum contains anti-MA antibodies, it blocks binding by the FITC anti-MA antibodies.

Another major problem with this method is that of binding of the anti-human FITC conjugate to cell surface immunoglobulin. Because of their complexity and the problems associated with these methods, they are not routinely used.

Anti-complement immunofluorescence (ACIF): ACIF is the method most commonly used for detecting antibodies to the EBNA complex. The method was originally developed by Reedman and Klein [24], and an adaptation of this method using Raji cells is described below.

a) Wash the cells twice in complement fixation buffer (CFB) and resuspend in 0.95% trisodium citrate.
b) Cytospin approximately 5×10^4 cells on to glass slides and allow to air dry (the cells can be added to the glass slide and allowed to air

121

dry without spinning; however, cytospin preparations give clearer results).

c) Fix in acetone and methanol (1:1) for 3 minutes at −20°C and store at −20°C in an airtight container if not being used immediately.

d) Dilute sera in CFB containing 10% complement. Obtain complement from EBV seronegative donors as follows: allow peripheral blood to clot at room temperature and place at 4°C for 2 hours; remove the serum, aliquot into suitable volumes and store at −70°C. It should not be refrozen, and should be removed from the freezer immediately prior to use.

e) Add 25µl diluted sera to the slides and incubate at 37°C for 60 minutes in a moist chamber.

f) Wash for 30 minutes in CFB with two changes of CFB.

g) Add 25µl FITC-conjugated anti-human C3 diluted in CFB and incubate for a further 60 minutes at room temperature. The optimal dilution for the conjugate must be determined empirically for each batch.

h) Wash as before, mount in PBS/glycerol (1:1) and examine under a UV microscope at ×400 magnification. A typical intense granular nuclear staining should be present with positive sera (Figure 4).

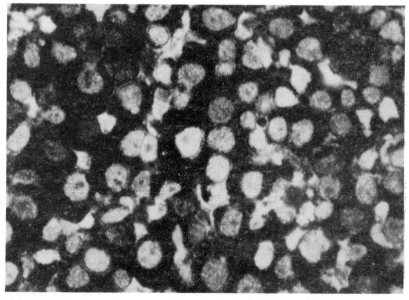

Figure 4 Raji cells stained for EBV nuclear antigen by anti-complement immunofluorescence (×400 magnification)

Known titre positive and negative control sera should be included with each batch of tests. In addition, duplicate slides using an EBV negative BL cell line, such as Ramos, should be prepared, and each serum tested on these cells to exclude non-specific nuclear staining.

A modification of this method [37] employs a complement enhancement step between the first and second layers to increase sensitivity. After the first wash CFB containing 10% complement is added to the slides, which are incubated for 20 minutes at 37°C. The slides are rewashed in CFB and anti-human C3 added as above.

Antibody titres to EBNA-1 and EBNA-2 can be differentiated by IF using an EBV-negative Burkitt lymphoma cell line transfected with the relevant EBNA coding region of the virus [38].

Enzyme linked immunosorbent assay (ELISA): An ELISA has been developed for detecting antibodies to VCA, EA and EBNA using purified protein from P3HR-1 and Raji cell lines. Results indicated a high degree of correlation between ELISA and IF titres [39]. However, in a limited trial of a commercially-available ELISA, a lack of sensitivity compared to indirect IF was noted [40]. As yet ELISAs are not in widespread use in the UK.

A sensitive ELISA assay to detect antibodies to the gp350 component of the MA complex has also been developed [41]. It has proved a thousand-fold more sensitive than conventional IF techniques, but as yet has only been used for research purposes to measure gp300/350 antibody levels in cotton-top tamarins following immunisation with the putative gp300/350 subunit vaccine [31].

Immunoblotting: Antibodies to the six EBNA proteins can be detected by immunoblotting, following separation of protein from EBV-infected cells by sodium dodecyl sulphate-polyacrylamide gel electrophoresis (SDS-PAGE) [42] and transfer to nitrocellulose by western blotting [43]. Once the protein has been transferred the nitrocellulose is cut into strips and incubated for 30 minutes with PBS, powdered milk (5%) and Tween 20 (0.05%) to prevent non-specific binding of protein to the nitrocellulose. The strips are incubated with patients' sera diluted in PBS/Tween for 4–5 hours before being washed for 60 minutes with frequent changes of PBS/Tween. Following washing, the strips are incubated with horseradish peroxidase conjugated anti-human IgG for 30 minutes before being washed and developed with diaminobenzidine tetrahydrochloride (DAB) and hydrogen peroxide (0.012%). The EBNA proteins are visualised as discrete bands on the nitrocellulose (Figure 5). The B95-8 cell line [44] is commonly used in immunoblotting, as all six EBNAs are expressed in this line. As with IF techniques an EBV negative cell line should be used as a negative control.

Antibodies to LMP, which are not detectable by IF, can only be

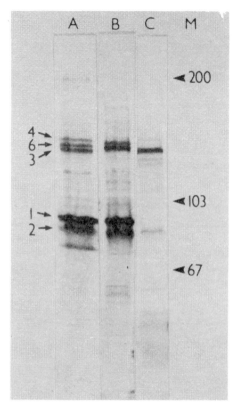

Figure 5 Immunoblot of an EBV immortalised lymphoblastoid cell line probed for EBV nuclear antigen with three different human sera. Serum A has antibodies to EBV nuclear antigens 1, 2, 3, 4 and 6; serum B to EBV nuclear antigens 1, 2, 4 and 6; serum C (from a chronic IM patient) to EBV nuclear antigens 2 and 3 only. EBV nuclear antigen-5 is not present in this cell line, which is unusual in that EBV nuclear antigen-1 has a higher molecular weight than EBV nuclear antigen-2. M indicates positions of molecular weight markers.

reliably detected by first immunoprecipitating LMP from EBV-infected LCLs, followed by immunoblotting as for the EBNA proteins [45].

In addition to being time-consuming and requiring specialist equipment, a major drawback of immunoblotting is that it is not very quantitative. As a consequence this method is generally used for research purposes only.

Heterophile antibody test: Heterophile antibodies (HA), which are not strictly speaking EBV-specific, are antibodies, usually of the IgM glass, which agglutinate red blood cells of species other than human. They are typically produced in IM. The rapid slide test, of which a number are commercially available, is commonly used to detect HA and is an adaptation of the Paul-Bunnell (PB) test.

The original PB test involved the incubation of serial dilutions of heat-inactivated sera with sheep erythrocytes [46]. Positive reactions showed macroscopic agglutination. A problem with this method was of non-specific agglutination caused by Forssman type antibodies. This problem was overcome in an adaptation of the original PB test which used guinea-pig kidney suspension to absorb out Forssman type antibodies, and ox erythrocytes to absorb out IM-specific HA providing a differential diagnostic test [47].

It is this differential test which forms the basis of the rapid slide tests, some of which use horse erythrocytes instead of sheep, which are a more sensitive indicator of agglutination [48].

Rapid ELISA: Recently a rapid ELISA (Monolert, Ortho Diagnostics, UK), has been developed. This test, for diagnosing acute IM, detects the presence of IgG and IgM class antibodies to a peptide containing the glycine-alanine repeat sequence of ENBA-1. These antibodies arise early in acute IM and are undetectable by conventional IF techniques [49]. The test is carried out on a peptide coated plastic spoon and takes less than ten minutes to perform.

Studies using this peptide in a conventional ELISA using microtitre plates demonstrated high specificity and sensitivity [50]. The rapid ELISA is claimed to be as highly specific and sensitive.

NON-SEROLOGICAL METHODS OF DETECTION

Direct detection of antigen:
1 Immunofluorescence. The ACIF technique described in the previous section can also be used to detect the presence of EBNA-positive cells in biopsy or post mortem material, where involvement of EBV is suspected. Tissue is snap frozen, thin-sectioned and fixed onto glass slides with acetone and methanol (1:1) at −20°C. The method for staining is as previously described, using known EBNA-positive and negative control sera as the first layer. A modification of the method uses a horseradish peroxidase conjugated anti-human C3 antibody instead of an FITC conjugate, which allows for better preservation of the tissue histology [51] (Figure 6).
2 Immunoblotting. Biopsy material can also be analysed by immuno-blotting as previously described, probing with known EBNA antibody positive and negative control sera. This method is a relatively insensitive technique and requires a large number of cells, compared to immunocytochemical and *in situ* techniques.

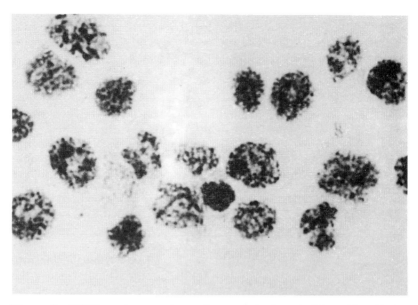

Figure 6 EBV nuclear antigen detected in lymphoblastoid cells by anti-complement immunoperoxidase staining (×600 magnification)

Virus isolation: EBV shedding from pharyngeal and/or salivary gland epithelial cells occurs either continuously at low levels or intermittently in most normal seropositive individuals [7]. However, the methods of detection are insensitive and time consuming, involving the outgrowth of LCL from cord blood lymphocytes, after incubation with throat washing or saliva samples. Random samples give 10–20% positivity in this assay [52] and this can be increased to around 60% if virus is first concentrated by ultracentrifugation. Studies using this method indicate that levels of shedding remain relatively constant with time, and normal seropositive individuals can be divided into groups of high, low or non-shedders [7].

Since most seropositive individuals shed virus, its isolation is not diagnostic of any specific EBV associated disease. However, high titre virus can be recovered from acute IM patients and from immuno-suppressed individuals, such as renal transplant recipients [10]. High titre virus shedding in immunosuppressed patients is sometimes referred to as a 'reactivated' type of EBV infection, although it occurs without clinical symptoms.

Detection of viral DNA:
1   Southern blotting and hybridisation. DNA from biopsy material, or cell lines, can be prepared using standard methods [53]. The ex-

tracted DNA is digested with restriction endonucleases and separated on an agarose gel by electrophoresis. The separated DNA is transferred to nitrocellulose by southern blotting [54] and hybridised with radiolabelled EBV probes. This technique is widely used in research laboratories and is highly specific and sensitive. However, it is a time-consuming procedure, requiring expensive equipment and using radioisotopes. For these reasons it is not routinely used in diagnostic laboratories.

2   *In situ* hybridisation. The method of *in situ* hybridisation can be used to detect genomic sequences of EBV in LCL and BL cell lines. The cells are fixed in paraformaldehyde and probed with radio-labelled or biotinylated EBV-specific EBV probes [55]. However, the method is not as sensitive as immunocytochemical methods, and has not been used on BL and NPC biopsy material with any great success.

3   Polymerase chain reaction (PCR). PCR is a recently developed, highly sensitive technique for amplifying and detecting specific DNA sequences in biological material [56]. PCR is capable of detecting very low abundance DNA, as low as one copy per cell, and has great potential for use in diagnosis of viral infection and the detection of viral DNA in biopsy material.

PCR has been used to detect viral sequences in a variety of situations including the detection of HIV in blood samples from HIV seropositive individuals [57]. However, the technique has yet to be applied routinely to the detection of EBV.

Diagnosis of EBV-associated disease

SEROLOGY IN THE NORMAL SEROPOSITIVE INDIVIDUAL

In normal healthy individuals, previously infected with EBV, IgG class antibodies to VCA are invariably present, with titres normally ranging from 8 to 512. Antibodies to EA are usually low (<32) or absent. Antibody titres to EBNA-1 range from 10 to 160 in normal individuals, with levels anti-EBNA-2 being low or absent [58]. Antibodies to MA are detectable in most seropositive individuals [36]. It should be noted that considerable variation in antibody titres of identical sera have been reported when tested by different laboratories [59].

INFECTIOUS MONONUCLEOSIS

EBV is known to be the causative agent of IM [2], a usually benign, self-limiting lymphoproliferative disease common in adolescence. Signs and symptoms of IM include: pyrexia, fatigue, lymphadenopathy, pharyngitis and splenomegaly, with large numbers of atypical lymphocytes present in the peripheral blood. In the majority of cases, symptoms resolve in 1–6 months, but occasional complications include hepatitis and pneumonitis and, very rarely, IM may be fatal. In a small minority of patients, symptoms may persist for more than a year leading to chronic IM, a sometimes severe and debilitating illness.

The diagnostic test most commonly used for IM is the rapid slide test for HA mentioned earlier. Heterophile antibodies arise early in infection and usually persist for 1–6 months. A positive result can be regarded as diagnostic in cases of acute onset when backed up by a typical clinical syndrome and the presence of circulating atypical lymphocytes. Occasionally HA can persist for longer than six months, and positive results have been observed in cases of chronic IM [60] and post-viral fatigue syndrome (PVFS) [61]. While a positive test in acute illness can usually be regarded as diagnostic, a negative result does not exclude a diagnosis of IM, as HA-negative cases of IM are not uncommon, particularly in children under the age of 12 years. In these cases, EBV specific serology can be of value. Detection of anti-VCA IgM can usually be considered diagnostic, although anti-VCA IgM can occasionally be detected in chronic IM patients [62]. Anti-VCA IgG antibodies are of little value in the diagnosis of IM, unless a seroconversion can be demonstrated, a rare event as most patients do not present early enough in the disease. A rising antibody titre to VCA IgG may be observed in IM, but again the titre has usually reached maximum levels by the time of presentation. Antibodies to EA-D are usually present in acute infection and fall to low and undetectable levels in convalescence [58]. Their diagnostic value is limited as raised anti-EA antibody levels are found in a variety of EBV-associated and non-associated conditions. At best they provide supportive evidence for diagnosis, and probably reflect increased viral shedding. IgG class antibodies to EBNA-1 do not usually arise until convalescence, whereas antibodies to EBNA-2 arise earlier in the illness and fall to low or undetectable levels in convalescence [58]. Absence of EBNA-1 antibodies should not be regarded as diagnostic as they are often low or absent in chronic IM, nor should the presence of anti-EBNA-1 exclude the diagnosis of IM, as in a proportion of cases antibodies arise soon after onset [58]. Despite the battery of serological methods available the rapid slide test for HA, coupled with the haematological and clinical details, remains the simplest and least expensive method for diagnosis of IM. In HA-negative cases where IM is suspected, particularly in young children, detection of IgM class antibodies to VCA is a valuable aid to diagnosis.

CHRONIC IM

Signs and symptoms of chronic IM range from fever, pharyngitis, malaise, myalgia and lymphadenopathy to potentially life-threatening problems including anaemia, thrombocytopaenia, hypogamma-globulinaemia and pneumonitis [63]. Onset of chronic IM usually follows acute IM, and may be as a consequence of impaired cell-mediated immune response to the virus [60].

To meet the criteria for a diagnosis of chronic IM three conditions need to be satisfied. Firstly, symptoms should have persisted for at least 12 months. Secondly, onset of persistent symptoms should have been preceded by a proven case of IM, and thirdly, evidence of active EBV infection should be present. Patients suffering from chronic IM frequently have significantly raised (>1024) VCA and EA antibody titres. In addition abnormal EBNA serology has been reported in this group of patients. EBNA-2 antibody levels are frequently higher than those to EBNA-1, the reverse of the situation found in normal seropositive individuals, and similar to that found in acute IM [58]. A positive HA result may be observed [60] and, more rarely, anti-VCA IgM may be detected [62]. Similar antibody profiles have been observed in patients diagnosed as suffering from PVFS [61]. It may be that some PVFS patients are in fact suffering from chronic IM, but often details of acute onset of symptoms are incomplete, and it is difficult to make such a diagnosis in the absence of a proven case of IM preceding onset of chronic symptoms.

BURKITT'S LYMPHOMA

Burkitt's lymphoma (BL) is a monoclonal B cell tumour, characterised by a chromosome translocation resulting in the juxtaposition of the *c-myc* oncogene with regions of the immunoglobulin genes [64]. There are two distinct forms of BL: sporadic and endemic. Sporadic BL occurs at low incidence throughout the world affecting all age groups and has no known viral association. Endemic BL occurs at relatively high incidence in certain geographically-restricted areas of the world, in particular Central Africa and Papua New Guinea, affecting children between the age of two and 12 years. In addition to the translocation involving *c-myc*, an apparently invariant feature of endemic BL is the finding of EBV DNA in the tumour cells, suggesting a possible role for the virus in the development of BL [65]. It is widely recognised that other co-factors – for example, malaria – must also be involved in the evolution of the tumour.

The presence of EBV in tumour cells from cases of endemic BL can be demonstrated by southern blotting and probing for viral DNA, or by direct detection of EBNA-1. However, both these methods are reliant on the collection and correct preparation of biopsy material, and are technically demanding, requiring specialist equipment. Usually a diagnosis of BL is made purely on clinical and histological grounds. The involvement, or otherwise, of EBV can be demonstrated using EBV serology. Greatly increased IgG antibody titres to VCA, MA and EA-R are generally observed [36], often a tenfold or greater difference in anti-VCA IgG between BL and control sera is found [66]. EBNA titres may also be raised but this is not always the case. A prospective serological study in Uganda followed the EBV serological status of 42,000 children over a number of years [66]. Twelve of these children developed BL, and several months prior to development of the tumour all had appreciably raised antibody titres. This study indicates that EBV serology might conceivably be of value in the early diagnosis of BL. However, in practical and financial terms, such monitoring is not feasible. A more selective use of EBV serology may be more feasible, particularly in the monitoring of treatment. It has been observed that antibody titres, particularly to EA-R, decrease following treatment and in remission [32]. Correspondingly, a consistently high, or rising, titre in remission indicates a poor prognosis, commonly preceding relapse. Thus, careful monitoring of patients in remission may help in the early detection and treatment of tumour relapse.

NASOPHARYNGEAL CARCINOMA

EBV has also been implicated in the development of undifferentiated nasopharyngeal carcinoma (NPC). NPC is a tumour of the epithelial cells of the nasopharynx and is found predominantly in adults over the age of 35 years. It has a worldwide distribution, with areas of high incidence, for example, Southern China, where it is the most common tumour in male adults and the second most common in females. The evidence linking EBV with NPC is similar to that linking the virus with BL, in that virus genome and viral antigens can be detected in the tumour cells [67] with patients demonstrating greatly raised EBV antibody levels [68].

As with BL, it is likely that other co-factors are involved in addition to EBV. Although NPC occurs with high incidence in certain areas, it appears that the tumour is more closely associated with certain ethnic groups rather than a particular geographic location [69]. This suggests a genetic predisposition or an environmental factor such as diet.

Diagnosis of NPC is usually made on histological grounds. However, the presence of EBV DNA in tumour material can readily be demonstrated [67]. The only nuclear antigen to be expressed in all biopsies is EBNA-1, although LMP can be detected in approximately 50% of biopsy samples [70]. Significantly raised antibody titres to VCA, MA and EA-D can be found in most patients [68]. Of particular diagnostic importance are the IgA class antibodies to VCA and EA found in patients' sera and saliva, the presence of which in the latter is considered diagnostic.

Monitoring IgA antibody titres to EA and, to a lesser extent, VCA can provide a guide to prognosis for patients with remission following treatment. Patients with rising antibody titres have a higher risk of relapse than those with stable or falling antibody titres [71,72]. In addition, various prospective serological studies have been able to identify the group of individuals most likely to develop NPC on the basis of elevated IgA class antibodies to VCA and EA [69]. In areas of very high incidence, serological monitoring of the population may well prove beneficial, as the tumour is often not clinically apparent until metastasis has occurred, at which stage the prognosis is poor.

X-LINKED LYMPHOPROLIFERATIVE SYNDROME

X-linked lymphoproliferative syndrome (XLPS), also known as Duncan's syndrome, accounts for approximately 50% of fatal IM cases in the USA. It is a rare X-linked condition in which primary EBV infection can lead to a rapidly fulminant, and usually fatal, case of IM, with B-cell infiltration and necrosis at multiple organ sites. Alternatively a fatal B-cell lymphoproliferation or lymphoma may develop. The precise mechanism by which this occurs is not known, but probably involves an abnormal cell-mediated response to the virus infection [73].

Diagnosis of XLPS is much the same as for IM, and in addition, EBNA-positive infiltrating lymphocytes can often be detected in post mortem or biopsy material. Serological studies of XLPS families commonly reveals a carrier state in healthy female relatives as evidenced by elevated anti-EA and/or anti-VCA titres. In such cases genetic counselling can be given. The reason for such elevated titres is not known, but it may indicate an impairment in immunological control of the virus.

POST TRANSPLANT LYMPHOPROLIFERATIVE DISEASE

A relatively frequent complication in patients with profound immuno-suppression following organ transplantation is the development of poly-clonal lymphoproliferative lesions, which may progress to malignant monoclonal B-cell tumours. Studies monitoring the incidence of lymphoproliferative disease in these patients report incidences of 1–13% in renal transplant patients and 7.3% in cardiac transplant patients. Post transplant lymphomas have also been reported in bone marrow and heart/lung transplant recipients [74].

It is possible to detect the presence of EBV DNA [75] and EBNA [10] in most of these lesions, and it is thought that immunosuppression results in a loss of virus-specific immunosurveillance leading to an uncontrolled proliferation of EBV-infected B-cells, with the outgrowth of a malignant monoclonal population in a proportion of cases [10]. Withdrawal of immunosuppressive therapy can result in the regression of the early polyclonal lesions [76]. Similar lymphoproliferative disease is found in patients suffering from acquired immune deficiency syndrome (AIDS) and other T cell immunodeficiencies [77]. In addition, AIDS patients have a relatively high incidence of BL, a proportion of which are EBV-associated.

The majority of EBV-associated post transplant lymphomas and lymphoproliferative lesions appear to occur following primary EBV infection, although less frequently may occur in patients who were seropositive prior to transplant [78]. Serological diagnosis of primary infection is usually made retrospectively, on sera taken for other purposes, as these patients rarely exhibit symptoms of IM. Diagnosis of primary infection should be made on the basis of a positive HA test, or the presence of anti-VCA IgM. It is often possible to demonstrate a seroconversion in these patients using stored sera, taken at frequent intervals for other reasons. It is not unusual to find low IgG titres to VCA and EA in primary infection, hence a low IgG titre to VCA does not preclude acute IM in these patients. Patients who were seropositive prior to transplant may have an antibody profile suggestive of reactivation, with high antibody titres to VCA and EA.

As has been mentioned previously, EBNA-positive cells can often be detected in biopsy or post mortem tissue, and this is a more suitable method for demonstrating the involvement of EBV in these lesions. Unlike BL, expression of latent EBV genes is not restricted with all the EBNAs and LMP being detectable by immunoblotting [79].

ORAL HAIRY LEUKOPLAKIA

Oral hairy leukoplakia (OHL) is a recently-described epithelial lesion of the tongue commonly found in patients infected with HIV [80]. Diagnosis is usually made on clinical grounds and development of this lesion is considered to indicate a high risk of progression to AIDS.

Greenspan *et al* [81] detected EBV DNA in 13/13 cases and VCA in 19/20 cases of OHL. In addition, *in situ* hybridisation has been used to detect the presence of EBV DNA in the epithelial cells [82]. Serology is of little value in the diagnosis of OHL, as AIDS patients frequently have raised EBV antibody levels due to immunosuppression and increased viral replication.

The presence of EBV in these lesions suggests an aetiological role for the virus. This view is further supported by the observation that high dose acyclovir, which is known to suppress EBV replication *in vitro* and *in vivo*, results in regression of OHL [83].

ABNORMAL EBV SEROLOGY

Abnormally-elevated antibody titres to EBV-specific antigens are commonly found in a variety of conditions, including rheumatoid arthritis, Hodgkin's lymphoma, chronic lymphocytic leukaemia and ataxia telangiectasia [84]. These findings have given rise to speculation that EBV may play a role in the development of these diseases. However, there is little evidence to suggest that EBV is an aetiological agent. More probably, the raised antibody titres are due to increased shedding of virus following immune function disturbance.

Problems associated with EBV serology

A common problem encountered when using indirect IF is non-specific immunofluorescent staining; the usual cause of this is the presence of autoantibodies, which makes EBV serology a particular problem in autoimmune disease. Usually the non-specific staining is distinguishable from EBV-specific staining, but it may be so intense as to potentially mask EBV-specific staining. It is a wise precaution to include an

EBV-negative B-cell line, such as Ramos, to act as a control for non-specific staining. In addition, cells which were no longer viable when fixed will also stain non-specifically. It is important to ensure that the cells are viable when prepared for slides, and that all washing and fixing steps are carried out as rapidly as possible.

The presence of rheumatoid factor (RF) in sera can cause false positive anti-VCA IgM results [85]. RF is an IgM molecule which can bind IgG, hence the presence of RF in a patient's sera can result in anti-VCA IgG being bound to the anti-human IgM conjugate, giving a false positive result. IgM-positive samples should be screened for the presence of RF, which can be absorbed out, using IgG bound latex beads prior to testing if necessary.

Further problems can occur in determining titration end points. It is usual to take the end point as the dilution at which the intensity of fluorescence begins to diminish [86]. However, in some sera, the diminishing intensity is a very gradual process, making accurate end point readings difficult. Other factors which influence titration readings are the power of the objective lens on the microscope and the operator. Identical sera screened in different laboratories can give very variable results [59]; and, as yet, there are no reference sera available, nor any form of quality control to ensure standardisation of results. It is essential that known titre positive and negative control sera are included in each run. If control antibody titres vary by more than one doubling dilution, the tests should be repeated. Control sera should be aliquoted in suitable volumes and stored at $-20°C$. They should not be freeze-thawed repeatedly as this may lead to a decrease in titre. Where multiple sera from the same patient are being tested, it is advisable to screen them in parallel.

Conclusions

To conclude, there are cases where EBV serology can be of great value in making a diagnosis (Table 1), particularly in cases of IM where HA are absent, yet IM is suspected on clinical grounds. EBV serology may also be of value in determining the nature of persistent symptoms following IM, but a diagnosis of chronic IM is very dependent on good clinical data on the acute onset of illness being available. In cases of BL and NPC, where EBV is implicated as a co-factor in tumorigenesis, serology can be of value in confirming the involvement of the virus. Serology can also provide a guide to prognosis and an early warning of relapse, particularly in NPC. More importantly, it may be possible to pinpoint the subsection of the population at risk from NPC in areas of high incidence, as evidenced by raised EBV antibody titres, allowing careful monitoring and early treatment.

EBV serology is, however, of little value in immunosuppressed patients, who frequently display abberent antibody responses. In pa-

Table 1 Diseases associated with EBV and the methods which provide, or are supportive of, such a diagnosis

Disease	Diagnostic	Supportive
IM	HA test	anti-VCA IgG
	anti-VCA IgM	anti-EA IgG
Endemic BL	Direct detection	anti-VCA IgG
	of EBNA-1	anti-EA-R IgG
NPC	anti-VCA IgA	anti-VCA IgG
	anti-EA IgA	anti-EA-D IgG
	Direct detection	
	of EBNA-1	
XLPS	Direct detection	HA test
	of EBNA-1	anti-VCA IgM
Post transplant lympho-	Direct detection	None
proliferative disease	of EBNA-1	
Chronic IM	None	EBNA1/EBNA2
		antibody ratio
		anti-VCA IgG
		anti-EA IgG
OHL	Direct detection of VCA	None

tients with post transplant lymphoproliferative disease, it may be possible to demonstrate retrospectively a primary EBV infection prior to development of the lesion, but confirmation of the involvement of EBV relies on the direct detection of antigen in biopsy or *post mortem* tissue.

Acknowledgements

We would like to thank Dr MJ Allday and Dr JA Thomas for providing the photographs for Figures 5 and 6 respectively.

References

1   Henle G, Henle W. Immunofluorescence in cells derived from Burkitts lymphoma. *J Bacteriol* 1966, **91**, 1248–56.
2   Niederman JC, Evans AS, Subrahmanyan L, McCollum RW. Prevalence, incidence and persistence of EB virus antibody in young adults. *N Engl J Med* 1970, **282**, 361–5.
3   Sixbey JW, Nedrud JG, Raab-Traub N, Hanes RA, Pagano JS. Epstein-Barr virus replication in oropharyngeal epithelial cells. *N Engl J Med* 1984, **310**, 1225–30.
4   Klein G, Svedmyr E, Jondal M, Persson PO. EBV-determined nuclear antigen (EBNA)-positive cells in the peripheral blood of infectious mononucleosis patients. *Int J Cancer* 1976, **17**, 21–6.
5   Rickinson AB, Yao QY, Wallace LE. The Epstein-Barr virus as a model of virus-host interactions. *Brit Med Bull* 1985, **41**, 75–9.

6 Epstein MA, Achong BG. Various forms of Epstein-Barr virus infection in man: established facts and a general concept. *Lancet* 1973, **ii**, 836–9.

7 Yao QY, Rickinson AB, Epstein MA. A re-examination of the Epstein-Barr virus carrier state in healthy seropositive individuals. *Int J Cancer* 1985, **35**, 35–42.

8 Nilsson K, Klein G, Henle W, Henle G. The establishment of lymphoblastoid lines from adult and fetal human lymphoid tissue and its dependence on EBV. *Int J Cancer* 1971, **8**, 443–50.

9 Rickinson AB. Cellular immunological responses to the virus infection. In: Epstein MA, Achong BG, eds. The Epstein-Barr virus: recent advances. London: Heinemann, 1986, 75–125.

10 Crawford DH, Edwards JMB, Sweny P, Hoffbrand AV, Janossy G. Studies on long term T-cell-mediated immunity to Epstein-Barr virus in immunosuppressed renal allograft recipients. *Int J Cancer* 1981, **28**, 705–9.

11 Crawford DH, Thomas JA, Janossy G *et al.* Epstein-Barr virus nuclear antigen-positive lymphoma after cyclosporin A treatment in patient with renal allograft. *Lancet* 1980, **1**, 1355–6.

12 Dambaugh T, Hennessy K, Fennewald S, Kieff E. The virus genome and its expression in latent infection. In: Epstein MA, Achong BG, eds. The Epstein-Barr virus: recent advances. London: Heinemann, 1986, 13–45.

13 Dambaugh T, Hennessy K, Chamnankit L, Kieff E. U2 region of Epstein-Barr virus DNA may encode Epstein-Barr nuclear antigen 2. *Proc Natl Acad Sci USA* 1984, **81**, 7632–6.

14 Fingeroth JD, Weis JJ, Tedder TF, Strominger JL, Biro PA, Fearon DT. Epstein-Barr virus receptor of human B lymphocytes is the C3d receptor CR2. *Proc Natl Acad Sci USA* 1984, **81**, 4510–4.

15 Pope JH, Horne MK, Scott W. Transformation of foetal human leukocytes *in vitro* by filtrates of a human leukaemic cell line containing herpes-like virus. *Int J Cancer* 1968, **3**, 857–66.

16 Sixbey JW, Davis DS, Young LS, Hutt-Fletcher L, Tedder TF, Rickinson AB. Human epithelial cell expression of an Epstein-Barr virus receptor. *J Gen Virol* 1987, **68**, 805–11.

17 Allday MJ, Crawford DH, Griffin BE. Epstein-Barr virus latent gene expression during the initiation of B-cell immortalization. *J Gen Virol* 1989, **70**, 1755–64.

18 Hurley EA, Thorley-Lawson DA. B cell activation and the establishment of Epstein-Barr virus latency. *J Exp Med* 1988, **168**, 2059–75.

19 Sugden B, Phelps M, Domoradzki J. Epstein-Barr virus DNA is amplified in transformed lymphocytes. *J Virol* 1979, **31**, 590–5.

20 Rowe M, Rowe DT, Gregory CD *et al.* Differences in B cell growth phenotype reflect novel patterns of Epstein-Barr virus latent gene

expression in Burkitt's lymphoma cells. *EMBO J* 1987, **6**, 2743–51.

21 Hinuma Y, Konn M, Yamaguchi J, Wudarski DJ, Blakeslee JR, Grace JT. Immunofluorescence and herpes-type virus particles in the P3HR-1 Burkitt lymphoma cell line. *J Virol* 1967, **1**, 1045–51.

22 Epstein MA, Achong BG, Barr YM *et al.* Morphological and virological investigations on cultured Burkitt tumor lymphoblasts (strain Raji). *J Natl Cancer Inst* 1966, **37**, 547–59.

23 Klein G, Giovanella B, Westman A, Stehlin JS, Mumford D. An EBV-genome-negative cell line established from an American Burkitt lymphoma; receptor characteristics. EBV infectibility and permanent conversion into EBV positive sublines by *in vitro* infection. *Intervirology* 1975, **5**, 319–34.

24 Reedman BM, Klein G. Cellular localization of an Epstein-Barr virus (EBV)-associated complement-fixing antigen in producer and non-producer lymphoblastoid cell lines. *Int J Cancer* 1973, **11**, 499–520.

25 Yates JL, Warren N, Sugden B. Stable replication of plasmids derived from Epstein-Barr virus in various mammalian cells. *Nature* 1985, **313**, 812–5.

26 Hammerschmidt W, Sugden B. Analysis of immortalizing functions of Epstein-Barr virus in human B lymphocytes. *Nature* 1989, **340**, 393–7.

27 Rickinson AB, Young LS, Rowe M. Influence of the Epstein-Barr virus nuclear antigen EBNA2 on the growth phenotype of virus-transformed B cells. *J Virol* 1987, **61**, 1310–7.

28 Wang D, Liebowitz D, Kieff E. An EBV membrane protein expressed in immortalized lymphocytes transforms established rodent cells. *Cell* 1985, **43**, 831–40.

29 Thorley-Lawson DA, Israelsohn ES. Generation of specific cytotoxic T cells with a fragment of the Epstein-Barr virus-encoded p63/latent membrane protein. *Proc Natl Acad Sci USA* 1987, **84**, 5384–8.

30 Pearson GR, Luka J. Characterisation of the virus-determined antigens. In: Epstein MA, Achong BG, eds. Epstein-Barr virus: recent advances. London: Heinemann, 1986, 47–73.

31 Epstein MA, Morgan AJ. Progress with subunit vaccines againt the virus. In: Epstein MA, Achong BG, eds. The Epstein-Barr virus: recent advances. London: Heinemann, 1986, 271–89.

32 Henle G, Henle W, Klein G. Demonstration of two distinct components in the early antigen complex of Epstein-Barr virus infected cells. *Int J Cancer* 1971, **8**, 272–82.

33 Zur Hausen H, O'Neill FJ, Freeze UK, Hecker E. Persisting oncogenic herpesvirus induced by the tumour promter TPA. *Nature* 1978, **272**, 373–5.

34 Henle W, Henle G, Zajac BA, Pearson G, Waubke R, Scriba M. Differential reactivity of human serums with early antigens induced by Epstein-Barr virus. *Science* 1970, **169**, 188–90.

35 Klein G, Clifford P, Klein E *et al*. Membrane immunofluorescence reactions of Burkitt lymphoma cells from biopsy specimens and tissue cultures. *J Natl Cancer Inst* 1967, **39**, 1027–44.

36 Klein G, Pearson G, Henle G, Henle W, Goldstein G, Clifford P. Relation between Epstein-Barr viral and cell membrane immuno-fluorescence in Burkitt tumor cells. III. Comparison of blocking of direct membrane immunofluorescence and anti-EBV reactivities of different sera. *J Exp Med* 1969, **129**, 697–705.

37 Klein G, Svedmyr E, Jondal U, Persson PO. EBV-determined nuclear antigen (EBNA) positive cells in the peripheral blood of infectious mononucleosis patients. *Int J Cancer* 1976, **17**, 21–6.

38 Welinder C, Larsson NG, Szigeti R *et al*. Stable transfection of a human lymphoma line by sub-genomic fragments of Epstein-Barr virus DNA to measure humoral and cellular immunity to the corresponding proteins. *Int J Cancer* 1987, **40**, 389–95.

39 Luka J, Chase RC, Pearson GR. A sensitive Enzyme-Linked Immunosorbent Assay (ELISA) against the major EBV-associated antigens. I. Correlation between ELISA and immunofluorescence titers using purified antigens. *J Immunol Methods* 1984, **67**, 145–56.

40 Brown K, Dept of Virology, The London Hospital (personal communication).

41 Randle BJ, Epstein MA. A highly sensitive enzyme-linked immunosorbent assay to quantitate antibodies to Epstein-Barr virus membrane antigen gp340. *J Virol Methods* 1984, **9**, 201–8.

42 Laemmli UK. Cleavage of structural proteins during the assembly of the head of bacteriophage T4. *Nature* 1970, **227**, 680–5.

43 Burnette WN. 'Western blotting': electrophoretic transfer of pro-teins from sodium dodecyl sulfate-polyacrylamide gels to un-modified nitrocellulose and radiographic detection with antibody and radioiodinated protein A. *Anal Biochem* 1981, **112**, 195–203.

44 Miller G, Shope T, Lisco H, Stitt D, Lipman M. Epstein-Barr virus: transformation, cytopathic changes and viral antigens in squirrel monkey and marmoset leukocytes. *Proc Natl Acad Sci USA* 1972, **69**, 383–7.

45 Rowe M, Finke J, Szigeti R, Klein G. Characterization of the serological response in man to the latent membrane protein and the six nuclear antigens encoded by Epstein-Barr virus. *J Gen Virol* 1988, **69**, 1217–28.

46 Paul JR, Bunnell WW. The presence of heterophile antibodies in infectious mononucleosis. *Am J Med Sci* 1932, **183**, 90–104.

47 Davidsohn I, Stern K, Kashiwagi C. The differential test for infectious mononucleosis. *Am J Clin Pathol* 1951, **21**, 1101–13.

48 Lee CL, Davidsohn I, Slaby R. Horse agglutinins in infectious mononucleosis. *Am J Clin Pathol* 1968, **49**, 3–11.

49 Smith RS, Rhodes G, Vaughan JH, Horwitz CA, Geltosky JE, Whalley AS. A synthetic peptide for detecting antibodies to Epstein-Barr virus nuclear antigen in sera from patients with infectious mononucleosis. *J Infect Dis* 1986, **154**, 885–9.

50 Geltosky JE, Smith RS, Whalley A, Rhodes G. Use of a synthetic peptide based ELISA for the diagnosis of infectious mononucleosis and other diseases. *J Clin Lab Analysis* 1987, **1**, 153–62.

51 Pi G, Zeng Y, Zhao W, Zhang Q. Development of an anti-complement immunoenzyme test for detection of EB virus nuclear antigen (EBNA) and antibody to EBNA. *J Immunol Methods* 1981, **44**, 73–8.

52 Golden HD, Chang RS, Prescott W, Simpson E, Cooper TY. Leukocyte-transforming agent: prolonged excretion by patients with mononucleosis and excretion by normal individuals. *J Infect Dis* 1973, **127**, 471–3.

53 Maniatis T, Fritsch EF, Sambrook J. Molecular cloning: a laboratory manual. Cold Spring Harbor: Cold Spring Harbor Laboratory Press, 1982.

54 Southern EM. Detection of specific sequences among DNA fragments separated by gel electrophoresis. *J Mol Biol* 1975, **98**, 503–17.

55 Teo CG, Griffin BE. Epstein-Barr virus genomes in lymphoid cells: Activation in mitosis and chromosomal location. *Proc Natl Acad Sci USA* 1987, **84**, 8473–7.

56 Saiki RK, Gelfand DH, Stoffel S *et al*. Primer-directed enzymatic amplification of DNA with a thermostable DNA polymerase. *Science* 1988, **239**, 487–91.

57 Loche M, Mach B. Identification of HIV-infected seronegative individuals by a direct diagnostic test based on hybridisation to amplified viral DNA. *Lancet* 1988, **2**, 418–21.

58 Henle W, Henle G, Andersson J *et al*. Antibody responses to Epstein-Barr virus-determined nuclear antigen (EBNA)-1 and EBNA-2 in acute and chronic Epstein-Barr virus infection. *Proc Natl Acad Sci USA* 1987, **84**, 570–4.

59 Holmes GP, Kaplan JE, Steward JA, Hunt B, Pinsky PF, Schonberger LB. A cluster of patients with a chronic mononucleosis-like syndrome: is Epstein-Barr virus the cause? *JAMA* 1987, **257**, 2297–302.

60 Borysiewicz LK, Hawroth SJ, Cohen J, Mundin J, Rickinson A, Sissons JGP. Epstein-Barr virus-specific immune defects in patients with persistent symptoms following infectious mononucleosis. *Q J Med* 1986, **58**, 111–21.

61 Hotchin NA, Read R, Smith DG, Crawford DH. Active Epstein-

Barr virus infection in post-viral fatigue symdrome. *J Infect* 1989, **18**, 143–50.

62　Tosato G, Straus S, Henle W, Pike SE, Blaese RM. Characteristic T cell dysfunction in patients with chronic active Epstein-Barr virus infection (chronic infectious mononucleosis). *J Immunol* 1985, **134**, 3082–8.

63　Schooley RT, Carey RW, Miller G, *et al*. Chronic Epstein-Barr virus infection associated with fever and interstitial pneumonitis: clinical and serological features and response to antiviral chemotherapy. *Ann Intern Med* 1986, **104**, 636–43.

64　Lenoir GM. Role of the virus, chromosomal translocations and cellular oncogenes in the aetiology of Burkitt's lymphoma. In: Epstein MA, Achong BG, eds. The Epstein-Barr virus: recent advances. London: Heinemann, 1986, 183–205.

65　Geser A, Lenoir GM, Anvret M, *et al*. Epstein-Barr virus markers in a series of Burkitt's lymphomas from the West Nile District, Uganda. *Eur J Cancer Clin Oncol* 1983, **19**, 1393–404.

66　de-Thé G, Geser A, Day NE, *et al*. Epidemiological evidence for causal relationship between Epstein-Barr virus and Burkitt's lymphoma from Ugandan prospective study. *Nature* 1978, **274**, 756–61.

67　Zur Hausen H, Schulte-Holthausen H, Klein G, *et al*. EBV DNA in biopsies of Burkitt tumours and anaplastic carcinomas of the nasopharynx. *Nature* 1970, **228**, 1056–8.

68　Henle W, Henle G, Ho JHC, *et al*. Antibodies to Epstein-Barr virus in nasopharyngeal carcinoma and other head and neck neoplasms and control groups. *J Natl Cancer Inst* 1970, **44**, 225–31.

69　de-Thé G, Zeng Y. Population screening for EBV markers: toward improvement of nasopharyngeal carcinoma control. In: Epstein MA, Achong BG, eds. The Epstein-Barr virus: recent advances. London: Heinemann, 1986, 237–49.

70　Young LS, Dawson CW, Clark D, *et al*. Epstein-Barr virus gene expression in nasopharyngeal carcinoma. *J Gen Virol* 1988, **69**, 1051–65.

71　Henle W, Ho JH, Henle G, Chan JC, Kwan HC. Nasopharyngeal carcinoma: significance of changes in Epstein-Barr virus-related antibody patterns following therapy. *Int J Cancer* 1977, **20**, 663–72.

72　de-Vathaire F, Sancho-Garnier H, de-Thé H, *et al*. Prognostic value of EBV markers in the clinical management of nasopharyngeal carcinoma (NPC): a multicenter follow-up study. *Int J Cancer* 1988, **42**, 176–81.

73　Ando I, Morgan G, Levinsky RJ, Crawford DH. A family study of the X-linked lymphoproliferative syndrome: evidence for a B cell defect contributing to the immunodeficiency. *Clin Exp Immunol* 1986, **63**, 271–9.

74　Cleary ML, Dorfman RF, Sklar J. Failure in immunological control of the virus infection: post-transplant lymphomas. In Epstein MA, Achong BG, eds. The Epstein-Barr virus: recent advances. London: Heinemann, 1986, 163–81.

75 Hanto DW, Frizzera G, Gajl-Peczalska KJ, *et al.* Epstein-Barr virus-induced B-cell lymphoma after renal transplantation: acyclovir therapy and transition from polyclonal to monoclonal B-cell proliferation. *N Engl J Med* 1982, **306**, 913–8.

76 Starzl TE, Porter KA, Iwatsuki S, *et al.* Reversibility of lymphomas and lymphoproliferative lesions developing under cyclosporin-steroid therapy. *Lancet* 1984, **i**, 583–7.

77 Purtilo DT. Opportunistic cancers in patients with immunodeficiency syndromes. *Arch Pathol Lab Med* 1987, **111**, 1123–9.

78 Ho M, Miller G, Atchison RW *et al.* Epstein-Barr virus infections and DNA hybridization studies in posttransplantation lymphoma and lymphoproliferative lesions: the role of primary infection. *J Infect Dis* 1985, **152**, 876–86.

79 Thomas JA, Hotchin NA, Allday MJ *et al.* Immunohistology of Epstein-Barr virus associated antigens in B cell disorders from immunocompromised individuals. *Transplantation* 1990, **49**, 944–53.

80 Greenspan D, Greenspan JS, Conant M, Petersen V, Silverman S, De Souza Y. Oral 'hairy' leucoplakia in male homosexuals: evidence of association with both papillomavirus and a herpes-group virus. *Lancet* 1984, **2**, 831–4.

81 Greenspan JS, Greenspan D, Lennette ET *et al.* Replication of Epstein-Barr virus within the epithelial cells of oral 'hairy' leukoplakia, an AIDS-associated lesion. *N Engl J Med* 1985, **313**, 1564–71.

82 Löning T, Henke RP, Reichart P, Becker J. In situ hybridisation to detect Epstein-Barr virus DNA in oral tissues of HIV-infected patients. *Virchows Arch [A]* 1987, **412**, 127–33.

83 Friedman-Kien AE. Viral origin of hairy leukoplakia. *Lancet* 1986, **2**, 694–5.

84 Henle W, Henle G. Epstein-Barr virus-specific serology in immunologically compromised individuals. *Cancer Res* 1981, **41**, 4222–5.

85 Henle G, Lennette ET, Alspaugh MA, Henle W. Rheumatoid factor as a cause of positive reactions in tests for Epstein-Barr virus-specific IgM antibodies. *Clin Exp Immunol* 1979, **36**, 415–22.

86 Edwards JMB. Antibodies to Epstein-Barr virus: IgG, IgM and IgA. In: Edwards JMB, Taylor CED, Tomlinson AH, eds. Immunofluorescence techniques in diagnostic microbiology, London: HMSO, 1982: 65–90 (PHLS monograph series No. 18).

Chapter 7

Rabies – recent developments in research and human prophylaxis

S Gardner and A King

There are few viruses, if any, as successful as rabies virus: it has survived for millennia; it thrives in most parts of the world; it can replicate in all warm-blooded animals, within which it secretes itself from their immunological defences; almost invariably the resultant disease is fatal, yet the virus ensures its own survival by causing the afflicted, when close to death, to find another host and victim. Even so, throughout history the virus has caused epizootics of disease which have appeared and then inexplicably disappeared, only to reappear at a later date. For example, rabies disappeared from central Europe at the beginning of this century, only for the present epizootic to arise in wildlife in the mid-1940s.

These peculiar modulations suggest that either the virus is able to change its properties to adapt to new environments (eg species) or some change in the host itself permits the virus to succeed without changing. Historical disease syndromes such as 'Oulo-fato' and Derriengue were suggestive of some virus variation, but the similarity of the clinical disease throughout the world and in such a wide variety of species led to the generally held view that it was caused by a single virus type, and that variations in the epidemiological pattern of disease were due to differences in species susceptibility. Not until the discovery of Lagos bat, Mokola and Duvenhage viruses, the rabies-related viruses, which could be distinguished from each other and from classical rabies virus on the basis of cross-protection tests in mice, was this view seriously challenged. This review will explore these recent developments and discuss how they can be used to control and prevent rabies infections in animals and man.

Identification of rabies virus

THE VIRUS

Rabies viruses are members of the Lyssavirus genus of the Rhabdoviridae and, in common with other Rhabdoviruses, possess an unsegmented negative strand genome which codes for five proteins: G, M, N, L and NS. The virion can be divided into two structural and functional units, the envelope and the nucleocapsid core [1]. The envelope contains glycoprotein (G) part of which is exposed in the form of 'spikes' on the virion surface, and the membrane protein (M) which is not exposed. The ribonucleocapsid core contains N (nucleo-), L (large) and NS (non-structural, but this is a misnomer) proteins and the RNA. Cross-reactions amongst the N proteins may be determined by indirect immunofluorescence and by complement fixation. The G protein is specific and several serotypes are now recognised by neutralisation and cross-protection studies (Table 1).

ANTIGENIC VARIATION

Antigenic differences amongst the rabies viruses can be detected by use of monoclonal antibody (Mab) techniques [2]. Mabs directed against the G-protein, which is responsible for both the attachment of virus to host cell and for the induction of humoral antibody within the host (Mab-Gs) are used in neutralisation tests. Those directed against the core proteins (Mab-Ns) are used in indirect tests on fixed material, usually after virus adaptation to cell culture. Since during infection core proteins are more abundantly produced, and fewer manipulations are required to detect them, Mab-N profiles of rabies viruses are more frequently studied.

Groups or panels of Mabs can be used for a variety of purposes. For example, crude polyclonal antisera for diagnostic techniques can be replaced by a pool of wide spectrum Mabs prepared in the laboratory and without further recourse to animals. Mabs can also be used for the

Table 1 Lyssavirus genus

Serotype	Virus	Host
1	Rabies	Warm-blood animals
2	Lagos bat	Bat (frugivorous), cat
3	Mokola	Shrew, dog, cat, man and rodent
4	Duvenhage	Bat (insectivorous), man
–	Kotonkan	Midge
–	Obodhiang	Mosquito

differentiation of rabies viruses: they can be used in the surveillance of the rabies viruses (street or vaccine) causing death in oral vaccination areas (see p. 146); and antigenic differences in isolates from geographically-separate outbreaks can be identified. This latter use, combined with epidemiological surveillance, has been instrumental in improving the understanding of the complexities of intra- and inter-species transmission of disease [3]; this understanding is essential if control programmes are to be initiated.

Epidemiology

Rabies is a zoonosis which continues to spread, especially in wildlife. The prevalent type of rabies in mainland Europe is fox rabies, whereas in Asia it is canine rabies. The wolf still plays an important role as vector along with the dog in some regions of Iraq, Iran, Afghanistan and some republics of the Soviet Union. In recent decades the raccoon dog, transported westwards from Siberia for fur and hunting purposes, has become a prolific vector of the disease in the western Soviet Union and in Poland.

The rabies type currently prevalent in Africa is canine, but in southern Africa sylvatic rabies independent of the canine cycle is found in the yellow mongoose and in the herbivorous kudu antelope. Antelope rabies became established in Namibia in 1977 [4] and claimed the lives of 50,000 antelope annually until 1983, since when the epidemic has spontaneously regressed [5].

In North America, 40 years of vaccination and pet animal control has greatly reduced canine rabies, but the disease is enzootic in foxes, skunks and raccoons. Within these species, compartmentation [6] occurs; that is, the disease is reported in one major host species in certain geographical areas while it is reported much less frequently in the same species in other areas of endemic rabies. By their elegant studies combining surveillance, inter-species transmission experiments and Mabs analyses, Smith and Baer [3] have shown that host susceptibility, pathogenicity and virus variation are all important factors in single-species involvement of disease, and they have drawn attention to the need for further research in these areas if control programmes are to be successful. Rabies in insectivorous bats accounts for some 15% of all rabies cases in the USA and Smith and Baer have also shown that spillover of disease from bats to terrestrial animals occurs rather more frequently than was at one time thought, although there is no suggestion that at present these bat rabies viruses cause cycles of disease in terrestrial animals.

In central and South America, canine rabies is the cause of many human deaths and vampire bat rabies is responsible for severe economic losses in cattle. Little information is available on rabies in the terrestrial wildlife of this vast area.

BAT RABIES IN EUROPE

Only since the 1950s has the extent of rabies in bats other than vampires been recognised, although in 1931 the disease was diagnosed in frugivorous and insectivorous species in addition to vampires in Trinidad [7]. All of the 31 European bat species are insectivorous, and rabies was confirmed in European bats for the first time in 1954 [8]. Throughout the following 30 years only 14 cases were recorded [9], but since 1985 a further 330 cases have been identified (Figure 1), mostly from serotine bats, *Eptesicus serotinus*, although representatives from at least five other species have also been found positive. Mabs analyses have shown the serotine isolates to be serotype 4 viruses as distinct from the serotype 1 viruses which cause terrestrial animal rabies. The virus which led to the death from rabies, believed to be of bat origin, of a biologist in Finland is distinct from all other rabies viruses so far examined, and is closely related to isolates from two *Myotis dasycneme* bats made by Dr J Haagsma in Holland [10]. These are the only virus isolates to have been made from bats other than serotines.

In each of the eight countries with recent infection in bats, the disease came to light by the accidental involvement of a member of the public. Although these contacts have resulted in three human deaths, including two in the USSR, present indications are that current post-exposure treatment available in Europe prevents human infection by the European bat rabies viruses. Three people bitten by confirmed rabid bats and treated with human diploid cell rabies vaccine and human anti-rabies immunoglobulin have remained well [11]. Protection studies in animals using the Finnish and *Myotis dasycneme* isolates have recently commenced at the CDC Atlanta, USA.

Of over 500 bats found sick or dead in the past three years, from 11 of the 15 indigenous British species submitted to Weybridge for rabies immunofluorescence examination, none have been found positive [12]. Representatives of three of the species (*E serotinus*, *Pipistrellus pipistrellus* and *M daubentoni*) found positive in Europe also reside in the UK and have been included in the survey.

Rabies control

URBAN

Canine rabies accounts for more than 99% of all human rabies cases and over 90% of all human post-exposure treatments recorded worldwide. Yet in the past, even before Pasteur, Scandinavian countries were able to rid themselves of it by sanitary measures including stray dog control. Other countries, including the UK, have used these techniques allied with quarantine and/or vaccination to eradicate and then maintain freedom from the disease. Currently the importation of mammals into the UK is controlled by the Rabies (Importation of Dogs, Cats and

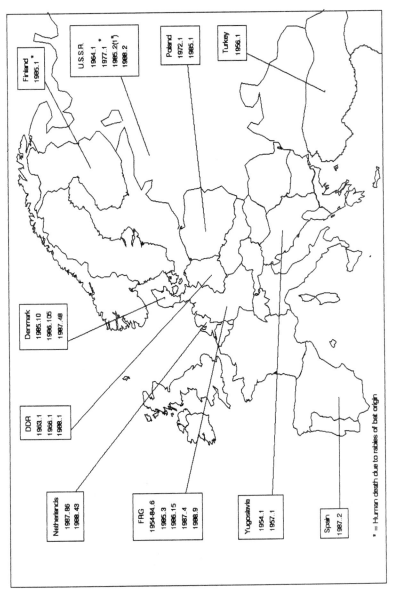

Figure 1   Bat rabies in Europe

Other Mammals) Order 1974 (as amended). It applies to a wide range of mammals but excludes farm livestock including horses, which are covered by separate regulations. These exempted animals may become subject to controls under the Importation Order if they have been in contact with any of those species subject to quarantine regulations. Under the Order, the animals must be licensed in advance to authorised quarantine premises and must be landed at a port or airport designated in the Order and transported to the quarantine premises by an authorised carrying agent, where they must be detained for a period of six months (life in the case of vampire bats). Dogs and cats must be vaccinated on arrival at the quarantine station. In North America canine rabies has been reduced from about 8000 cases a year shortly after the last war to fewer than 200 cases a year for the past few years, and human deaths from rabies have concomitantly declined from 94 in the four-year period 1946–49 to fewer than 10 in the 1980s [13–15]. In greater Buenos Aires, with a population of over seven million, an urban rabies control programme initiated in 1976 reduced canine rabies cases from over 4000 in the first year to zero by 1985. At the same time the number of human deaths from rabies fell from 10 in the first year to only one between 1980 and 1985 and the need for post-exposure treatment fell from over 45,000 in the first year to less than 16,000 in 1985, and continues to fall [16]. Guidelines for dog rabies control have been developed by WHO [17] and it was hoped that national programmes would be organised by all member states and a real effort made to eliminate the disease. Effective animal vaccines are available and if used to vaccinate 80% of the dog population would result in a substantial reduction of human rabies. Cat rabies may also be a problem in some countries but it has been found that when canine rabies is eliminated from urban areas cat rabies generally disappears. Thus the technology to control canine rabies by parenteral vaccination exists; all that is required is political will, education of the public and sufficient resources.

WILDLIFE

Canine rabies can be controlled because in general, dogs live in close association with man and are therefore within physical reach. This is rarely the case with wildlife. The reduction of stray dog and wildlife populations in order to break the chain of natural transmission has not succeeded, chiefly because wildlife species are able to repair their losses at a fast rate.

The findings that primary infection of experimental mice could take place in the buccal and lingual mucosa [18] and that live attenuated rabies virus given orally could immunise foxes [19,20] together formed the seed from which has blossomed the present-day programmes of wildlife rabies control. In the meantime research has been directed towards the development of vaccines efficacious for target species (the fox in Europe); safety for target and non-target species (vaccine was tested in all the rodent species indigenous to central Europe [21]); and effective measures to deliver the vaccine to the target species in the form of a bait.

Despite the satisfactory nature of all the proving trials, it was a courageous decision to disseminate live attenuated rabies virus in an attempt to create an immune barrier at the entrance to the Rhone Valley in 1978. The vaccine virus was contained in small plastic blister packages fixed under the skin of chicken heads used as bait, and 4,050 of these were distributed by hand over an area of approximately 335km^2 [22]. Rabies did not advance along the Rhone Valley.

With continued field trials, Switzerland has been freed of rabies except for incursions from highly endemic areas along its borders. The Federal Republic of Germany, following their introduction of a machine-manufactured bait, has cleared the disease from large parts of the country. Italy, Austria, Belgium, Luxembourg and France have been collaborating in other field trials. Pilot field trials of a similar nature, but with variations in immunising virus and bait presentation form, are also under way in Canada [23]. The next step forward may involve the use of recombinant vaccines: several of these are under construction and one has been extensively tested and shown to be efficacious in a variety of species [24].

Progress has also been made in the prevention or rabies in man independently of the control of rabies in dogs and wildlife.

Human vaccines

HUMAN DIPLOID CELL VACCINE (HDCV)

In 1976 a major turning point was reached in post-exposure treatment. For the first time a potent and safe vaccine was available to treat patients who had been bitten or otherwise exposed to potentially rabid animals. Ever since the time of Pasteur, the most widely used vaccine had been of nervous tissue origin. Not only was this type associated with serious neurological side effects, it was also often ineffective in preventing rabies. To overcome neuroparalytic accidents, other kinds of vaccines had been produced. One was made in the brains of newborn rats, mice or rabbits and the other, duck-embryo vaccine, in avian tissues. Reactions to these newer vaccines still occurred, although less frequently, but neither type was sufficiently antigenic for the number of doses to be reduced. Progress was made in the early 1960s when Wiktor and his colleagues succeeded in adapting rabies virus to grow in the human diploid cell strain WI–38 [25], which they did by subculturing suspensions of trypsinised, infected cells with uninfected cells. It took a further 10 years' research and development to produce a vaccine and to establish its immunogenicity and safety before it was acceptable for human use.

The greater potency of the human diploid cell rabies vaccine meant that fewer doses were needed. Instead of the traditional treatment, 14–21 daily doses of rabies vaccine followed by boosters, a satisfactory antibody response was achieved with just five or six spaced injections. The current vaccination schedule, recommended by the World Health Organization for vaccines with a minimum potency 2.5IU/ml, is for five or six doses to be given on days 0, 3, 7, 14, 30 and 90 [26]. The last dose is optional but is generally advised in the UK. The vaccine (1ml) should be given by deep subcutaneous or intramuscular inoculation into the upper arm.

Between June 1975 and January 1976 this HDCV post-exposure schedule was used for the first time to treat 45 people bitten, many severely, by confirmed rabid dogs and wolves in Iran [27]. Forty-four of the patients were also passively immunised with a mule anti-rabies serum. None of the people treated developed rabies.

The UK was one of the first countries to license HDCV. The vaccine was used from November 1975 for pre-exposure vaccination of people at occupational risk and from May 1976 for post-exposure treatment; whereas in the USA it was not licensed until June 1980. The UK was able to take early advantage of the new vaccine because the Medical Research Council had already tested HDCV in volunteers using a pre-exposure schedule of two doses given four weeks apart [28]. All the vaccinees studied produced satisfactory levels of neutralising antibody and none experienced serious side effects.

During the 13 years since HDCV was licensed here more than 80,000 doses have been given to people in England and Wales. The number of patients treated annually following exposure to possible rabies is shown in Table 2.

In the UK and other industrialised countries, post-exposure treatment is now more comfortable for the patient and easy for the doctor to administer. It is reassuring for both to know that treatment failures with HDCV are exceptional. However HDCV is not available to developing countries because of its high cost. Furthermore it is in those countries, where the dog is the principal vector and reservoir of the virus, that rabies is a major public health problem. The only vaccine available to the majority of the population in these places is still the traditional Semple-type nervous tissue vaccine made from the brains of infected sheep, goats and rabbits. For this reason, research has been directed to try to reduce the cost of post-exposure treatment and to make potent, safe vaccines more accessible to those where the risk from rabies is greatest.

Table 2   Post-exposure treatment in England and Wales, 1975–88

Year	Number of people treated
1975	126
1976*	181
1977	213
1978	216
1979	336
1980	306
1981	930
1982	757
1983	586
1984	548
1985	463
1986	709
1987	908
1988	806

*Human diploid cell rabies vaccine used from May 1976

SECOND-GENERATION CELL CULTURE VACCINES

A number of second-generation cell culture vaccines have been developed in recent years using cells more easily grown in bulk than are human diploid cells. Different virus strains have been used, and also alternative methods of purification and concentration. This has allowed the cost of vaccine to be reduced. Many of these vaccines are commercially available worldwide and, not infrequently, people return to the UK having started a course of treatment with one of them. At present only HDCV, made by Institut Mérieux in France, is licensed for human use in the UK, so treatment is continued with this vaccine. No problems have arisen from a change of vaccine mid-course and rabies antibody titres have been satisfactory in those people who were given more than one type of cell culture vaccine. Rabies vaccines with a similar potency to HDCV have been produced in fetal bovine cells, dog kidney cells, chick embryo fibroblasts and in a continuous heteroploid cell line of monkey origin (Vero), which has been approved as a suitable substrate for vaccine production. Clinical trials have been carried out both in exposed and unexposed people.

The two most widely used of these new rabies vaccines, apart from HDCV, are Rabipur®, a purified chick embryo cell vaccine (PCEC, Behring, West Germany), and Verorab® or Imovax®, made in Vero cells by Institut Mérieux, France. Behring also make Rabivac® which is the only HDCV where continuous zonal centrifugation is used to purify and concentrate rabies virus. Antigens purified by this method give fewer side effects.

PURIFIED CHICK EMBRYO CELL VACCINE (PCEC)

The rabies virus used in this vaccine is Flury–LEP (low egg passage) strain which was originally grown in chick embryos. The virus was adapted to grow in primary hamster kidney and then passaged in human diploid cells and finally adapted to primary chick fibroblasts [29]. The antigen was inactivated with beta propiolactone (BPL) then purified and concentrated by zonal centrifugation. Trials of PCEC in veterinary students showed that satisfactory levels of rabies-neutralising antibody were produced; the response was similar to that achieved with HDCV. Both general and local side effects were reported by the students but IgE antibody to avian proteins was not induced. However, Interferon, which may help prevent rabies from developing, was present within the first 30 hours after vaccination [30]. Each dose was adjusted to >2.5IU/ml.

Sixty-nine patients, 25 of them children, were treated in Thailand following exposure to laboratory-confirmed rabies in dogs and cats [31]. All were given 1ml PCEC on days 0, 3, 7, 14, 30 and 90 as recommended by WHO; about half of them were also given human rabies immunoglobulin (HRIG). No one developed rabies and there were few side effects.

PCEC rabies vaccine is as immunogenic as HDCV and side effects are similar. Urticarial reactions have been reported in atopic individuals. Each dose of PCEC costs about one third the price of HDCV [32].

VERO CELL VACCINE (PVRV)

Vero cells developed from *Cercopithecus aethiops* kidney cells were used as cell substrate and grown on microcarriers in large tanks [33]. The rabies virus strain was similar to that used to prepare HDCV and was inoculated during the growth phase of the cells. Infectious virus titres of $10^{6.6}$ $TCID_{50}$/ml were achieved and multiple harvests collected. Concentration was obtained by ultrafiltration and inactivation was with BPL. The vaccine was then purified by zonal centrifugation and stabilised with human albumin. Each dose contained 2.5IU or more of rabies antigen in 0.5ml [33].

Vero cell rabies vaccine was used to vaccinate both unexposed and exposed people in France [34]. A three-dose pre-exposure course (days 0, 7 and 28) produced a better antibody response at six weeks than a two-dose schedule (days 0 and 28). Ninety people exposed to suspected rabid dogs, cats and bovines were treated with the vaccine and remained well during a two-year follow-up period.

PVRV vaccine also underwent trials in Thailand [35]. One hundred and six patients bitten by laboratory-confirmed rabid animals were given six doses of vaccine, according to the WHO-recommended schedule.

Forty-seven also received human rabies immunoglobulin. Rabies antibody was detected in all patients on day 14 and persisted for one year, but again none developed rabies.

Alternative post-exposure regimens

The development of second-generation cell culture vaccines has caused some reduction in the cost per dose. Nevertheless, these newer vaccines are still too expensive for developing countries to use routinely. Various regimens have therefore been investigated in an attempt to either reduce the number of doses given after exposure or to give smaller volumes by the intradermal (ID) route, or both.

The ID route was shown in 1975 to be as effective in inducing a rabies-neutralising antibody response as the intramuscular route [28]. Furthermore Turner and his colleagues [36] reported that seven of ten volunteers given 0.1ml HDCV ID in four different sites developed rabies-neutralising antibody by seven days after inoculation. In a later study Nicholson and his colleagues [37] also showed that when 0.1ml HDCV was given ID in eight sites, not only was a high level of rabies antibody produced but there was a cell-mediated immune response. These results were important in designing suitable regimens for post-exposure treatment.

Optimum therapy as recommended by WHO [26] consists of the combined use of antirabies serum or HRIG and vaccine. Passive immunisation in most developing countries is achieved by the use of an animal serum. This is not always available and if given may be associated with hypersensitivity reactions. An early antibody response is therefore desirable when vaccine alone is given. The aim of post-exposure treatment is to prevent rabies virus becoming irreversibly attached to nerves and inaccessible to host defences. This is achieved firstly by thoroughly washing all wounds with plenty of soap and water to eliminate as much virus as possible and secondly by stimulating an immune response. Neutralising antibody, cell mediated immunity and interferon may all have a role in preventing rabies.

A rapid antibody response is achieved in a person previously vaccinated by giving booster doses of an immunogenic vaccine. A booster response is seen even after an interval of several years from the primary course of vaccination [38]. A rise in antibody from undetectable or detectable levels occurred within two to seven days in the majority of those studied [38–42]. Moreover, the rabies antibody rise was faster in those previously immunised than in those undergoing a primary course of vaccination either pre- or post-exposure, maximum titres being reached between 8 and 18 days compared with 28 and 38 days. Clearly the delay in the production of rabies-neutralising antibody may be of greater consequence in those previously unvaccinated.

Several regimens of vaccination have been shown to induce an early antibody response and some have been used effectively in field trials. Warrell and her colleagues in 1983 [43] investigated three economical regimens and compared these with the schedule recommended by WHO for cell culture and nervous tissue derived vaccines. They also studied the effect of adjuvant and concurrent use of HRIG. Their results showed that although the WHO-recommended six-dose schedule for cell culture vaccines was generally superior, a satisfactory antibody response could be achieved by giving one tenth of the dose ID or subcutaneously with adjuvant. However, only if 0.1ml HDCV was given ID in eight different sites on day 0 could an early antibody response be induced by day 7. This was confirmed in a later study [44] and it was suggested that the safest and most economical regimen was to give 0.1ml HDCV ID into eight sites on day 0, four sites on day 7 and one site on days 28 and 91. This regimen uses only 1.7 doses of vaccine, compared with six doses given on days 0, 3, 7, 14, 30 and 90 in the standard WHO schedule recommended for vaccine with an antigenic potency >2.5IU/ml. Also the number of clinic visits is reduced from six to four. Seventy-eight exposed patients in Thailand were treated with this regimen [45]. In 36, the exposure was severe and the patients were also given equine anti-rabies serum. All had been bitten by confirmed rabid dogs or cats but none developed rabies. This regimen has been used successfully with HDCV and also the new Vero cell rabies vaccine [35].

The proportion of people developing antibody by day 7 with this eight-site regimen has varied in different studies from 20–100% (43–47); generally only small numbers of vaccinees were studied. However, only between 10% and 33% developed antibody levels >0.5IU/ml, which is the arbitrary level recommended by WHO as giving adequate levels of protection. In one study [46], none reached this level. An improvement in the numbers of people developing antibody early was achieved by Ubol and Phanuphak [48]. They gave 0.1ml HDCV ID at four sites on days 0, 3 and 7 and 0.1ml ID at single sites on days 28 and 91. With this regimen, 50% of patients had >0.5IU/ml rabies antibody by day 7. This regimen also induced prompt cell-mediated immunity. Only 1.5 doses of vaccine were used, so this may be a better schedule than the eight-site one when passive immunisation is unavailable.

In non-exposed volunteers, a multi-site ID regimen for post-exposure treatment has also been shown to be satisfactory with the new PCEC vaccine [47].

Although these multiple-site ID methods have advantages in the reduction of cost and clinic visits they are still more expensive than regimens using Semple-type nervous tissue vaccines.

Haverson and Wasi [49] reduced the cost further by giving just 0.1ml HDCV on day 0, 3, 7, 14 and 28 for a cost of £13 for five injections. This schedule was used in field studies in Thailand and 219 exposed persons were treated successfully. No case of rabies occurred and at one year after treatment with four or five doses, all 31 patients studied had residual rabies antibody; in 23 (74%) the titre was 0.5IU/ml or more. Although this regimen is not ideal, it is to be preferred when money is restricted and the only alternative is nervous tissue vaccine.

In Yugoslavia a different regimen has been investigated [50]. Vodopija and his colleagues gave four doses of vaccine in a 2–1–1 post-exposure schedule. Two doses of vaccine, 1ml intramuscularly into each arm, were given on day 0 and single doses on days 7 and 21. With this regimen HDCV, PCEC and PVRV induced antibody by day 7 in 65%, 83% and 78% respectively of vaccinees and in all of them antibody was still present on day 90.

Treatment failures

HDCV has been used to treat many thousands of people exposed to possible rabid animals during the past 12 years and its efficacy has been outstanding. However it would be surprising if no vaccination failures had occurred with cell culture vaccines, as most of those treated would not have been previously immunised and vaccination therefore was started after the person had been infected.

The first treatment failure occurred in 1981 [51]. An American citizen living in Rwanda was attacked by a dog and bitten on the right foot and possibly also the left hand. She was given the first dose of HDCV on the same day as the incident and received three further doses on day 3, 7 and 14. Twenty days after the bites she developed fever and weakness of the left arm and eventually flaccid paralysis of all four limbs and respiratory failure. She died six weeks after the onset of the illness despite intensive medical care. Although rabies virus was not isolated from the patient, high levels of rabies antibody in both serum and CSF, together with pathological findings of acute encephalitis, supported the diagnosis of rabies.

The very short incubation period of 20 days probably meant that virus invaded nerve endings shortly after exposure and before antibody had had time to develop despite prompt vaccination. She was not, however, given passive immunisation and therefore had not received optimum treatment.

Since this case was reported there have been others. At least 16 people treated with cell culture vaccines after exposure have died of rabies [52]. Vaccination failures have occurred with HDCV, PCEC and PVRV. All of the patients had major exposures and in the majority the incubation period was short, 21 days or less. Treatment was frequently not started promptly within 24 hours of the biting incident and only half received combined serum and vaccine.

At least one person has died despite optimum treatment including local infiltration of the wound with HRIG [53]. Unfortunately, this patient was given the vaccine into the gluteal region and not into the deltoid and only low levels of rabies antibody were present at onset of symptoms 21 days after he was bitten on the finger by a yellow mongoose. There were several explanations for failure to develop a satisfactory immune response. These included interference of active immunisation by concurrent use of HRIG and also the possibility that this particular patient may have been a poor responder to vaccines. It is essential, therefore, that rabies vaccine is administered in the best site for achieving an antibody response. When the standard WHO schedule for cell culture vaccines is used, the site preferred is the deltoid region. A different site should be used for concomitant administration of HRIG or anti-rabies serum.

Correct post-exposure treatment for all major exposures in a previously unvaccinated person consists of:

1 Immediate, thorough washing of wound(s) with soap and water.
2 HRIG 20IU/kg body weight or heterologous antirabies serum 40IU/kg body weight – up to half the dose is infiltrated into the wound area and the remainder is given intramuscularly into the gluteal region or lateral thigh.
3 A potent cell culture vaccine – 1ml given on day 0, 3, 7, 14, 30 and 90 into the deltoid region.
4 Rabies antibody test on day 30 to ensure the patient has developed a satisfactory antibody response.

A person who has had rabies vaccine before exposure should receive, as soon as possible after exposure, one to four booster doses of a vaccine with an antigenic potency >2.5IU on days 0, 3, 7 and 90. The number of doses given depends on the clinical details and previous vaccination history. Systemic passive immunisation should not be given [26].

Pre-exposure prophylaxis

Now that potent, safe rabies vaccines are available, they are increasingly being given not only to people at occupational risk but also to those travelling or working in endemic regions.

Several different schedules have been used. WHO recommends that three doses of rabies vaccine with a potency >2.5IU/ml are given on days 0, 7 and 28 or 0, 28 and 56, with further booster doses every one to three years while the person is at risk [26].

The schedule recommended in the UK is a primary course of two doses of 1ml HDCV with a four-week interval between each dose; a reinforcing dose is given after 12 months. Further doses are given every one to three years depending on risk, and after any possible exposure to rabies virus. A serum is tested for rabies antibody three to four weeks after the primary course. With this schedule only 1–2% of people fail to sero-convert and require additional doses [54].

The ID route may also be used. As discussed above, it is effective providing the inoculation is done properly. Where groups of people require vaccination at the same time, the cost is reduced. Several clinics in various parts of the country are now offering this service to travellers [55]. For those at greatest risk the intramuscular route is advised [54].

People entitled to be offered rabies vaccine free of charge are listed in *Immunization against Infectious Disease* published by the Department of Health, Welsh Office and Scottish Home and Health Department (1988). People in most of the categories are at risk in the UK because they come into contact with imported animals. The UK is a rabies-free country with strict quarantine regulations and rabies vaccine is therefore not intended for anyone handling native animals with the exception of bats.

Rabies vaccine is also available free of charge for people working with animals in countries where rabies is endemic. For others travelling or working in endemic areas vaccine should be obtained from commercial sources.

Rabies vaccine reactions

The cell culture rabies vaccines have an excellent safety record and millions of doses have been given worldwide. Nevertheless, as with most vaccines, side effects have been reported, and in the UK have occurred in about 20% of those given a prophylactic pre-exposure course of HDCV [54]. Local reactions were present in 15% of those vaccinated and pain, erythema and swelling at site of inoculation were the most common complaints.

Systemic symptoms were less frequent but influenza-like illness was reported by 2% and headache, fever, fatigue and malaise were typical responses. It is possible that all these symptoms were caused by interferon production.

Hypersensitivity reactions developed in about 1% of vaccinees and in some instances were severe [54]; clinically the majority of these were Type III hypersensitivity reactions and consisted of urticaria, pruritic rashes and oedema. Only one case of anaphylaxis was reported and this person had also been given typhoid vaccine.

Hypersensitivity reactions to HDCV have also been reported in the USA [56]. One hundred and eight clinical reports of allergic reactions were analysed by the Centers for Disease Control (CDC), Atlanta, USA. Eighty-seven cases were Type III, twelve were indeterminate type and nine were Type 1 reactions. Type III reactions occurred between two and 21 days after HDCV, and in the majority were associated with booster doses.

Such Type III reactions have been investigated by two groups to determine the antigen and antibody responsible [57,58]. Their results suggested that BPL, which is used to inactivate rabies virus, modifies non-viral human proteins, eg human albumin (HA) present in HDCV.

Individuals sensitised at primary immunisation developed IgE antibody to the BPL–HA complex with resulting urticarial reactions after subsequent booster doses.

Neuroparalytic reactions following immunisation with HDCV have been reported in two children at least temporally associated with the use of HDCV [59,60]. Both children had a Guillain-Barré-like syndrome from which they recovered completely.

Peripheral neuropathy has been reported in two patients in the UK, following vaccination with HDCV [54] and in another patient in France given fetal bovine cell rabies vaccine [61]. The latter patient recovered after two plasma exchanges. The risk of possible neurological complications, albeit low, and hypersensitivity reactions should be considered when advising rabies vaccination either pre- or post-exposure for those at minimal or no risk of exposure to rabies.

Epidemiology of post-exposure treatment in the United Kingdom

Currently about 800 people are given anti-rabies treatment annually in the UK (Table 2). The majority have either been bitten by an animal or had a non-bite exposure in a country where rabies is endemic. Some patients are bitten by confirmed rabid animals but for others the risk from rabies is small. Unfortunately there is often insufficient information available to reassure that there is no risk, as the dog or cat involved in the biting incidents has not been observed for a period of 10 to 14 days. Many people may therefore be treated unnecessarily for want of information. The animals involved in incidents which resulted in post-exposure treatment in 1988 are given in Table 3. Sixty-six per cent

Table 3   Animal responsible for post-exposure treatment in 1988

Animal	Number of exposures
Dog	505 (66%)
Cat	154 (20%)
Other domesticated species	
Horse (6); Donkey (2); Camel (2); Cow (1)	11 (1%)
Wild mammals	
Monkey (69); Rodent (9); Bat (5); Squirrel (5); Hyrax (3); Otter (2); Potto (2); Leopard (1); Lion (1); Mongoose (1); Deer (1); Fox (1)	100 (13%)
Human	16
Not recorded (probably all dogs)	20
Total	806

of people who reported incidents were exposed to dogs and 20% to cats. Wild mammal encounters accounted for 13%. There was 5% increase in wild mammal bites abroad in 1988 compared with those reported in 1982 [54], with monkeys accounting for 9% of bites compared with 5% in 1982.

Dog bites are extremely common everywhere. The British dog population was estimated to be six million in 1979 and there were possibly over 200,000 dog bites [62]. In the USA every year, at least one million people are bitten [63]. Fortunately the UK is rabies-free and has strict quarantine regulations, and therefore anti-rabies treatment is only considered if the animal is known to have been imported into the UK. Not many countries enjoy our favourable position.

The geographical region where the exposures took place is shown in Table 4. The same pattern is seen as was found in 1982 [54]. Approximately half the incidents occurred in Europe (which includes Turkey). During 1988 there were 16,075 cases of animal rabies in Europe: 81.7% in wild mammals and 18.2% in domesticated species [64]. Altogether there were 1,462 rabid dogs and cats reported to the WHO Collaborating Centre for Rabies, with Turkey notifying one third (560) of them. Seventeen per cent of British citizens bitten in Europe during 1988 were exposed in Turkey. Overall 52% of all patients given anti-rabies treatment were at genuine risk, having been bitten in countries where the dog is the principal vector and reservoir and human deaths from rabies are not uncommon.

Human rabies cases in the United Kingdom

There have been 22 human rabies cases in the UK this century. The last indigenous case was in Wales in 1902; all the others were infected abroad. Eleven people have died of rabies since 1975 and details of these patients are given in Table 5. None were given post-exposure treatment. Eight of those bitten were adults and three were children. The biting animal was known to be a dog in 10 instances. The site of the

Table 4 Geographical region where biting incident occurred in 1988

Region	Number of exposures
Europe	453 (57%)
Asia	182 (23%)
Africa	121 (15%)
America	33 (4%)
Not recorded	17
Total	806

Table 5   Human deaths from rabies in the United Kingdom, 1975–88

Case no	Year	Age (years)	Country where bitten	Vector	Site of bite	Incubation period
1	1975	30	The Gambia	Dog	Lip	6 weeks
2	1975	22	India or Nepal	Dog	Not known	Not known
3	1976	53	Bangladesh	Not known	Not known	>14 months
4	1977	11	Pakistan	Dog	Thigh	2 months
5	1977	4	India	Dog	Knee	2 months
6	1978	55	India	Dog	Hand	2½ months
7	1981	23	India	Dog	Leg	2 months
8	1986	45	Zambia	Dog	Arm	2 months
9	1987	8	India	Dog	Leg	2 months
10	1988	42	Bangladesh	Dog	Arm	6 months
11	1988	34	Pakistan	Dog	Not known	>6 months

bite where known varied; four were on a lower limb, three on an upper limb and one on the lip. The incubation period was also variable: the shortest was six weeks and the longest more than 14 months.

The incubation period in humans is usually 20–60 days, and this was so for seven patients. Three patients had an unusually long incubation period. The patient with the longest incubation period was thought not to have been outside the UK for 14 months prior to onset of symptoms. There was no doubt about the diagnosis in this case, as rhabdovirus particles were seen by electron microscopy (Dr AM Field, personal communication).

All eleven patients listed in Table 5 were bitten in countries where rabies is a serious problem. These deaths might have been prevented by prompt post-exposure treatment. Animal bites in such countries should not be ignored.

Future research

Today, a little over a century since Pasteur developed the first rabies vaccine, the basic technology to control and eradicate the disease is known. Yet there are surprising gaps in our knowledge awaiting future research. How does the virus gain access to the central nervous system? Where in the body and in what form is the virus during an incubation period which may last weeks or months and sometimes even years? How is the virus able to alter the behavioural mechanisms of aggression? What is the role played by its ribonucleoprotein in the induction of immunity?

Much of the credit for improvements in the general knowledge of rabies belongs to the Veterinary Public Health Unit of the WHO. It has provided the platform for consultations and international meetings on such diverse aspects as surveillance and control in carnivores, post-exposure treatment, potency testing of human and veterinary vaccines and European bat rabies.

Future wildlife rabies control leading to eradication requires further research into vaccines (modified live virus and/or recombinant), bait presentation and delivery systems which will reach target species other than the fox. The ultimate goal of canine rabies eradication will not be achieved unless research reveals an orally-applied immunising agent, which can be packaged to whet the appetite of the dog, but which cannot infect the dog or its best friend, be it man or animal.

In the human rabies field the most pressing problem is one of logistics: how to meet vaccine requirements in canine rabies-endemic areas at affordable cost. A corollary is the production of antiserum for use in post-exposure treatment – could further research into human cell-derived Mabs to replace animal-derived polyclonal antiserum be rewarding?

References

1   Tordo N, Poch O. Structure of rabies virus. In: Campbell JB, Charlton KM (eds) Rabies. Boston: Kluwer, 1988, 25–45.

2   Wiktor TJ, Koprowski H. Monoclonal antibodies against rabies virus produced by somatic cell hybridization: Detection of antigenic variants. *Proc Natl Acad Sci USA* 1978, **75**, 3938–42.

3   Smith JS, Baer GM. Epizootiology of rabies: the Americas. In: Campbell JB, Charlton KM (eds) Rabies. Boston: Kluwer, 1988, 267–99.

4   Barnard BJ, Hassel RH. Rabies in kudus (Tragelaphus strepsiceros) in South West Africa/Namibia. *J South Afr Vet Ass* 1981, **52**, 309–14.

5   Schneider HP. Rabies in South West Africa/Namibia. In: Kuwert E, Merieux C, Koprowski H, Bogel K (eds) Rabies in the Tropics. Berlin: Springer-Verlag, 1985, 520–35.

6   Bisseru B. Rabies. London: Heinemann, 1972, 155.

7   Pawan JL. The transmission of paralytic rabies in Trinidad by the vampire bat (Desmodus rotundus murinus Wagner, 1840). *Ann Trop Med Parasitol* 1936, **30**, 101–30.

8   Mohr W. Die Tollwut (Rabies). *Med Klin Berl* 1957, **52**, 1057–60.

9   Bat rabies cases 1984–86. *Rabies Bull Europe* 1986, **10(4)**, 12, 14.

10 Haagsma J, King AA. Unpublished data.

11 King A, Crick J. Rabies-related viruses. In: Campbell JB, Charlton KM (eds) Rabies. Boston: Kluwer, 1988, 177–99.

12 King AA, Davies PK. Unpublished data.

13 Centers for Disease Control. *Rabies Surveillance Annual Summary 1980–82*, August 1983.

14 Centers for Disease Control, Rabies surveillance 1986. Appendices *MMWR* 1987, **36**, No. 3S 15S–27S.

15 Centers for Disease Control, Rabies surveillance, United States, 1987. *MMWR* 1988, **37**, No. SS4.

16 Larghi OP, Arrosi JC, Nakajata-a J, Villa-Nova A. Control of urban rabies. In: Campbell JB, Charlton KM (eds) Rabies. Boston: Kluwer, 1988, 407–22.

17 WHO Guidelines for dog rabies control. Geneva, 1984, VPH/ 83.43.

18 Correa-Giron EP, Allen R, Sulkin SE. The infectivity and pathogenesis of rabiesvirus administered orally. *Am J Epidem* 1970, **91**, 203–15.

19 Black JG, Lawson KF. Sylvatic rabies studies in the silver fox (Vulpes vulpes). Susceptibility and immune response. *Can J Comp Med* 1970, **34**, 309–11.

20 Baer GM, Abelseth MK, Debbie JG. Oral vaccination of foxes against rabies. *Am J Epidem* 1971, **93**, 487–90.

21 Wandeler AI, Bauder W, Prochaska S, Steck F. Small mammal studies in a SAD baiting area. *Comp Immun Mic and Inf Dis* 1982, **5**, 173–6.

22 Wandeler AI. Control of wildlife rabies: Europe. In: Campbell JB, Charlton KM (eds) Rabies. Boston: Kluwer, 1988, 365–80.

23 (Cited by) Perry BD. The oral immunization of animals against rabies. In: Grunsell CSG, Raw M-E, Hill FWG (eds) The Veterinary Annual. London: Butterworth 1989, 37–47.

24 Rupprecht CE, Kieny M-P. Development of a vaccinia-rabies glycoprotein recombinant virus vaccine. In: Campbell JB, Charlton KM (eds) Rabies. Boston: Kluwer, 1988, 335–64.

25 Wiktor TJ, Fernandes MV, Koprowski H. Cultivation of rabies virus in human diploid cell strain WI-38. *J Immun* 1964, **93**, 353–66.

26 WHO Expert Committee on Rabies seventh report. Geneva: WHO, 1984, 27–34. (Technical Report series 709.)

27 Bahmanyar M, Fayaz A, Nour-Salehi S, Mohammadi M, Koprowski H. Successful protection of humans exposed to rabies infection. Postexposure treatment with the new human diploid cell rabies vaccine and antirabies serum. *JAMA* 1976, **236**, 2751–4.

28 Aoki FY, Tyrrell DAJ, Hill LE, Turner GS. Immunogenicity and acceptability of a human diploid-cell culture rabies vaccine in volunteers. *Lancet* 1975, **i**, 660–2.

29 Barth R, Gruschkau H, Bijok U, *et al.* A new inactivated tissue culture rabies vaccine for use in man. Evaluation of PCEC-vaccine by laboratory tests. *J Biol Standard* 1984, **12**, 29–46.

30 Scheiermann N, Baer J, Hilfenhaus J, Marcus I, Zoulek G. Reactogenicity and immunogenicity of the newly developd purified chick embryo cell (PCEC) – rabies vaccine in man. *Zbl Bakt Hyg A* 1987, **265**, 439–50.

31 Wasi C, Chaiprasithikul P, Chavanich L, Puthavathana P, Thong-

charoen P, Trishanananda M. Purified chick embryo cell rabies vaccine. *Lancet* 1986, **i**, 40.

32 Nicholson KG, Farrow PR, Bijok U, Barth R. Pre-exposure studies with purified chick embryo cell culture rabies vaccine and human diploid cell vaccine: serological and clinical responses in man. *Vaccine* 1987, **5**, 208–10.

33 Roumiantzeff M, Ajjan N, Branche R *et al*. Rabies vaccine produced in cell culture: production control and clinical results. In: Kurstak E (ed.) Applied virology. Orlando: Academic Press, 1984, 241–96.

34 Dureux B, Canton P, Gerard A *et al*. Rabies vaccine for human use, cultivated on Vero cells. *Lancet* 1986, **ii**, 98.

35 Suntharasamai P, Warrell MJ, Warrell DA *et al*. New purified Vero-cell vaccine prevents rabies in patients bitten by rabid animals. *Lancet* 1986, **ii**, 129–31.

36 Turner GS, Aoki FY, Nicholson KG, Tyrrell DAJ, Hill LE. Human diploid cell strain rabies vaccine. Rapid prophylactic immunisation of volunteers with small doses. *Lancet* 1976, **i**, 1379–81.

37 Nicholson KG, Prestage H, Cole PJ, Turner GS, Bauer SP. Multi-site intradermal antirabies vaccination. Immune responses in man and protection of rabbits against death from street virus by post-exposure administration of human diploid-cell-strain rabies vaccine. *Lancet* 1981, **ii**, 915–18.

38 Rodrigues FM, Mandke VB, Roumiantzeff M *et al*. Persistence of rabies antibody 5 years after pre-exposure prophylaxis with human diploid cell antirabies vaccine and antibody response to a single booster dose. *Epidemiol Infect* 1987, **99**, 91–5.

39 Berlin BS, Goswick C. Rapidity of booster response to rabies vaccine produced in cell culture. *J Inf Dis* 1984, **150**, 785.

40 Kuwert EK, Marcus I, Werner J *et al*. Post-exposure use of human diploid cell culture rabies vaccine. *Develop Biol Standard* 1977, **37**, 273–86.

41 Turner GS, Nicholson KG, Tyrrell DAJ, Aoki FY. Evaluation of a human diploid cell strain rabies vaccine: final report of a three year study of pre-exposure immunization. *J Hyg Camb* 1982, **89**, 101–10.

42 Fishbein DB, Bernard KW, Miller KD *et al*. The early kinetics of the neutralizing antibody response after booster immunizations with human diploid cell rabies vaccine. *Am J Trop Med Hyg* 1986, **35**, 663–70.

43 Warrell MJ, Warrell DA, Suntharasamai P *et al*. An economical regimen of human diploid cell strain anti-rabies vaccine for post-exposure prophylaxis. *Lancet* 1983, **ii**, 301–4.

44 Warrell MJ, Suntharasamai P, Nicholson KG *et al*. Multi-site intradermal and multi-site subcutaneous rabies vaccination: im-

proved economical regimens. *Lancet* 1984, **i**, 874–6.

45 Warrell MJ, Nicholson KG, Warrell DA *et al.* Economical multi-ple-site intradermal immunisation with human diploid-cell-strain vaccine is effective for post-exposure rabies prophylaxis. *Lancet* 1985, **i**, 1059–62.

46 Suntharasamai P, Warrell MJ, Warrell DA *et al.* Early antibody responses to rabies post-exposure vaccine regimens. *Am J Trop Med Hyg* 1987, **36**, 160–5.

47 Suntharasamai P, Warrell MJ, Viravan C, *et al.* Purified chick embryo cell rabies vaccine: economical multisite intradermal reg-imen for post-exposure prophylaxis. *Epidemiol Infect* 1987, **99**, 755–65.

48 Ubol S, Phanuphak P. An effective economical intradermal regim-en of human diploid cell rabies vaccination for post-exposure treatment. *Clin Exp Immunol* 1986, **63**, 491–7.

49 Harverson G, Wasi C. Use of post-exposure intradermal rabies vaccination in a rural mission hospital. *Lancet* 1984, **ii**, 313–15.

50 Vodopija I, Sureau P, Lafon M, *et al.* An evaluation of second generation tissue culture rabies vaccines for use in man: a four-vaccine comparative immunogenicity study using a pre-exposure vaccination schedule and an abbreviated 2-1-1 post-exposure sche-dule. *Vaccine* 1986, **4**, 245–8.

51 Devriendt J, Staroukine M, Costy F, Vanderhaeghen J-J. Fatal encephalitis apparently due to rabies. Occurrence after treatment with human diploid cell vaccine but not rabies immune globulin. *JAMA* 1982, **248**, 2304–6.

52 Rabies vaccine failures. Editorial: *Lancet* 1988, **i**, 917–8.

53 Shill M, Baynes RD, Miller SD. Fatal rabies encephalitis despite appropriate post-exposure prophylaxis. A case report. *N Engl J Med* 1987, **316**, 1257–8.

54 Gardner SD. Prevention of rabies in man in England and Wales. In: Pattison JR (ed.) Rabies, a growing threat. Wokingham: Van Nostrand Reinhold, 1983, 39–49.

55 Furlong J, Lea G. Rabies prophylaxis simplified. *Lancet* 1981, **i**, 1311.

56 Schnurrenberger P, Dreesen D, Brown J, *et al.* Systemic allergic reactions following immunization with human diploid cell rabies vaccine. *MMWR* 1984, **33**, 185–7.

57 Swanson MC, Rosanoff E, Gurwith M, Deitch M, Schnurrenberger P, Reed CE. IgE and IgG antibodies to β-propiolactone and human serum albumin associated with urticarial reactions to rabies vac-cine. *J Infect Dis* 1987, **155**, 909–13.

58 Anderson MC, Baer H, Frazier DJ, Quinnan GV. The role of specific IgE and beta-propiolactone in reactions resulting from

booster doses of human diploid cell rabies vaccine. *J Allergy Clin Immunol* 1987, **80**, 861–8.

59 Bøe E, Nyland H. Guillain-Barré syndrome after vaccination with human diploid cell rabies vaccine. *Scand J Infect Dis* 1980, **12**, 231–2.

60 Bernard KW, Smith PW, Kader FJ, Moran MJ. Neuroparalytic illness and human diploid cell rabies vaccine. *JAMA* 1982, **248**, 3136–8.

61 Courrier A, Stenbach G, Simonnet PH *et al*. Peripheral neuropathy following fetal bovine cell rabies vaccine. *Lancet* 1986, i, 1273.

62 Bewley BR. Medical hazards from dogs. *BMJ* 1985, **291**, 760–1.

63 Callaham M. Dog bite wounds. *JAMA* 1980, **244**, 2327–8.

64 WHO Collaborating Centre for rabies surveillance and research. Rabies in Europe 4th quarter, 1988. *Rabies Bull Europe* 1988, **12** (**4**) 2, 23.

Chapter 8

Treatment of cytomegalovirus infections

A Webster and PD Griffiths

For the vast majority of previously healthy individuals, cytomegalovirus (CMV) infection is trivial. Primary infection occurs frequently during childhood and adolescence, and exposure to virus continues throughout life. In the UK, 70% of individuals have experienced infection by age 40 [1]. Transmission occurs by many different routes: infants may acquire the infection *in utero*, or perinatally from infected maternal cervical secretions or breast milk; infection may also be transmitted via saliva, or by sexual contact; transfusion of infected blood or blood products, or transplantation of a solid organ from a CMV-seropositive donor may also transmit infection. In common with other herpesviruses, CMV is not eliminated following primary infection, but persists in a latent form from which it may subsequently reactivate. Reinfection may also occur on subsequent exposure to other viral strains.

Severe CMV disease is seen in patients with impaired cell-mediated immunity (CMI), for example in those patients receiving immuno-suppressive therapy following organ transplantation, or following bone marrow transplantation, or in patients with acquired immune deficiency syndrome (AIDS). Severe CMV disease also occurs in infants infected *in utero*.

CMV disease may occur as a result of primary infection, reactivation or reinfection and may be caused directly by viral replication, termed 'lytic' infection, or be immunologically mediated. Infection with CMV is usually disseminated; virus may be isolated from many tissues, but may not cause disease at all sites.

Clinical manifestations

For the vast majority of previously healthy individuals, infection is entirely asymptomatic, but may rarely produce a mononucleosis syndrome, with fever, tonsilitis, lymphadenopathy and hepato-splenomegaly, which follows a benign, self-limiting course. Immunocompromised patients may experience non-specific signs and symptoms of infection, such as fever, which is characteristically high and swinging, and malaise. These symptoms may persist for many days, but do not adversely affect the patients' prognosis. However, they may prolong the period of in-patient care.

Despite the disseminated nature of the infection, disease is often limited to a single organ, and individual patient groups appear to experience characteristic manifestations of CMV disease, presumably due to differences in their particular underlying immune dysfunction.

RETINITIS

Retinitis is seen most often in patients with AIDS, and occasionally in other patient groups. Patients present with loss of visual acuity or visual fields, or with blurring of vision. Fundoscopy typically shows fluffy white exudates together with haemorrhages with a perivascular distribution. Untreated, it is relentlessly progressive, and may result in blindness. The typical histological appearance of retinal necrosis, with swollen retinal cells containing intranuclear and intracytoplasmic inclusions [2], probably results directly from viral replication. CMV infection is virtually universal in AIDS patients [3], but established retinitis is only a late event in the course of disease progression in AIDS, and as such, is probably a marker of profound immunosuppression: in one study, none of eight untreated patients survived more than six weeks from the diagnosis of CMV retinitis [2]. However, with increasing awareness of CMV infection in AIDS patients, asymptomatic CMV infections may be diagnosed earlier so that potentially sight-preserving therapy may be started promptly. Nevertheless, successful treatment may leave retinal atrophy and scarring, which predisposes to later retinal detachment, so close clinical follow-up is mandatory.

GASTROINTESTINAL DISEASE

Oesophagitis, gastritis and colitis are also associated with CMV infection, and probably occur as a result of lytic infection. An aetiological role for CMV is supported by evidence of CMV vasculitis in the GI tract [4], and by the symptomatic response following antiviral therapy (see p. 172). Gastrointestinal disease appears to be a particular problem in AIDS patients. Patients with colitis typically present with diarrhoea and abdominal pain; oesophagitis or gastritis usually present with dysphagia and epigastric pain. Endoscopy may reveal patchy erythema, oedema and mucosal erosions, haemorrhage and ulceration. Untreated, symptoms may be persistent and progressive, resulting in debility and weight loss [4].

PNEUMONITIS

CMV pneumonitis has been most extensively studied in organ transplant recipients, in whom it frequently causes respiratory failure and death. Presenting features include dyspnoea and fever, and radiological examination of the chest may show patchy shadowing. It is associated with an interstitial cellular infiltrate, and is thought to be due to a cell-mediated cytotoxic response to infected lung tissue, rather than as a result of viral replication *per se* [5]. Evidence for an immunological component in this disease is provided by the observed association of pneumonitis with graft versus host disease (GVHD) in BMT recipients [6]. CMV has also frequently been isolated from broncho-alveolar lavage fluid in AIDS patients with pneumonia. However, the majority of these patients usually had coinfection with *Pneumocystis carinii* [7], and therefore it is difficult to assess whether CMV acts as a pathogen under these circumstances. There is no evidence that the presence of CMV adversely affects the outcome of treatment of *P carinii* pneumonia.

CONGENITAL INFECTION
A small proportion of infants infected with CMV *in utero* may be severely affected. These infants have multisystem disease, and may present with jaundice, hepato-splenomegaly, encephalitis, thrombo-cytopenia or growth retardation. Mortality in these rare cases is high and most survivors are left with neurological sequelae.

OTHER MANIFESTATIONS
Hepatitis, manifested by elevated levels of hepatic transaminases or alkaline phosphatase, has been associated with CMV infection [8,9]. Leukopenia has also occurred in association with CMV infection [10] and may be due to viral replication in the bone marrow. In the absence of any more severe manifestation of CMV disease, slow resolution appears to be the rule. Histological findings typical of CMV infection have been described in the brains of AIDS patients with encephalitis [11]. However, the evidence for a causal role for CMV in encephalitis is lacking, since other opportunistic pathogens may also be present, and HIV itself infects the central nervous system.

Many other clinical syndromes have been ascribed to CMV infection, such as sclerosing cholangitis [12], including the 'vanishing bile duct' syndrome [13], and adrenalitis [14], but there is no evidence that CMV is the aetiological agent.

Diagnosis
Management of severe CMV infection depends upon achieving an accurate diagnosis, which rests upon detection of virus, or the typical histological changes in tissue. Conventional cell culture is limited in its usefulness, since the virus may take three weeks to produce the characteristic cytopathic effect, by which time the patient may well have succumbed to his infection. Fortunately, more rapid tests are available

for detection of CMV: these include direct immunofluorescence for viral antigens [15]; detection of early antigens by immunofluorescence following overnight culture of infected cell monolayers [16]; or alternatively, DNA–DNA hybridisation may be used [17]. The presence of typical intranuclear inclusions in histological sections may suggest the diagnosis, but this should be confirmed by a specific test, since other viral infections may produce a similar appearance. Because of the frequency with which CMV may be isolated in the absence of symptoms, even in severely immunocompromised patients, it is necessary to demonstrate virus within a particular organ by taking a biopsy of affected tissue. This can pose particular problems in patients with haematological abnormalities, such as thrombocytopenia or haemophilia. Broncho-alveolar lavage is a useful alternative in the diagnosis of respiratory infections. Since the retina is not accessible for biopsy, diagnosis of retinitis rests upon the typical fundoscopic appearance, in conjunction with the isolation of virus from other sites. In consequence, the diagnosis of CMV disease is often based on circumstantial evidence and/or 'clinical impression', and is therefore often presumptive. This makes evaluation of the outcome of natural infection particularly difficult, and assessment of the efficacy of antiviral drugs even more controversial.

Management of infection

The use of blood and blood products from CMV-seronegative donors may reduce the incidence of primary CMV infection in seronegative recipients, or may reduce the risk of reinfection in seropositive recipients. Transplanted organs may also be a source of infection, and therefore selection of a CMV-seronegative donor may benefit the patient. However, a choice of donors is seldom available, and in any individual the overall benefits of transplantation must be weighed against the risks of CMV infection.

Interruption or reduction of therapy in patients receiving immunosuppressive agents may be effective in the treatment of CMV disease. This measure alone may be successful in the management of less severe infection. However, this may predispose to graft rejection, which may be a high price to pay for the successful treatment of CMV. Obviously, this option is not available for patients whose immunodeficiency is not iatrogenic.

There is a limited choice of specific antiviral agents available for the treatment of CMV infection. These drugs have serious side effects (see p. 169), and should therefore be reserved for patients whose infection has a poor prognosis if left untreated. Unfortunately, few of the agents to be reviewed have been subjected to the vigorous evaluation demanded by a placebo-controlled randomised trial. A role for specific antiviral therapy has been established for AIDS patients with retinitis in order to preserve sight. However, this complication appears to coincide with an inexorable decline in immune function, and retinitis in this group of patients frequently relapses on cessation of treatment. Severe gastro-intestinal disease in these patients also appears to respond to specific therapy (see p. 172).

Specific antiviral therapy is used in an attempt to reduce the high morbidity and mortality associated with pneumonitis in transplant recipients. It is ethically impossible to withhold treatment from these patients, although the benefits of therapy have not been demonstrated convincingly.

Patients with other types of immune dysfunction may suffer from severe CMV disease, for example those with congenital immuno-deficiencies, or occasionally as a result of chemotherapy for other malignant diseases. The effects of antiviral therapy in these groups have been little studied, if at all. Similarly, the potential benefits of specific antiviral therapy in congenitally- or perinatally-infected infants are unknown. Less severe CMV infections may pass undiagnosed, and the prognosis for these patients, and any possible role for antiviral therapy, remain unclear.

Antiviral agents

Many antiviral agents have been assessed for the treatment of CMV infections, mostly with disappointing results. Overall, the prognosis for severe, established CMV infection remains poor, although treating the disease at an early stage may potentially improve the response rate.

Initially, agents such as vidarabine, interferon and acyclovir, alone and in combination, were used in the treatment of CMV disease with no effect [18–27]. More recently, attention has been focused on two drugs with improved activity against CMV.

GANCICLOVIR

Ganciclovir (9–[2–hydroxy–1–(hydroxymethyl) ethoxymethyl] guano-sine) is an acyclic nucleoside structurally related to acyclovir, and is phosphorylated by host thymidine kinase to form a triphosphate which is a potent inhibitor of viral DNA polymerase. It has excellent *in vitro* activity against all the human herpesviruses, including CMV [28]. It is, however, associated with a dose-dependent myelotoxicity. This can be particularly troublesome in AIDS patients concomitantly receiving zido-vudine, which is also myelotoxic. Severe neutropenia often necessitates withdrawal of one or other drug. This can also be problematic in patients undergoing engraftment of bone marrow. In preclinical studies it had a cytostatic effect on the testis which was irreversible. Patients requiring treatment should therefore be counselled that the drug may render them permanently sterile.

Ganciclovir, which has recently been licensed in the United Kingdom, must be administered by intravenous infusion; no orally-effective formulation is available.

FOSCARNET

Foscarnet (trisodium phosphonoformate hexahydrate), a pyrophosphate analogue, is also effective *in vitro* against herpesviruses by its inhibitory action on viral DNA polymerase [29]. It has recently been licensed but, because of its short half-life and instability at low pH, must be given as a continuous intravenous infusion. Major side effects associated with this drug are altered calcium metabolism, anaemia and renal toxicity.

Treatment of congenital infection

Many workers have been motivated to treat congenitally-infected infants by the poor prognosis of those born with symptoms, despite the known toxicity of the drugs, and their unknown long-term effects on growing children: fluorodeoxyuridine, idoxyuridine, cytosine arabinoside and adenine arabinoside have all been tried. Despite some improvements in virological status, little appreciable beneficial effect on the clinical course of these patients has been described, although toxicity has been marked.

Treatment of transplant recipients

Signs and symptoms of CMV infection characteristically appear in the second and third months post transplant, and interstitial pneumonitis is the most worrying manifestation of this. Treatment of this complication has produced disappointing results. Ganciclovir has been largely unsuccessful in the treatment of pneumonitis in BMT recipients, with many patients succumbing to their infection as a result of respiratory failure [30,31] (see Table 1). Response rates in patients with solid organ transplants (see Table 2) and pneumonitis are reportedly better [32,33], as are those in patients with gastrointestinal infection [30]. Foscarnet has also been used in BMT and renal transplant recipients with pneumonitis; the outcome in the former group was poor, since all patients died. Renal transplant recipients with pneumonitis fared better, with 4/6 patients surviving. Those with other manifestations of CMV infection responded more favourably [34]. Overall, the response to antiviral therapy among transplant recipients appears to be influenced by the nature of the underlying immunodeficiency, with BMT recipients faring badly, and by the type of infection, with pneumonitis responding less well than lytic infections.

The mode of action of immunoglobulin preparations from CMV immune donors is far from clear. Initially it was assumed that the antibodies neutralised CMV found in body fluids, but this seems unlikely since it has been shown that CMV *in vivo* is coated with a host

Table 1 Antiviral agents in the treatment of CMV disease in BMT recipients

Drug	Organ affected	Criterion	Result	Relapse	Ref
Ganciclovir	Lung	Improved	5/11	2/5	[30]
	Retina		1/1		
	GI		3/3	0/3	
Ganciclovir	Lung	Improved	1/10		[31]
Foscarnet	Lung	Survived	0/9		[34]
CMV Ig	Lung	Survived	9/18		[56]
CMV Ig+ ganciclovir	Lung	Survived	7/10		[36]
CMV Ig+ ganciclovir	Lung	Survived	13/25		[37]

Table 2 Antiviral agents in the treatment of CMV disease in recipients of transplants other than bone marrow

Drug	Organ transpl	Organ affected	Criterion	Result	Relapse	Ref
Ganciclovir	Kidney, liver	GI, lung, retina, liver	Improved	5/6	0/4	[30]
Ganciclovir	Kidney	Lung	Improved	5/5	1/5	[33]
Ganciclovir	Heart, heart-lung	Lung	Improved	4/7		[32]
Foscarnet	Kidney	Lung	Survived	4/6		[34]

GI = gastrointestinal tract

protein, beta$_2$ microglobulin, which protects the virus 'from neutralisation [35]. Perhaps the antibodies enhance antibody-dependent cellular cytotoxicity against CMV-infected cells. Alternatively, they might block a putative CMI target on lung cells produced by CMV infections [5]. Combinations of immunoglobulin and ganciclovir have recently been reported to give better survival [36,37].

Treatment of patients with AIDS
In contrast to transplant recipients, AIDS patients do not appear to be particularly susceptible to CMV pneumonitis, but suffer more frequently from retinal or gastrointestinal disease. Overall response rates

of the order of 80% have been reported using ganciclovir in the treatment of CMV retinitis and gastrointestinal disease [38–41] (see Table 3). However, relapse was reported in almost two thirds of these patients when treatment was stopped. Thus, in order to prevent progression of retinitis and eventual blindness, treatment of this complication should be continued indefinitely. Because of the associated toxicity of ganciclovir, which may preclude the simultaneous use of zidovudine, low dose maintenance regimens have been recommended. Intravitreous administration has been suggested as a possible alternative method of treatment which avoids systemic side effects [42]. However, this mode of administration may be associated with local side effects.

Encouraging results have recently been reported using foscarnet in the treatment of retinitis, although relapse remained a major problem [43]. However, this drug might be a less toxic alternative in the treatment of these patients. Farthing *et al* [44] report beneficial effects with this drug in the treatment of eight AIDS patients with pneumonitis. However, four of these patients were concurrently receiving high dose co-trimoxazole for confirmed or suspected *P carinii* pneumonia.

Prophylaxis for CMV infections
Due to the toxicity of antiviral agents effective against CMV, and the poor prognosis of established disease in certain patients, attention has been directed towards the use of prophylactic antiviral regimens. CMV hyperimmune immunoglobulin, interferon and acyclovir have been used to this end, with variable results (Table 4). Immunoglobulin from CMV seropositive donors has been used in several studies in an attempt to reduce the incidence and severity of CMV infections in transplant

Table 3 Antiviral agents used for the treatment of CMV disease in AIDS patients

Drug	Organ affected	Criterion	Result	Relapse	Ref
Ganciclovir	Lung, GI, retina, liver	Improved	0/6		[30]
Ganciclovir	Retina	Improved	6/7	5/5	[40]
Ganciclovir	Lung, retina, GI	Improved or stabilised	17/22	11/14	[41]
Ganciclovir	Lung, retina, GI	Improved	12/14	9/11	[38]
Ganciclovir	GI	Improved	30/41	13/33	[39]
Foscarnet	Lung	Improved	8/8		[44]
Foscarnet	Retina	Improved	13/13	9/11	[43]

GI = gastrointestinal tract

Table 4 Prophylaxis for CMV infections in BMT recipients

Drug schedule	Duration (d)	Design	Criterion	Result treated	Result control	p value	Ref
IFN every 3rd day median start = d18	80	CMV− or CMV+ patients	Infection Pneumonitis	15/39 6/15	20/40 8/20		[51]
CMV + Ig d −5, −1, 6, 20, 34, 48, 62	62	CMV− patients*	Infection Disease	6/41 2/41	11/44 4/44		[47]
CMV + Ig weekly	120	CMV− or CMV+ (low) patients	Infection Pneumonitis	18/38 6/38	21/37 12/37	0.02	[48]
CMV + Ig d25, 50, 75	75	CMV− or CMV+ patients†	Infection** Pneumonitis	0/17 0/17	16/38 5/38	0.01	[46]
Acyclovir (high dose) start d5	30	CMV+ patients††	Infection Pneumonitis	51/86 16/86	49/65 20/65	0.04	[52]
Acyclovir (high dose) start d0	84	all patients	Disease Pneumonitis	4/53 1/53	15/51 9/51	0.002	[53]

*Some infections diagnosed by serology alone
All trials are randomised, with control groups receiving no treatment except:
**Control groups given either non-immune immunoglobulin, or no immunoglobulin
†Treated group CMV+ HSV+, control group CMV+ HSV−
††Patients also randomised to receive CMV−, or unscreened blood products
Ig = immunoglobulin; CMV+ = CMV seropositive; CMV− = CMV seronegative; IFN = interferon

recipients. Direct comparisons between studies are difficult because of differences in the preparations used and in the study populations.

A reduction in the incidence of CMV infections and pneumonitis has been reported in CMV-seronegative renal transplant recipients [45]. In BMT recipients, results are conflicting: Condie *et al* [46] report a decrease in infections and pneumonitis in CMV-seronegative and seropositive recipients; Bowden *et al* [47], however, did not observe a protective effect in giving CMV immunoglobulin to CMV-seronegative recipients over and above that of giving CMV-seronegative blood products. In a population of BMT recipients seronegative, or with low levels of antibody to CMV, Winston *et al* [48] report a reduction in incidence of CMV infection and pneumonitis.

Two placebo-controlled randomised trials of prophylactic interferon administration to renal transplant recipients have been conducted in Boston. The first studied seropositive or seronegative patients and showed a significantly prolonged time to CMV excretion and a significant reduction in CMV viraemia, but no clinical effect [49]. The second studied seropositive patients only, and showed, as before, a significantly prolonged time to CMV excretion, but no effect on viraemia, and a significant reduction in overall CMV disease [50]. These results are encouraging, but further reports from other centres will be required to show which effects of prophylactic interferon can be reproduced predictably.

Interferon, given after marrow engraftment (median 18 days post-transplant), had no effect on the frequency or severity of CMV infections in CMV-seropositive or seronegative BMT recipients [51].

In a recent study, Meyers *et al* [52] reported a beneficial effect of high dose acyclovir on CMV infection and pneumonitis following BMT. However, patients in the treatment and control groups were selected on the basis of herpes simplex virus (HSV) seropositivity, since it was considered unethical to withhold acyclovir from HSV-seropositive patients. Thus the two groups are not strictly comparable, especially since the acquisition of CMV infection is linked statistically to the prior acquisition of HSV [1], so that the results must be viewed with caution. Nevertheless, a recent report of beneficial effect from high-dose oral acyclovir in renal allograft recipients suggests that this drug may indeed be an effective prophylactic, though not therapeutic, agent [53].

Clearly, the role for antiviral agents in CMV prophylaxis has yet to be elucidated; standardisation of treatment protocols and study populations would aid this process.

Conclusions

CMV is a ubiquitous virus; infection is systemic, affecting many organs, but is entirely asymptomatic for the vast majority of individuals. In patients with impaired immunity, several disease syndromes have been recognised, including pneumonitis, retinitis, gastrointestinal disease and hepatitis, as well as non-specific signs, such as fever. The prognosis for these syndromes is variable, depending to some extent upon the underlying immune dysfunction. Non-specific measures, such as modification of the immunosuppressive regimen, may be effective in management of some cases. Changes in the management of the underlying disease may also significantly alter the incidence of CMV disease in patients iatrogenically immunosuppressed. For example, measures to reduce the incidence of GVHD in BMT recipients, such as T-cell depletion, may also reduce the incidence of CMV pneumonitis. Conversely, the use of anti-thymocyte globulin to counteract rejection of transplanted organs may predispose to CMV disease [54,55]. Similar effects may also result from modifications of dosage and duration of immunosuppressive therapy. It is conceivable that the development of highly selective immunosuppressive regimens might entirely eliminate CMV disease as a complication in some patient groups, but there is no sign of this as yet. Currently-available drugs with activity against CMV are toxic and difficult to administer, and there are as yet few instances where demonstrated benefits outweigh the risks associated with their use. These include pneumonitis in transplant recipients, and retinitis and gastrointestinal disease in AIDS patients. Other diseases may be relative indications for their use, and each case must be assessed individually. Early recognition and treatment of potentially-severe disease is clearly desirable, since the prognosis for established disease is poor. Prophylactic regimens have been employed in an attempt to eliminate CMV disease as a complication of transplantation, but the benefits of this have yet to be proven.

References

1   Berry NJ, MacDonald Burns D, Wannamethee G, *et al.* Sero-epidemiologic studies on the acquisition of antibodies to cytomegalovirus, herpes simplex virus, and human immunodeficiency virus among general hospital patients and those attending a clinic for sexually transmitted diseases. *J Med Virol* 1988, **24**, 385–93.

2   Holland GN, Pepose JS, Pettit TH, Gottlieb MS, Yee RD, Foos RY. Acquired immune deficiency syndrome, ocular manifestations. *Ophthalmology* 1983, **90**, 859–73.

3   Mintz L. Drew WL, Miner RC, Braff EH. Cytomegalovirus infections in homosexual men. An epidemiological study. *Ann Intern Med* 1983, **99**, 326–9.

4   Meiselman MS, Cello JP, Margaretten W. Cytomegalovirus colitis. Report of the clinical endoscopic, and pathologic findings in two patients with the acquired immune deficiency syndrome. *Gastroenterology* 1985, **88**, 171–5.

5   Grundy JE, Shanley JD, Griffiths PD. Is cytomegalovirus inter-
     stitial pneumonitis in transplant recipients an immunopathological
     condition? *Lancet* 1987, **ii**, 996–9.
6   Weiner RS. Interstitial pneumonia following bone marrow trans-
     plantation. In: Gale RP, Champlin R eds. Progress in bone marrow
     transplantation. New York: Liss, 1987, 507–23. (UCLA symposia
     on molecular and cellular biology series; vol 53.)
7   Murray JF, Felton CP, Garay SM, *et al*. Pulmonary complications
     of the acquired immunodeficiency syndrome: report of a National
     Heart, Lung and Blood Institute workshop. *N Engl J Med* 1984,
     310, 1682–8.
8   Reichert CM, O'Leary TJ, Levens DL, Simrell CR, Macher AM.
     Autopsy pathology in the acquired immune deficiency syndrome.
     Am J Pathol 1983, **112**, 357–82.
9   Luby JPL, Brunett W, Hull AR, Ware AJ, Shorey JW, Peters PC.
     Relationship between cytomegalovirus and hepatic function
     abnormalities in the period after renal transplant. *J Infect Dis* 1974,
     129, 511–8.
10   Marker SC, Howard RJ, Simmons RL, *et al*. Cytomegalovirus
     infection: a quantitative prospective study of 320 consecutive renal
     transplants. *Surgery* 1981, **89**, 660–71.
11   Post MJ, Hensley GT, Moskowitz LB, Fischl M. Cytomegalic
     inclusion virus encephalitis in patients with AIDS: CT, clinical and
     pathologic correlation. *AJR Am J Roentgenol* 1986, **146**, 1229–34.
12   Schneiderman DJ, Cello JP, Laing FC. Papillary stenosis and
     sclerosing cholangitis in the acquired immunodeficiency syndrome.
     Ann Intern Med 1987, **106**, 546–9.
13   O'Grady JG, Alexander GJM, Sutherland S, *et al*. Cytomegalo-
     virus infection and donir/recipient HLA antigens: interdependent
     co-factors in pathogenesis of vanishing bileduct syndrome after
     liver transplantation. *Lancet* 1988, **ii**, 302–5.
14   Greene LW, Cole W, Greene JB, *et al*. Adrenal insufficiency as a
     complication of the acquired immunodeficiency syndrome. *Ann
     Intern Med* 1984, **101**, 497–8.
15   Volpi A, Whitley RJ, Ceballos R, Stagno S, Pereira L. Rapid
     diagnosis of pneumonia due to cytomegalovirus with specific mono-
     clonal antibodies. *J Infect Dis* 1983, **147**, 1119–20.
16   Griffiths PD, Panjwani DD, Stirk PR, *et al*. Rapid diagnosis of
     cytomegalovirus infection in immunocompromised patients by de-
     tection of early antigen fluorescent foci. *Lancet* 1984, **ii**, 1242–5.
17   Chou S, Merigan TC. Rapid detection and quantitation of human
     cytomegalovirus in urine through DNA hybridization. *N Engl J
     Med* 1983, **308**, 921–5.
18   Meyers JD, McGuffin RW, Neiman PE, Singer JW, Thomas ED.
     Toxicity and efficacy of human leukocyte interferon for treatment

of cytomegalovirus pneumonia after marrow transplantation. *J Infect Dis* 1980, **141**, 555–62.

19 Meyers JD, McGuffin RW, Bryson YJ, Cantell K, Thomas ED. Treatment of cytomegalovirus pneumonia after marrow transplant with combined vidarabine and human leukocyte interferon. *J Infect Dis* 1982, **146**, 80–4.

20 Wade JC, Hintz M, McGuffin RW, Springmeyer SC, Connor JD, Meyers JD. Treatment of cytomegalovirus pneumonia with high-dose acyclovir. *Am J Med* 1982, **73** (suppl 1A), 249–56.

21 Wade JC, McGuffin RW, Springmeyer SC, Newton B, Singer JW, Meyers JD. Treatment of cytomegalovirus pneumonia with high-dose acyclovir and human leukocyte interferon. *J Infect Dis* 1983, **148**, 557–62.

22 Winston DJ, Ho WG, Schroff RW, Champlin RE, Gale RP. Safety and tolerance of recombinant leukocyte A interferon in bone marrow transplant recipients. *Antimicrob Agents Chemother* 1983, **23**, 846–51.

23 Meyers JD, Day LM, Lum LG, Sullivan KM. Recombinant leukocyte A interferon for the treatment of serious viral infections after marrow transplant: a phase I study. *J Infect Dis* 1983, **148**, 551–6.

24 Shepp DH, Newton BA, Meyers JD. Intravenous lymphoblastoid interferon and acyclovir for treatment of cytomegaloviral pneumonia. *J Infect Dis* 1984, **150**, 776–7.

25 Pollard RB, Egbert PR, Gallagher JG, Merigan TC. Cytomegalovirus retinitis in immunosuppressed hosts. I. Natural history and effects of treatment with adenine arabinoside. *Ann Intern Med* 1980, **93**, 655–64.

26 Ch'ien LT, Cannon NJ, Whitley RJ *et al.* Effect of adenine arabinoside on cytomegalovirus infections. *J Infect Dis* 1974, **130**, 32–9.

27 Balfour HH, Bean B, Mitchell CD, Sachs GW, Boen JR, Edelman CK. Acyclovir in immunocompromised patients with cytomegalovirus disease: a controlled trial at one institution. *Am J Med* 1982, **73 (suppl 1A)**, 241–8.

28 Mar E-C, Cheng Y-C, Huang E-S. Effect of 9-(1,3-dihydroxy-2-propoxymethyl)guanine on human cytomegalovirus replication *in vitro. Antimicrob Agents Chemother* 1983, **24**, 518–21.

29 Oberg B. Antiviral effects of phosphonoformate (PFA, foscarnet sodium) *Pharmacol Ther* 1982, **19**, 387–415.

30 Erice A, Jordan C, Chace BA, Fletcher C, Chinnock BJ, Balfour HH. Ganciclovir treatment of cytomegalovirus disease in transplant recipients and other immunocompromised hosts. *JAMA* 1987, **257**, 3082–7.

31 Shepp DH, Dandliker PS, de Miranda P *et al.* Activity of 8-[2-hydroxy-1-(hydroxymethyl)ethoxymethyl]guanine in the treatment of cytomegalovirus pneumonia. *Ann Intern Med* 1985, **103**, 368–73.

32 Keay S, Bissett J, Merigan TC. Ganciclovir treatment of cyto-megalovirus infections in iatrogenically immunocompromised patients. *J Infect Dis* 1987, **156**, 1016–21.

33 Rostoker G, Ben Maadi A, Buisson C, Deforge L, Weil B, Lang P. Ganciclovir for severe cytomegalovirus infection in transplant recipients. *Lancet* 1988, **ii**, 1137–8.

34 Ringden O, Wilczek H, Lonnqvist B, Gahrton G, Wahren B, Lernestedt J-O. Foscarnet for cytomegalovirus infections. *Lancet* 1985, **i**, 1503–4.

35 McKeating JA, Griffiths PD, Grundy JE. Cytomegalovirus in urine specimens has host beta$_2$microglobulin bound to the viral envelope: a mechanism of evading the host immune response? *J Gen Virol* 1987, **68**, 785–92.

36 Emanuel D, Cunningham I, Jules-Elysee K *et al.* Cytomegalovirus pneumonia after bone marrow transplantation successfully treated with the combination of ganciclovir and high-dose intravenous immune globulin. *Ann Int Med* 1988, **109**, 777–82.

37 Reed EC, Bowden RA, Dandliker PS, Lilleby KE, Meyers JD. Treatment of cytomegalovirus pneumonia with ganciclovir and intravenous immunoglobulin in patients with bone marrow transplants. *Ann Int Med* 1988, **109**, 783–7.

38 Laskin OL, Stahl-Bayliss CM, Kalman CM, Rosecan LR. Use of ganciclovir to treat serious cytomegalovirus infections in patients with AIDS. *J Infect Dis* 1987, **155**, 323–7.

39 Chachoua A, Dieterich D, Krasinski K, *et al.* 9-(1,3-dihydroxy-2-propoxymethyl) guanine (ganciclovir) in the treatment of cyto-megalovirus gastrointestinal disease with the acquired immuno-deficiency syndrome. *Ann Intern Med* 1987, **107**, 133–7.

40 Masur H, Lane HC, Palestine A, *et al.* Effect of 9-(1,3-dihydroxy-2-propoxymethyl) guanine on serious cytomegalovirus disease in eight immunosuppressed homosexual men. *Ann Intern Med* 1986, **104**, 41–4.

41 Collaborative DHPG Treatment Study Group. Treatment of serious cytomegalovirus infections with 9-(1,3-dihydroxy-2-propoxymethyl)guanine in patients with AIDS and other immuno-deficiencies. *N Engl J Med* 1986, **314**, 801–5.

42 Ussery FM, Conklin R, Stool E, *et al.* Ganciclovir by intravitreal injection in the treatment of AIDS-associated cytomegalovirus (CMV) retinitis. Stockholm: IVth International Conference on AIDS, 1988, [Abstract 7178].

43 Walmsley SL, Chew E, Read SE, *et al.* Treatment of cyto-megalovirus retinitis with trisodium phosphonoformate hexa-hydrate (foscarnet). *J Infect Dis* 1988, **157**, 569–72.

44 Farthing C, Anderson MG, Ellis ME, Gazzard BG, Chanas AC. Treatment of cytomegalovirus pneumonitis with foscarnet (tri-

sodium phosphonoformate) in patients with AIDS. *J Med Virol* 1987, **22**, 157–62.

45 Snydman DR, Werner BG, Heinze-Lacey B, *et al.* Use of cyto-megalovirus immune globulin to prevent cytomegalovirus disease in renal-transplant recipients. *N Engl J Med* 1987, **317**, 1049–54.

46 Condie RM, O'Reilly RJ. Prevention of cytomegalovirus infection by prophylaxis with an intravenous, hyperimmune, native, unmod-ified cytomegalovirus globulin. Randomized trial in bone marrow transplant recipients. *Am J Med* 1984, **76 (suppl 3A)**, 134–41.

47 Bowden RA, Sayers M, Flournoy N, *et al.* Cytomegalovirus im-mune globulin and seronegative blood products to prevent primary cytomegalovirus infection after marrow transplantation. *N Engl J Med* 1986, **314**, 1006–10.

48 Winston DJ, Ho WG, Lin C-H, *et al.* Intravenous immune globulin for prevention of cytomegalovirus infection and interstitial pneumonia after bone marrow transplantation. *Ann Intern Med* 1987, **106**, 12–18.

49 Cheeseman SH, Rubin RH, Stewart JA *et al.* Controlled clinical trial of prophylactic human-leukocyte interferon in renal trans-plantation. *N Engl J Med* 1979, **300**, 1345–9.

50 Hirsch MS, Schooley RT, Cosimi AB, *et al.* Effects of interferon-alpha on cytomegalovirus reactivation syndromes in renal-transplant recipients. *N Engl J Med* 1983, **308**, 1489–93.

51 Meyers JD, Flournoy N, Sanders JE, *et al.* Prophylactic use of human leukocyte interferon after allogeneic marrow trans-plantation. *Ann Intern Med* 1987, **107**, 809–16.

52 Meyers JD, Reed EC, Shepp DH, *et al.* Acyclovir for prevention of cytomegalovirus infection and disease after allogeneic bone mar-row transplantation. *N Engl J Med* 1988, **318**, 70–5.

53 Fryd DS. A randomized, placebo-controlled trial of oral acyclovir for the prevention of cytomegalovirus disease in recipients of renal allografts. *N Eng J Med* 1989, **320**, 1381–7.

54 Pass RF, Whitley RJ, Diethelm AG, Welchel JD, Reynolds DW, Alford CA. Cytomegalovirus infection in patients with renal trans-plants: Potentiation by antithymocyte globulin and an incompatible graft. *J Infect Dis* 1980, **142**, 9–17.

55 Rubin RH, Cosimi AB, Hirsch MS, Herrin JT, Russell PS, Tol-koff-Rubin NE. Effects of antithymocyte globulin on cyto-megalovirus infection in renal transplant recipients. *Trans-plantation* 1981, **31**, 143–5.

56 Griffiths PD, Stirk PR, Blacklock HA, Milburn HJ, du Bois RM, Prentice HG. Rapid diagnosis and treatment of cytomegalovirus pneumonitis. In: Gale RP, Champlin R eds. Progress in bone marrow transplantation. New York: Liss, 1987, 583–7. (UCLA symposia on molecular and cellular biology series; vol 53.)

Chapter 9

Hantavirus

G Lloyd

Interest in hantaviruses has been stimulated by a marked increase in hantavirus infection being identified throughout Europe, together with other recognised endemic areas of the world, namely eastern Russia, China, Korea, Japan and Scandinavia. Significant progress has been made only within the past decade in elucidating the aetiology and epidemiology of this disease, resulting in many different hantavirus isolates being obtained from human and rodent sources.

The abundant literature of earlier years has already been amply reviewed [1–4], and this chapter focuses on the significant points of importance which enhance our overall understanding of the disease within the context of recent developments in hantavirus research.

Historical perspective

Hantaviruses are responsible for causing a range of clinical manifestations collectively known as haemorrhagic fever with renal syndrome (HFRS) [5] or, more recently, hantavirus disease (HVD) [6].

With increased understanding of the nature of HVD, its epidemiology and viral aetiology, there is a realisation that hantaviruses have affected man for centuries.

Its first major recorded impact on the western world was between 1951 and 1954, during the war in the Korean peninsular. During this outbreak, over 3,000 United Nations troops developed a disease resulting in a mortality rate of between 5–10%. The disease was characterised by fever, headache, backache, haemorrhagic manifestations, acute renal failure and shock [5]. After extensive and detailed reporting of this outbreak, it became apparent that HVD was not new, and has probably been known since 1919 in the far east of the USSR. Some reports suggest haemorrhagic fever accompanied by renal dysfunction is found in Chinese medical records 1,000 years old!

Records from major conflicts over the last century also indicate HVD is a problem within the circumstances war creates, particularly in prolonged trench warfare [7]. Common clinical features now recognised as HVD were described and recorded in many of the following wars: 'general dropsy with renal involvement' during the American Civil War of 1861–66; 'trench nephritis' amongst British, French, American and German soldiers in 1915–16 [8–11]; 'Far Eastern Nephroso-nephritis' among Russian soldiers in the Far East during 1932; Songo fever among Japanese soldiers during the war with China, 1934 [12–13]; and 'feldnephritis' among German soldiers during 1941–42 in Russia and Finland [14].

During the US Army Korean War investigation, Gajdusek [5] and Tamuru [16] observed that the Far Eastern haemorrhagic fever bore close clinical and epidemiological resemblance to an acute renal disease in Scandinavia (Nephropathia epidemica). They hypothesized that both diseases were aetiologically related. Twenty-five years later, Lee and colleagues [17] reported the isolation of the aetiological agent, Hantaan virus, making it possible to confirm the immunological identities of three geographically-separate diseases [18]. Subsequent immunological studies confirmed the serological relatedness of Russian, Korean, Manchurian, Japanese, Scandinavian and Chinese variants of HVD [19–21].

With the knowledge that similarly-described diseases throughout the world are related, several terms were suggested to standardise the enormous descriptive terminology (Table 1) under a single name. The World Health Organization Committee recommended that all related diseases be called hemorrhagic fever with renal syndrome (HFRS). In order to standardise both the terminology relating to the diseases and

Table 1   Examples of synonyms and acronyms used to describe hantavirus disease

---

Churilov's disease
Epidemic haemorrhagic fever (EHF)
Epidemic nephropathy
Epidemic nephroso-nephritis
Endemic nephropathy
Far Eastern nephrosonephritis
Haemorrhagic fever with renal syndrome (HFRS)
Haemorrhagic nephro-nephritis
Korean haemorrhagic fever
Manchurian fever
Nephropathia epidemica (NE)
Nidoko fever
Scandinavian epidemic nephropathy
Songo fever
Tayinshan disease
War nephritis

---

the numerous recent viral isolations, the 29th International Colloquium on Hantaviruses held in 1987 proposed hantavirus disease (HVD) and the hantaviruses respectively [6].

Clinical features of hantavirus infections

Prior to serodiagnosis, diagnosis of HVD was based on clinical judgment only and the course of the disease was documented from Russian, Chinese and Korean war publications. The severe form of HVD normally associated with the Far East follows a characteristic clinical course and is traditionally divided into five phases: febrile, hypotensive, oliguric, diuretic and convalescent.

Briefly, the incubation period of the disease is thought to be 2–3 weeks and the febrile phase begins abruptly, with fever, chills, prostration, headache, backache, anorexia, nausea and vomiting common. There is a typical erythematous flush of the face, neck, shoulders and upper thorax. The hypotensive and oliguric phases occur at approximately the fifth and ninth days of illness. These phases are marked by the development of thrombocytopaenia (100,000 platelet/mm^2), proteinuria, evidence of severe plasma loss through damaged capillary endothelium and moderate to severe acute renal failure in severe cases. Only the worst cases go into thermal shock [22]. Bleeding is mostly confined to petechiae, significant haemorrhage occurring in only about 10% of severe cases. The onset of the diuretic phase between days 12 and 14 suggests improvement. In a few cases, sudden diuresis in an already dehydrated patient has caused fluid balance problems. During convalesence some patients become anaemic and most will have lost weight. Recovery may be protracted but usually takes four months or less.

The principal complications and causes of death are associated with circulatory failure, shock, uraemia and complications of renal failure [23–24]. Shock is the principal or contributory cause of death in over three-quarters of cases, uraemia being listed as the immediate cause of death in half the cases. Since these complications occur chiefly in the hypotensive and oliguric phases, three quarters of all deaths occur within the first 10 days of illness.

The mild or moderate HVD of Scandinavia [25], and recently China [26], is characterised by sudden onset of fever, headache, nausea, myalgia and lumbar and abdominal pain. This is a self-limiting disease in which nephritis leads to moderate renal dysfunction and an uncomplicated recovery [27,28].

In most countries, local descriptions differ little in essentials and the distinction between the eastern, severe form of HVD with mortality rates of 5–10% and the more benign European HVD remains clear.

Pathology and pathophysiology

The multisystemic pathology of classical HVD is characterised by damage to capillaries and small blood vessels, resulting in vasodilation and congestion, with microscopic and gross haemorrhages [22,29–31]. Most deaths occur in the hypotensive and oliguric phases when vesicular damage is apparent by intense capillary engorgement, interstitial and retroperitoneal oedema. Haemorrhages are widespread but found most consistently in the subepicardium and subendocardium of the right atrium, medulla of the kidney and arterior lobe of the pituitary. In addition, the kidney tubules are dilated and tubular epithelial cells exhibit various degrees of degradation. Inflammatory, principally mononuclear, infiltrates are observed especially in the cortex. In renal biopsies of NE patients, the above tubulo-intersticial findings are present, although severity of haemorrhage is less marked [27,32,33].

Immunohistologic studies of renal biopsies reveal the deposition of IgG, IgM and IgA on both epithelial side and within the mesangium of glomeruli, suggesting circulation of immune complexes [34]. The origin of the immune complexes is unknown but it has been suggested that this may result from *in situ* reactions of circulating antibodies from peritubular capillaries with cellular and/or viral antigens from damaged tubular epithelial cells [35]. Since immune complexes are found early in the febrile phase of illness, immunopathogenetic mechanisms have been considered the basis of the disease. However, since there is a limited amount found in the kidney, an immune pathogenetic mechanism may play a less important role in the development of nephritis than viral cellular destruction [34].

Although there is widespread vascular endothelium damage resulting in abnormal vascular permeability, vasodilation, fluid transudation and oedema, and extravasation of red cells and haemorrhage, the pathogenesis is unknown. Primary capillary endothelial cell and peripheral blood mononuclear leukocyte hantavirus infection [36], endothelial cell dysfunction [37] and platelet dysfunction [38] have all been reported, but the sequence of events and their relationship to the characteristic alterations found in the kidney need further study.

Virus aetiology and characterisation

In 1940, Russian and Japanese workers established evidence of the viral aetiology of hantavirus infection, by producing a typical disease in human volunteers by intravenous injection of blood and urine of patients early in their illness. However, it was not until 1978 that a viral agent described as Hantaan virus was isolated from *Apodemus agrarius coreae* (Korean striped field mouse), which had been caught in the epizootic and epidemic regions of Korea [39]. In 1982, French and co-workers [40] were able to isolate and assay HV (76/118) in A549 (human lung carcinoma) and in Vero E6 (a clone of African green monkey kidney) continuous cell lines. The increased sensitivity and yield of these cells has allowed the isolation of additional virus strains and more advanced morphological and biochemical studies.

Using the E6 cell line many isolates have been derived from a variety of sources, demonstrating that hantaviruses are distributed worldwide and have specific rodent hosts. Hantaviruses have now been isolated in Vero E6 cells from many countries, including: Sweden, Finland and Belgium from lung and spleen of *Clethrionomys glareolus* (bank vole) [41,42]; China and Korea viruses from *Rattus norvegicus* (rat), *Apodemus species* (field mice) and human cases [43]; Japan [44], Korea [45], United Kingdom [46] from laboratory rats; USA from *Microtus pennsylvanicus* [47]; USSR from *Clethrionomys glareolus*, *Apodemus species*, *Microtus species* and human cases.

The virus was first considered to be a member of the Bunyaviridae (Table 2) on the basis of morphology [48] and morphogenesis [49] studies by electron microscopy. Hantaviruses have a surface structure composed of a square grid-like pattern [49] not previously described for animal viruses.

The hantavirus genus is a member of the Bunyaviridae family. The viral particle averages 98nm in diameter, is enveloped, has a density of 1.15–1.19g/ml, is heat labile and sensitive to lipid solvents. Hantaviruses have a tripartite, single-stranded, negative-polarity RNA genome, in which the three segments, L, M and S have molecular masses of approximately 2.7, 1.2 and 0.6Mda respectively. All three strands have a common 3′-terminal base sequence (AUCAUCAUCUG) which is not shared by any of the genera of Bunyaviridae. The viral envelope contains two specific glycoproteins, designated G1 and G2 [50–55].

Further characterisation of this genus has been determined at the antigenic level using polyvalent sera raised to specific hantavirus strains, and by monoclonal antibodies [56–60]. Plaque-reduction neutralisation assays and various immunoassays using polyvalent sera and monoclonal antibodies show at least four antigenic serotypes [57]: (1) those viruses

Table 2   The taxonomic status of hantavirus

Family Bunyaviridae

Genera	Prototype virus
Bunyavirus	Bunyamwera virus
Phlebovirus	Sandfly fever (Sicilian) virus
Uukuvirus	Uukuniemi virus
Nairovirus	Crimean/Congo haemorrhagic fever
Hantavirus	Hantaan virus

associated with *Apodemus* (prototype Hantaan 76/118); (2) those associated with *Rattus* (prototype Seoul); (3) those associated with *Clethrionomys* (prototype Puumala); and (4) those associated with *Microtus* (prototype Prospect Hill). To date both broad reacting and strain-specific antigenic sites have been found on both nucleocapsid and G2 antigens. Monoclonal antibodies to G1 have yet to be assessed but they are undoubtedly of strain-specific value. With the use of an increased panel of monoclonals, there is a tendency to increase the number of serotypes on the basis of serological reactions with one or two monoclonals [61]. Hence the subtyping within hantavirus serotypes will be an increasing observation as new isolates appear. At present, serotyping data subdivides hantaviruses into five broadly related types (Table 3).

Laboratory diagnosis

Serological diagnosis of HVD rests principally upon demonstrating a rise in titre of specific antibodies. Several methods have been developed and are currently in use (Figure 1), including indirect immunofluorescence (IF) [62], immunoperoxidase (IP) [63], enzyme-linked immunosorbent assay (ELISA) [64,65], all using Vero E6 cells as the antigen source (Table 2). Haemagglutination-inhibition (HI), using sucrose acetone extracted antigen prepared from brains of suckling mice infected with HV and erythrocytes from the Chinese grey goose (*Anser cygnoides*), is reported [66] to be equally or more specific than IF and IP techniques. The use of HI is specific within the hantavirus group and cross-reactions with other Bunyaviridae do not occur.

The immune adherence haemagglutination (IAHA) [67] and plaque reduction neutralisation tests [68,69] are useful in differentiating infections of closely-related hantaviruses. All the assays require a laboratory that has biosafety level 3 facilities at some stage of their protocols. In addition, protein products of various parts of the hantavirus genome [70] have also been expressed in a baculovirus system. These gene products are reactive with convalescent hantavirus immune sera for patients infected with severe HVD but demonstrate little reactivity with isolates from rodents in areas of mild HVD infection. However, with further developments using other hantavirus strains, gene products produced this way will reduce the reliance on laboratory facilities requiring biosafety level 3 and produce reagents of a group and type-specific nature, making differential diagnosis more precise.

Binding assays for specific IgG and IgM using enzyme immunoassay (ELISA) or IF are considered the procedure of choice for the early serological diagnosis of infection [71]. Specific IgM is detectable in 80–100% of cases within one week of onset [72]. Radial haemolysis complement fixation, an inexpensive and simple assay, is in widespread use in the Soviet Union and is reported as sensitive as IF [73].

Table 3 The serotype of hantaviruses and their relationship with rodent host, geographic distribution and human disease

Virus type	Subtype	Distribution	Host	Human disease
1 Hantaan	Hantaan	Korea		Severe (KHF)
	A9	China	*Ap agrarius*	Severe (KHF)
	23328	USSR		Severe (KHF)
	Fojnica	Yugoslavia	*A flavicollis*	Severe (KHF)
	Porogia	Greece	*A flavicollis*	Severe (KHF)
	Chen	China	Human	Severe (KHF)
	4605	USSR	Human	Severe (KHF)
2 Seoul	SR11	Japan	*R norvegicus*	Severe/moderate
	GB-B	UK	*R norvegicus*	Severe/moderate
	R22	China	*R norvegicus*	Moderate (EHF)
	80-39 (Seoul)	Korea	*R norvegicus*	Moderate (KHF)
	Tchoupitoulas	USA	*R norvegicus*	Unknown
	Girard Point	USA	*R norvegicus*	Unknown
3 Puumala	Puumala	Finland	*C glareolus*	Mild (NE)
	Sotkampo	Finland	*C glareolus*	Mild (NE)
	Hallnas	Sweden	*C glareolus*	Mild (NE)
	Cg1820	USSR (Baskiria)	*C glareolus*	Moderate (HFRS)
	Cg13891	Belgium	*C glareolus*	Unknown
4 Prospect Hill	–	USA	*M pennsylvanicus*	Unknown
5 Leaky	–	USA	*Mus musculus*	Unknown

NB This is by no means a complete list but it demonstrates worldwide distribution of hantavirus and the close serological relationship between isolates derived from a common host which by coincidence to the type of HVD normally expected. The only unexpected finding is the laboratory-based infection which in the case of the UK was imported and severely affected the laboratory personnel.

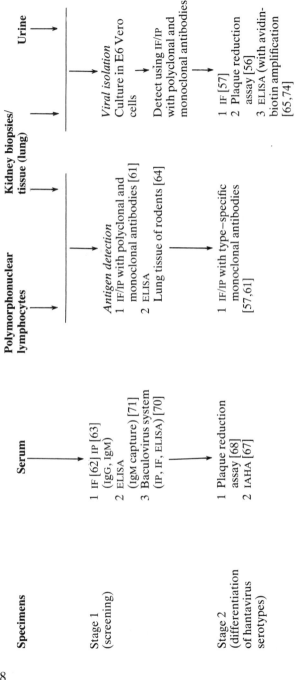

Figure 1 Routine laboratory diagnosis of hantavirus infection

IF = immunofluorescence test
IP = immunoperoxidase test

ELISA = enzyme linked immunosorbent assay
IAHA = immune adherence haemagglutination assay

Antigen detection systems, such as sandwich ELISA [74], appear to be more sensitive in the early diagnosis of the disease than are serologic tests. In the first week of illness, 87% of HVD patients were reported to be antigen positive, while 29% were IF antibody positive [75]. Demonstration of viral antigen on circulating platelets, mononuclear cells and neutrophils by IF has been reported as a successful approach to antigen detection [76].

Isolation of virus from peripheral blood or serum is very inconsistent. Isolation from urine is successful when obtained early in the infection [77]. However, primary isolation of virus from animal experiments has in the author's experience taken between 39–60 days after infection of E6 vero cells. Alternatively, RNA probes generated from cDNA clones of the M and S genome segments of Hantaan virus readily detect Far Eastern isolates, but have been less effective against the Scandinavian and European isolates [78].

Differential diagnosis depends to a great extent upon the stage in the patient's course of illness and the strain of infecting hantavirus. Early in the febrile phase, the infection cannot be differentiated from other viral fevers. An epidemiologic history should be sought. Exposure to rodents or other reservoir animals in forests in the spring and summer or to household or field rodents in autumn or winter are clues that suggest HVD. The temporal relationship of renal insufficiency with an acute febrile illness may also suggest leptospirosis, streptococcal or other bacterial causes of acute interstitial nephritis, as well as other viral aetiologies.

Treatment

There is no consensus that any specific therapy is of value. Attentive, supportive therapy remains the critical element in managing the three main stages of the disease as described by Katz [77]. The maintenance of circulation, management of fluid and electrolyte balance, and intervention with artificial dialysis may be life-saving. In the mild type of infection, a more conservative approach is usual, few needing parenteral fluid therapy or dialysis. Early indications of ribavirin used experimentally [79] and with human trials in China [80] have indicated a reduced mortality if treatment starts within five days of onset. Recombinant α-interferon [81] used in China failed to modify HVD significantly.

Epidemiological features of hantavirus

On serological evidence this newly-defined group of viruses has been reported within many rodent species and man worldwide [82]. Early clinical and epidemiological features of hantavirus-like diseases from various parts of the world have now been confirmed as being caused by an antigenically-related group of viruses. These agents also share similarities in their epidemiology and ecological characteristics, routes of transmission and mechanisms of virus maintenance.

Central to understanding the natural history of hantavirus is how the virus is maintained and transmitted within rodent colonies. Adult rodents show persistent infection without any clinical manifestations and secrete infectious virus for prolonged periods (see p. 193). After inoculation, the rodent is viraemic for about seven days. During the viraemic phase, virus is disseminated throughout the body and found primarily in the lungs, spleen and kidney for long periods, perhaps for the life of the rodent. Saliva contains virus for varying lengths of time after infection, dependent on the rodent, eg *Rattus* 13 days, *Apodemus* 40 days, and plays an important role in horizontal passage of the virus between rodents. This is more important in *Rattus* transmission since the virus shed in their saliva is reportedly the highest among the various rodent hosts. Infection of newborn rodents with hantavirus is fatal.

Humoral responses as measured by IF and neutralisation assays show antibodies persist for life, but do not have the capability of reducing the amount of antigen expressed in the organs of the body. At present the role of antibody in regulating the shedding of virus is little understood. T cell function is important for protection against and clearance of virus [83]. Cell-mediated studies have demonstrated that helper/inducer T cells are necessary to induce antibodies against hantavirus glycoproteins producing neutralising antibodies, and cytotoxic T cells are considered essential for viral clearance [84,85].

Many hantavirus isolates have been obtained from humans and rodent hosts and have been typed according to their serological cross-reactivity. The epidemiology of hantaviruses is intimately linked to the ecology of their principal host. The serological grouping also coincides with the main host and the severity of the disease (see Table 3).

Thus, there are currently five possible Hantavirus serotypes, each having many subtypes.

1 Severe HVD (Korean haemorrhagic fever type) associated with *Apodemus*-borne viruses – Hantaan and related viruses. There are well-defined human cases attributed to this group of viruses found in China [15], Korea [3], eastern USSR [86], Bulgaria [87] and more recently in Greece [88] and Yugoslavia [89]. Studies of the seasonality demonstrate that the time of increased virus transmission to man is linked to increased prevalence of hantavirus in *Apodemus* mice. Therefore, there are two peaks of human disease which occur in the spring and autumn. During these periods, *Apodemus* are reproductively more active, spend more time outside their burrows and become more associated with rural exposure.

2 Moderate HVD (hemorrhagic fever with renal syndrome) *Rattus-*

borne viruses, known as Seoul type. This group originally was associated with a form of HVD found in urban centres of Korea far from the endemic areas of severe HVD (KHF) [90]. In addition, related strains have been identified in China [15], western USSR [91], Japan [92], and in wharf rats in the USA and South America [93].

SR11 and GB-B type are closely related hantaviruses from infected laboratory rats. The problem of laboratory infections was first reported by Umenai and colleagues [94], when they described an outbreak of HVD among animal workers in Japan. Further research found it was widespread among research facilities in both Japan and Korea. These observations were extended to Belgium [95], the United Kingdom [96], and France [97]. An additional concern was that in the United Kingdom hantavirus isolates were obtained from rat immunocytomas in a non-microbiological environment derived from Belgium. Tumours stored for up to 10 years in liquid nitrogen yielded hantavirus isolates on culture. Concern was also directed to the study of rat origin cell lines or rat-mouse hybridomas as possible sources of hantavirus. Many cell lines of this nature held in the American Type Culture Collection were found negative on screening for hantavirus [98]. The associated clinical illness is of the moderate to severe type, yet the serological characterisation with type-specific monoclonals places the UK isolates as a subtype within the Seoul group.

3 Mild HVD (nephropathia epidemica) *Clethryonomys*-borne virus strains, known as the Puumala type. There are well-defined cases primarily in Scandinavia (Finland, Norway and Sweden) [40,61,99] and more recently in Belgium [100], France [101], the United Kingdom [102] and western USSR [103]. The disease is most often seen in men and is clearly associated with rural exposure. Commonly, patients are infected during the summer months in forest areas. Most cases occur in the autumn and winter. The virus has now been shown to be clearly distinct from the prototype Hantaan virus and the hantaviruses from domestic rats.

4 Prospect Hill type, isolated from *Microtus pennsylvanicus* in the USA [93]. Sera from *Microtus* reacts with the Korean *Apodemus* isolate, but not with *Clethryonomys* antigens. Alternatively, the antigen reacts more conclusively with sera from nephropathia patients. At present no human disease has been attributed to infection with Prospect Hill.

5 Leaky strain, isolated from *Mus musculus* from Texas, USA, which is currently unclassified. There is at present no associated human illness proven with this virus.

As outlined, five distinct virus types are now recognised within the genus and the total is likely to increase as investigations continue. Antibodies against hantaviruses, without disease, have already been extensively detected throughout the American continent (Alaska, Canada, USA, Brazil and Columbia); the Western Pacific and Southeast Asia (Taiwan, Hong Kong, Philippines, Malaysia, Thailand, India, Burma); in Africa (Egypt, Uganda, Gabon, Central African Republic, Nigeria); the Pacific Islands (Fiji, Hawaii); and Europe (Belgium, France, United Kingdom) [104].

It is worth pointing out that there is increased detection of imported cases of moderate HVD being imported into the United Kingdom [96] from endemic areas of, for example, Asia, emphasising that hantaviruses should be excluded in patients presenting a febrile illness after visiting endemic areas.

Seroepidemiological studies of human sera referred to the PHLS (1985–89) have shown evidence of hantavirus among patients suspected of leptospirosis (8%–12%), rickettsiae (2%–7%) and arboviral infection (3%–9%). All patients had a history of rural exposure or contact with water, or had an occupation which brought them into contact with rodents (water workers, agricultural employees, and so on). Examination of groups based on occupations considered high risk have revealed serological evidence of hantavirus (Table 4). The patients identified as having no international travel history serologically cross-react with the Puumala group and other European isolates. Together with the clinical history of a mild or asymptomatic hantavirus infection, contact with voles (*Clethrionomys* sp) and rural rats (*Rattus* sp) was considered the main point of contact. In 1987, several patients had a reported clinical illness compatible with HVD (Nephropathia epidemica) and leptospirosis. A significantly higher proportion of cases (8–12%) with suspected leptospirosis also have serological evidence of hantavirus. Because of the similarity in epidemiology and clinical presentation, it is

Table 4 Prevalence of hantavirus among defined occupations in the UK

Occupations	Number	% positive	Group serotype
Nature conservancy workers – Scotland	122	12.5	Puumala
Sewerage/water workers	96	4.3	Puumala
Animal laboratory personnel	560	18.9	Puumala/Seoul Hantaan
Farm workers	130	21.5	Puumala
Water sports activities	90	5.1	Puumala

recommended that hantavirus investigation should be undertaken in all cases of suspected leptospirosis.

In the case of laboratory workers, the additional exposure to rodents under their care cannot be ruled out. The figures assigned to laboratory personnel appear to be more prevalent in those with at least five years' experience. The majority seemed to have been exposed to hantavirus within the Puumala group, with no clinical evidence of infection.

Hantavirus antibodies can also be demonstrated in animals such as cats, dogs, guinea pigs and rabbits kept in animal rooms with sero-positive rodents [105]. It remains to be established whether they are good indicators of infection in rodents they are in contact with or whether they are of any risk to man.

Patients with an international travel history involving the Far East, China or Korea are being identified with HVD in the UK and demonstrate a serological reactivity with the Hantaan and Seoul group of viruses giving a mild to severe illness. From a limited, random UK seroepidemiological study of various animal populations it is evident that between 5–12% of rats, voles and rabbit species are hantavirus carriers within the UK. In addition, recent studies have shown that the cat family (wild and domestic) have a hantavirus seropositivity rate of between 5–15%. Since the inter-relationship between cats and rodents is well known and wounding within wild populations is a proven route of hantavirus transmission, then it should be of no surprise that the cat population should show serological evidence of hantavirus infection. The overall epidemiological significance of this data within the UK is difficult to interpret as far as human infection is concerned, since at present there are few cases reported or recognised within the UK.

Transmission characteristics

Evidence concerning the mode of spread of hantaviruses to man is derived principally from epidemiological observations. The consensus is that the virus spreads to humans from persistently infected rodents. Knowledge of the duration and level of infectious virus is essential to understanding the mechanism of viral spread but limited work has been published in this area. There is sufficient experimental evidence to suggest that infection with hantavirus of *Apodemus* sp [106], *Rattus* sp [107] and *Clethrionomys* sp [54, 108] leads to virus being isolated from tissue and body fluids up to a year after inoculation.

The mechanism of transmission indicates a principal role for respiratory infection via aerosols of infectious virus from rodents excreting the virus in the lungs, saliva, urine and faeces. Alternatively, incidental accounts of presumed transmission by rodent bites have been reported [91]. Saliva-mediated transmission plays an important role in horizontal passage of the virus between rodents, especially among *Rattus*, where the duration and level of virus shed is highest [109].

Horizontal transmission among humans has not been documented. Blood and urine are both infectious if obtained during the first five days of the disease. HVD only occurred if administered by the intravenous or intramuscular route. Direct inoculation of the upper respiratory tract did not result in infection, which contradicts the evidence produced using rodent material. Nosocomial spread was not documented during the Korean War experience, suggesting that secondary infection is low.

However, there has been a report of vertical transmission in Korea (Lee, personal communication) from two mothers to their fetuses, causing the abortion and death of the babies, but as yet this has not been confirmed as a major route of infection. Rodent students [110] have indicated that cross-placental transmission does not occur in the young when born to a persistently affected mother; subsequent infection occurs after birth through saliva and urine contact.

There has been little to suggest that, unlike other viruses of the family Bunyaviridae, arthropods serve as important vectors of these viruses [111]. Recent reports of hantavirus isolation from cats, insectivores including peridomestic species (*Suncus murinus*) and other domesticated animals have been reported but are unconfirmed [112]. However, recent isolation of hantavirus from *Trombicula scutellaris*, an ectoparasite of *A agrarius* in China, has renewed interest in the role of the chigger and gamasoid mites in HVD [113]. Further Chinese reports have indicated that hantavirus was recovered from *Haemolaelaps glasgowi* and *Eulaelaps stabularis* collected from *A agrarius* [113]. Experimental studies also demonstrated that all these mites and also *Ornithonysus bacoti* and *Haemolaelaps casalis* could be infected by feeding on infected rodents [114]. Whatever the role ectoparasites play in the transmission cycle of hantavirus in nature, proof that they are not necessary for inter-rodent transmission or transmission of the virus to man is clear. Aerosol spread is central to man's infection.

Many recorded cases of hantavirus among laboratory personnel associated with rodent contact have alerted the scientific community to the potential risk of working with laboratory species of *Rattus*, *Apodemus* and *Clethrionomys*. Reports from Japan [115,116], Korea [45], Belgium [117] and the United Kingdom [96] of laboratory infections confirm the aerosol route as a means of transmission to humans from *Rattus* species. In addition, the potential risk of processing rodent immunocytomas containing hantavirus has been shown to occur from persistently infected animals [96]. This supported an early observation in Japan, where tumours induced chemically with 4-nitroquinoline-N-oxide in Fisher rats proved to contain hantavirus [118].

Vaccines

The data would suggest that endemic hantavirus illness is primarily a disease of poverty and war. All the evidence recorded suggests that control should be directed at reducing the contact between human populations and rodents. Where social conditions cannot significantly comply, then there is a role for a vaccine. This would be directed towards clearly-defined populations in endemic areas, for example forest, water and agricultural workers, plus specific laboratory personnel.

Two approaches are currently being investigated. The first is centred in China, where it has been reported that an inactivated cell culture vaccine has given encouraging results in animal models [119]. The vaccine is based on formalin-inactivated hamster kidney cell cultures. It is reported that human trials are imminent within China.

Secondly, a cDNA clone containing the complete open reading frame of the Hantaan virus M genome segment has been cloned into vaccinia virus and forms the basis of a genetically-manipulated vaccine [120]. *In vitro* studies in mice have shown that the vaccinia construct raises neutralising antibodies against hantavirus. Vaccinia constructs expressing the S genome segment proteins fail to elicit neutralising antibodies. Although this work is at an early stage, results do hold out hope for a future vaccine. Use of the vaccinia virus expression system will also offer the opportunity to clone G1 and G2 separately, to determine which glycoprotein is responsible for stimulating neutralising antibodies.

Conclusion

During the past decade a considerable amount of new knowledge has advanced our understanding of hantaviruses and their related diseases. We now know that there are several different viruses capable of causing similar or identical disease; that the hantaviruses are maintained in chronically infected rodents; that there is a wider distribution than first suspected; that currently diagnosis is rapid and accurately diagnoses acute disease; and that antivirals such as ribavirin may have a place in treating cases if used early in the febrile phase.

Finally, with a better understanding of the molecular biology of hantaviruses, diagnostic reagents and vaccines derived from suitable expression vectors are now being produced. Their evaluation opens up new chapters in strain differentiation, diagnosis, host immunological responses and possible vaccines against the hantaviruses.

References

1   Cohen MS. Epidemic hemorrhagic fever revisited. *Rev Infect Dis* 1982, **4**, 992–7.

2   Gajdusek DC. Muroid virus nephropathies and muroid viruses of the Hantaan virus group. *Scand J Infect Dis Suppl* 1982, **36**, 96–108.

3   Lee HW. Korean hemorrhagic fever. *Prog Med Virol* 1982, **28**, 96–113.

4   Fisher-Hoch SP, McCormick JB. Haemorrhagic fever with renal syndrome: A review. *Abs Hyg Comm Dis* 1985, **60** (4), 1–20.

5   Gajdusek DC. Haemorrhagic fever with renal syndrome (Korean hemorrhagic fever, epidemic hemorrhagic fever, nephropathia epidemica): a newly recognised zoonotic plague of the Eurasian landmass with the possibility of related muroid nephropathies on other continents. In: Mackenzie JS, ed. Viral diseases in South-East Asia and the Western Pacific. Sydney: Academic Press, 1982, 576–94.

6   Desmyter J, van Ypersele de Strihou C, van der Groen G. Hantavirus disease or haemorrhagic fever with renal syndrome. *Lancet* 1984, **ii**, 158.

7   Clement J. Trench nephrities: past and present. 29th International Colloquium on Hantaviruses – Inaugural Session. Antwerp, Belgium, 1987, 10–11 December.

8   Abercrombie RG. Observation on the acute phase of five hundred cases of war nephritis. *J Roy Army Med Corps* 1916, **27**, 131–57.

9   Ameuille P. Du rôle de l'infection dans les néphrites de guerre. *Ann de Méd* 1915–16, **3**, 298–332.

10 Ullmann B. Über die diesem Kriege beobachtee neue Form akuter Nephritis. *Berl klin Wchnschr* 1916, **53**, 1046–9.

11 Grey H. Nephritis in fifty-six soldiers. *J Urol* 1919, **3**, 27–32.

12 Ishii S, Ando K, Watanabe N, *et al.* Studies on Song-go fever. *Jap Army Med J* 1942, **370**, 266–8.

13 Kitano M. A study of epidemic hemorrhagic fever. *Jap Army Med J* 1944, **370**, 269–83.

14 Myhrman G. Nephropathia epidemica: a new infectious disease in Northern Scandinavia. *Acta Med Scand* 1951, **140**, 52–6.

15 Chen HX, Qiu FX, Dong B-J, *et al.* Epidemiological studies on hemorrhagic fever with renal syndrome in China. *J Infect Dis* 1986, **154**, 394–8.

16 Tamura M. Occurence of epidemic hamorrhagic fever in Osaka City: First cases found in Japan with characteristic feature of marked proteinuria. *Biken J* 1964, **7**, 79–94.

17 Lee HW, Lee PW, Johnson KM. Isolation of the etiological agent of Korean hemorrhagic fever. *J Infect Dis* 1978, **137**, 298–308.

18 Johnson KM. Nephropathia epidemica and Korean hemorrhagic fever: The veil lifted? *J Infect Dis* 1980, **141**, 135–6.

19 Lee HW, Lee PW, Lähdevirta J, Brummer-Korvenkontio M. Aetiological relation between Korean haemorrhagic fever and nephropathia epidemica. *Lancet* 1979, **i**, 186–7.

20 Lee HW, Gajdusek DC, Gibbs CJ, Xu Z-Y. Aetiological relation between haemorrhagic fever with renal syndrome in People's

Republic of China. *Lancet* 1980, **i**, 819–20.

21   Lee PW, Gibbs CJ, Gajdusek DC, Hsiang CM, Hsiung GD. Identification of epidemic haemorrhagic fever with renal syndrome in China with Korean haemorrhagic fever. *Lancet* 1980, **i**, 1025–6.

22   Sheedy JA, Froeb HF, Batson HA, *et al.* The clinical course of epidemic hemorrhagic fever. *Am J Med* 1954, **16**, 619–38.

23   Penttinen K, Lähdevirta J, Kekomäki R, *et al.* Circulating immune complexes, immunoconglutinins, and rheumatoid factors in nephropathia epidemica. *J Infect Dis* 1981, **143**, 15–21.

24   Lähdevirta J, Collan Y, Jokinen EJ, Hiltunen R. Renal sequelae to nephropathia epidemica. *Acta Pathol Micrbiol Scand [A]* 1978, **86**, 265–71.

25   Lähdevirta J, Savola J, Brummer-Korvenkontio M, Berndt R, Illikainen R, Vaheri A. Clinical and serological diagnosis of Nephropathia epidemica, the mild type of haemmorrhagic fever with renal syndrome. *J Infect* 1984, **9**, 230–8.

26   Song G. Hang C-S, Liao H-X, Fu J-L. Antigenic comparison of virus strains of mild and classical types of epidemic haemorrhagic fever isolated in China and adaption of these to cultures of normal cells. *Lancet* 1984, **i**, 677–8.

27   Lähdevirta J. Nephropathia epidemica in Finland: a clinical histological and epidemiological study. *Ann Clin Res* 1971, **3**, 1–154.

28   Lähdevirta J. Clinical features of HFRS in Scandinavia as compared with East Asia. *Scand J Infect Dis Suppl* 1982, **36**, 93–5.

29   Trencseni T, Keleti B. Clinical aspects and epidemiology of hemorrhagic fever with renal syndrome. Analysis of clinical and epidemiological experiences in Hungary. Budapest: Akademiai Kiado, 1971, 1–237.

30   Hullighorst RL, Steer A. Pathology of epidemic hemorrhagic fever. *Ann Intern Med* 1953, **38**, 77–101.

31   Lukes RJ. The pathology of thirty-nine fatal cases of epidemic hemorrhagic fever. *Am J Med* 1954, **16**, 639–50.

32   Kuhlbäck B, Fortelius P, Tallgren LG. Renal histopathology in a case of nephropathia epidemica. Myhrman. A study of successive biopsies. *Acta Pathol Microbiol Scand* 1964, **60**, 323–33.

33   Collan Y, Lähdevirta J. Electron microscopy of nephropathia epidemica cell nuclei in kidney biopsies. *Acta Pathol Microbiol Scand [A]*, 1979, **87**, 71–7.

34   Jokinen EJ, Lähdevirta J, Collan Y. Nephropathia epidemica: Immunohistochemical study of pathogenesis. *Clin Nephrol* 1978, **9**, 1–5.

35   Tsai TF. Hemorrhagic fever with renal syndrome: Clinical aspects. *Lab Anim Sci* 1987, **37**, 419–27.

36   Takenka A, Gibbs CJ, Nakamura T, Gajdusek DC. Replication

of Hantaan virus in human leukocytes. In: Abstracts, Sixth International Congress of Virology, Sendai, Japan Sept 1–7, 1984, 274.

37   Kurata T, Aoyama Y, Yamamouchi T, *et al*. Dissemination of Hantaan or HFRS related virus in experimentally infected mice, and naturally infected wild and laboratory rats. Proceedings, 6th International Congress of Virology, Sendai. September 1984.

38   Xu Z-Y. Pathogenesis of epidemic haemorrhagic fever. *Chinese J Infect Dis* 1983, **8**, 127–30.

39   Lee HW, Lee PW. Korean hemorrhagic fever. I. Demonstration of causative antigen and antibodies. *Korean J Int Med* 1976, **19**, 371–83.

40   French GR, Foulke RS, Brand OA, Eddy GA, Lee HW, Lee PW. Korean hemorrhagic fever: propagation of the etiologic agent in a cell line of human origin. *Science* 1981, **211**, 1046–8.

41   Brummer-Korvenkontio M, Vaheri A, Hovi T, *et al*. Nephropathia epidemica: Detection of antigen in bank voles and serologic diagnosis of human infection. *J Infect Dis* 1980, **141**, 131–4.

42   Niklasson B, LeDuc J. Isolation of the Nephropathia epidemica agent in Sweden. *Lancet* 1984, **i**, 1012–3.

43   Song G, Qui XZ, Ni D-S, Zhao JN, Kong BX. Etiological studies of epidemic hemorrhagic fever. I. Isolation of EHF virus in *Apodemus agrarius* from non-endemic area of EHF and its characterization. *Acta Acad Med Sin* 1982, **4**, 73–77.

44   Kitamura K, Morita C, Komatsu T, *et al*. Isolation of virus causing hemorrhagic fever with renal syndrome (HFRS) through a cell culture system. *Jpn J Med Sci Biol* 1983, **36**, 17–25.

45   Lee PW, Amyx HL, Gibbs CJ, Gajdusek DC, Lee H-W. Propagation of Korean hemorrhagic fever virus in laboratory rats. *Infect Immun* 1981, **31**, 334–8.

46   Lloyd G, Jones N. Infection of laboratory workers with hantavirus acquired from immunocytomas propagated in laboratory rats. *J Infect* 1986, **12**, 117–25.

47   Lee PW, Amyx HL, Yanagihara R, Gajdusek DC, Goldgaber D, Gibbs CJ. Prospect Hill virus isolated from meadow voles in the United States. *J Infect Dis* 1985, **152**, 826–9.

48   Hung T, Xia S-M, Zhao TX, *et al*. Morphological evidence for identifying the viruses of hemorrhagic fever with renal syndrome as candidate members of the Bunyaviridae family. *Arch Virol* 1983, **78**, 137–44.

49   Tao H, Semao X, Zinyi C, Gan S, Yanagihara R. Morphology and morphogenesis of viruses of hemorrhagic fever with renal syndrome. II. Inclusion bodies – ultrastructural markers of Hantavirus-infected cells. *Intervirol* 1987, **27**, 45–52.

50   Schmaljohn CS, Dalrymple JM. Analysis of Hantaan virus RNA:

evidence for a new genus of Bunyaviridae. *Virology* 1983, **131**, 482–91.

51 Yoo D, Kang CY. Genomic comparison among members of Hantavirus group. In: Mahy B, Kolakofsky D, eds. The biology of negative strand viruses. Amsterdam: Elsevier, 1987, 424–31.

52 Schmaljohn CS, Hasty SE, Harrison SA, Dalrymple JM. Characterisation of Hantaan virions, the prototype virus of hemorrhagic fever with renal syndrome. *J Infect Dis* 1983, **148**, 1005–12.

53 Elliott LH, Kiley MP, McCormick JB. Hantaan virus: Identification of virion proteins. *J Gen Virol* 1984, **65**, 1285–93.

54 Schmaljohn CS, Schmaljohn AL, Dalrymple JM. Hantaan virus M RNA: coding strategy, nucleotide sequence and gene order. *Virology* 1987, **157**, 31–9.

55 Schmaljohn CS, Jennings GB, Haj J, Dalrymple JM. Coding strategy of the S genome segment of Hantaan virus. *Virology* 1987, **155**, 633–43.

56 Lee PW, Gibbs CJ, Gajdusek DC, Yanagihara R. Serotypic classification of Hantaviruses by indirect immunofluorescent antibody and plaque reduction neutralization tests. *J Clin Micro* 1985, **22**, 940–4.

57 Sugiyama K, Morikawa S, Matsuura Y, *et al.* Four serotypes of haemorrhagic fever with renal syndrome viruses identified by polyclonal and monoclonal antibodies. *J Gen Virol* 1987, **68**, 979–87.

58 Xu ZK, An XL, Wang MX. The antigenic analysis of haemorrhagic fever with renal syndrome viruses in China by monoclonal antibodies. *J Hyg* 1986, **97**, 369–75.

59 Xing Z, Song G, Li WM, Zheng XL, Liao HX, Fu JL. Antigenic comparison of viruses of hemorrhagic fever with renal syndrome isolated from China and Japan. *Chinese J Virol* 1987, **3**, 27–31.

60 van der Groen G, Beelaert G, Hoofd G, *et al.* Partial chracterisation of a hantavirus isolated from *Clethrionomys glareolus* captured in Belgium. *Acta Virol* 1987, **31**, 180–4.

61 van der Groen G, Yamanishi K, McCormick J, Lloyd G, Tkachenko EA. Characterisation of hantaviruses using monoclonal antibodies. *Acta Virol* 1987, **31**, 499–503.

62 Doo CD. Study on immunofluorescent antibodies against Hantaan, Seoul and Puumala viruses in sera from patients with hemorrhagic fever with renal syndrome in Korea. *Korea Univ Med J* 1983, **23**, 91–8.

63 van der Groen G, Beelaert G. Immunoperoxidase assay for the detection of specific IgG antibodies to Hantaan virus. *J Virol Methods* 1985, **10**, 53–8.

64 Gavrilovskaya IN, Apekina NS, Gorbachkova EA, Myasnikov YA, Rylceva EV, Chumakov MP. Detection by enzyme-linked immunosorbent assay of haemorrhagic fever with renal syndrome virus in lung tissue of rodents from European USSR. *Lancet* 1981, i, 1050.

65 Goldgaber D, Gibbs CJ, Gajdusek DC, Svermyr A. Definition of three serotypes of Hantaviruses by a double sandwich ELISA with biotin-avidin amplification system. *J Gen Virol* 1985, **66**, 1733–40.

66 Okuno Y, Yamianishi K, Takahashi Y, *et al*. Haemagglutination-inhibition test for haemorrhagic fever with renal syndrome using virus antigen prepared from infected tissue culture fluid. *J Gen Virol* 1986, **67**, 149–56.

67 Sugiyama K, Matsuura Y, Morita C, *et al*. An immune adherence assay for discrimination between etiologic agents of hemorrhagic fever with renal syndrome. *J Infect Dis* 1984, **149**, 67–73.

68 Kim Mj. Formation of neutralizing antibodies against Hantaan and Seoul viruses of sera from patients from haemorrhagic fever with renal syndrome patients in Korea. *Korea Univ Med J* 1986, **23**, 115–22.

69 Arikawa J, Takashima I, Hashimoto N. Cell fusion by haemorrhagic fever with renal syndrome (HFRS) viruses and its application for titration of virus infectivity and neutralizing antibody. *Arch Virol* 1985, **86**, 303–13.

70 Schmaljohn CS, Sugiyama K, Schmaljohn A, Bishop DHL. Baculovirus expression of the small genome segment of Hantaan virus and potential use of the expressed nucleocapsid protein as a diagnostic antigen. *J Gen Virol* 1988, **69**, 777–86.

71 Zheng X, Zhang D, Chang-sou H, *et al*. Studies on the experimental conditions of anti-μ-chain monoclonal antibody immunoglobulin M capturing enzyme linked immunosorbent assay (MaIgMcELISA) to detect the IgM antibody against epidemic hemorrhagic fever (EHF) virus. In: Abstract book of the 29th International Colloquium Hantaviruses, Antwerp, Belgium 10–11 December 1987.

72 Jiang YT, Li ZD, Song G. Seroepidemiological study of epidemic hemorrhagic fever with renal syndrome in China. *Chin Med J* 1981, **94**, 221–8.

73 Inouye S, Matsuno S, Kono R. Difference in antibody reactivity between complement fixation and immune adherence hemagglutination tests with virus antigens. *J Clin Microbiol* 1981, **14**, 241–6.

74 Xu ZK, Wang MX, Jiang SC, Ma WY, Wang HT. Detection of HFRS virus antigen and antibody by a McAb ELISA indirect sandwich method. In: 29th International Colloquium Hantaviruses, Antwerp, Belgium 10–11 December 1987.

75 Xue YI, Fang L. The observation of EHF virus in urine of EHF

patients. *J Xian Med Univ* 1987, **8**, 54–7.

76 Nagai T, Tanishita O, Takahashi Y, *et al*. Isolation of haemorrhagic fever with renal syndrome virus from leukocytes of rats and virus replication in cultures of rat and human macrophages. *J Gen Virol* 1985, **66**, 1271–8.

77 Katz S, Leedham CL, Kessler WH. Medical management of hemorrhagic fever. *J Am Med Assoc* 1952, **150**, 1363–6.

78 Schmaljohn CS, Lee HW, Dalrymple JM. Detection of Hantaviruses with RNA probes generated from recombinant DNA. *Arch Virol* 1987, **95**, 291–301.

79 Huggins JW, Kim GR, Brand OM, McKee KT. Ribavirin therapy for Hantaan virus infection in suckling mice. *J Infect Dis* 1986, **153**, 489–97.

80 Gui XE, Ho M, Cohen MS, Wang Q-L, Huang H-P, Xie Q-X. Hemorrhagic fever with renal syndrome: treatment with recombinant α interferon. *J Infect Dis* 1987, **155**, 1047–51.

81 Hsiang CM, Huggins JW, Guan MY, *et al*. Effective therapy of epidemic hemorrhagic fever with Ribavirin in a double-blind, random controlled trial. II. Improvement in some clinical signs of hematological, cardiology, renology and immunology. In: 29th International Colloquium Hantaviruses, Antwerp, Belgium 10–11 December 1987.

82 Lee HW, French GR, Lee PW, Baek LJ, Tsuchiya K, Foulke RS. Observations of rodents with the etiological agent of Korean hemorrhagic fever. *Am J Trop Med Hyg* 1981, **30**, 477–82.

83 Nakamura T, Yanagihara R, Gibbs CJ, Gajdusek DC. Immune spleen cell-mediated protection against fata Hantaan virus infection in infant mice. *J Infect Dis* 1985, **151**, 691–7.

84 Tanishita O, Takahashi Y, Okuno Y, *et al*. Persistent infection of rats with haemorrhagic fever with renal syndrome virus and their antibody responses. *J Gen Virol*, 1986, **67**, 2819–24.

85 Asada H, Tamura M, Kondo K, *et al*. Role of T lymphocyte subsets in protection and recovery from Hantaan virus infection in mice. *J Gen Virol* 1987, **68**, 1961–69.

86 Tkachenko EA, Ivanov A, Donets M, *et al*. Potential reservoir and vectors of haemorrhagic fever with renal syndrome (HFRS) in the USSR. *Ann Soc Belge Med Trop* 1983, **63**, 267–8.

87 Gavrilovskaja I, Vasilenk S, Chumakov M, Sindarov L. Hemorrhagic fever with renal syndrome in Bulgaria: Spreading and serologic proof. *Epidemiology Mikrobiology Infek Bol* 1984, **21**, 17–22.

88 Siamopoulos K, Antoniades A, Acritidis N, *et al*. Outbreak of haemorrhagic fever with renal syndrome in Greece. *Eur J Clin Micro* 1985, **4**, 132–4.

89 Gligic A, Obradovic M, Stojanovic R, *et al*. Hemorrhagic fever

with renal syndrome in Yugoslavia: Ten years investigation with main point on epidemic in 1986. In: 29th International Colloquium Hantaviruses, Antwerp, Belgium 10–11 December 1987.

90   Lee Hw, Baek LJ, Johnson KM. Isolation of Hantaan virus, the etiologic agent of Korean hemorrhagic fever, from wild urban rats. *J Infect Dis* 1982, **146**, 638–44.

91   Tkachenko EA, Bashkirtsev VN, van der Groen G, Dzagurova TK, Ivanov Ap, Ryl'Tseva EV. E6 cells of Hanta virus from *Clethrionomys glareolus* captured in the Bashkiria area of the USSR. *Ann Soc Belge Med Trop* 1984, **64**, 425–6.

92   Lee HW, Lee PW, Tamura M, Tamura T, Okuno Y. Etiological relation between Korean hemorrhagic fever and epidemic hemorrhagic fever in Japan. *Biken J* 1979, **22**, 41–5.

93   LeDuc JW, Smith GA, Johnson KM. Hataan-like viruses from domestic rats captured in the United States. *Am J Trop Med Hyg* 1984, **33**, 992–8.

94   Umenai T, Lee HW, Lee PW, *et al.* Korean haemorrhagic fever in staff in an animal laboratory. *Lancet* 1979, **i**, 1314–6.

95   Desmyter J, Le Duc JW, Johnson KM, Brasseur F, Deckers C, van Ypersele de Strihou C. Laboratory rat associated outbreak of haemorrhagic fever with renal syndrome due to Hantaan-like virus in Belgium. *Lancet* 1983, **ii**, 1445–8.

96   Lloyd G, Bowen ETW, Jones N, Pendry A. HFRS outbreak associated with laboratory rats in UK. *Lancet* 1984, **i**, 1175–8.

97   Douron E, Moriniere B, Matheron S, *et al.* HFRS after a wild rodent bite in the Haute-Savoie and risk of exposure to Hantaan-like virus in a Paris laboratory. *Lancet* 1984, **i**, 676–7.

98   LeDuc JW, Smith GA, Macey M, Hay RJ. Certified cell lines of rat origin appear free of infection with Hantavirus. *J Infect Dis* 1985, **152**, 1082–3.

99   Traavik T, Mehl R, Berdal BP, Lund S, Dalrymple JM. Nephropathia epidemica in Norway: Description of serological response in human disease and implication of rodent reservoirs. *Scand J Infect Dis* 1983, **15**, 11–16.

100   van Ypersele de Strihou C, van der Groen G, Desmyter J. Hantavirus nephropathy in Wester Europe: Ubiquity of hemorrhagic fevers with renal syndrome. *Adv Nephrol* 1986, **15**, 143–72.

101   Douron E, Rollin P, Assous M, Sureau AP. HFRS in France-Epidemic and clinical features. In: 29th International Colloquium Hantaviruses, Antwerp, Belgium. 10–11 December 1987.

102   Walker E, Pinkerton IW, Lloyd G. Scottish case of haemorrhagic fever with renal syndrome. *Lancet* 1984, **ii**, 982.

103   Chumakov MP, Gavrilovskaya IN, Boiko V, *et al.* Detection of hemorrhagic fever with renal syndrome (HFRS) virus in the lungs of bank voles *Clethrionomys glareolus* and redback voles *Clethrio-*

nomys rutilus trapped in HFRS foci in the European part of USSR, and serodiagnosis of this infection in man. *Arch Virol* 1981, **69**, 295–300.

104 LeDuc JW, Smith GA, Childs JE, *et al.* Global survey of antibody to Hantaan-related viruses among peridomestic rodents. *Bull WHO* 1986, **64**, 139–44.

105 France AJ, Burns SM. Hantavirus infection: a new imported viral haemorrhagic fever. *J Infect* 1988, **16**, 108–9.

106 Lee HW, Lee PW, Baek LJ, Song CK, Seong IW. Interspecific transmission of Hantaan virus, etiologic agent of Korean hemorrhagic fever in the rodent *Apodemus agrarius*. *Am J Trop Med Hyg* 1981, **30**, 1106–12.

107 Lee PW, Yanagihara R, Gibbs CJ, Gajdusek DC. Pathogenesis of experimental Hantaan virus infection in laboratory rats. *Arch Virol* 1986, **88**, 57–66.

108 Yanagihara R, Amyx HL, Gajdusek DC. Experimental infection with Puumala virus, the etiologic agent of nephropathia epidemica, in bank voles (*Clethrionomys glareolus*). *J Virol* 1985, **55**, 34–8.

109 Glass GE, Childs JE, Korch GW, LeDuc JW. Association of intraspecific wounding with hantaviral infection in wild rats (*Rattus norvegicus*). *Epidemiol Infect* 1988, **101**, 459–72.

110 Lloyd G, Jones N. Transmission of hantavirus during pregnancy in laboratory rats. In preparation.

111 Traub R, Hertig M, Lawrence WH, Harriss TP. Potential vectors and reservoirs of hemorrhagic fever in Korea. *Am J Hyg* 1954, **59**, 291–305.

112 Tang YW, Xu ZY, Zhu ZY, Tsai TS. Isolation of haemorrhagic fever with renal syndrome from *Suncus murinus*, an insectivore. *Lancet* 1985, **i**, 513–4.

113 Zhang Y, Bao MR, Sheng JZ. An investigation of haemorrhagic fever with renal syndrome virus infections in mites in epidemic areas. *Chin J Epid* 1986, **7**, 26–8.

114 Mong HY, Wu GH, Zhao XZ, *et al.* Experimental study on natural infection, biting and transoviral transmission of epidemic hemorrhagic fever virus in gamasoid mites. *Chin J Epid* 1986, **7**, 200–2.

115 Kawamata J. Studies on surveillance and control measures for zoonoses associated with animal experimentation, with special reference to haemorrhagic fever: a report to the Ministry of Education, Cultural Affairs and Science, Tokyo, 1981.

116 Nisioka S, Matsumoto H, Oku A, *et al.* Epidemic hemorrhagic fever occurring in Wakayama Medical College: A clinical and laboratory study of 7 patients. *Kansenshogaky Zasshi* 1981, **55**, 410–28.

117   Lee HW, Johnson KM. Laboratory-acquired infections with Hantaan virus, the etiologic agent for Korean hemorrhagic fever. *J Infect Dis* 1982, **146**, 645–51.

118   Yamanishi K, Dantas FJR, Takahasi M, *et al.* Isolation of haemorrhagic fever with renal syndrome (HFRS) virus from a tumor specimen in rat. *Biken J* 1983, **26**, 155–60.

119   Song G. Development of inactivated cell culture vaccine against hemorrhagic fever with renal syndrome (HFRS). In: 29th International Colloquium Hantaviruses, Antwerp, Belgium. 10–11 December 1987.

120   Pensiero MN, Jennings GB, Schmaljohn CS, Hay J. Expression of the Hantaan virus M genome segment by using a vaccinia virus recombinant. *J Virol* 1988, **62**, 696–702.

Chapter 10

Oculogenital *Chlamydia trachomatis* infections and their diagnosis

EO Caul

Until recently the genus Chlamydia was composed of two species – *C psittaci* and *C trachomatis*. With the recent recognition, identification and characterisation of the TWAR agent [1] as a potentially important cause of acute respiratory illness, a third member of the family has now been established. *C psittaci* is a zoonotic infection where avian species are the main natural hosts and man becomes incidentally infected. The TWAR agents were considered to be more closely related to *C psittaci* but now appear to be a single distinct species [2], with no recognised animal reservoir. Transmission probably occurs from person to person with man being the natural host. *C trachomatis* is the most common member of the Chlamydia genus known to infect man and is commonly transmitted by the sexual route [3,4]. In contrast to *C psittaci*, an animal reservoir is not recognised and persistence of the organism in man is facilitated by person to person contact. Since the early 1960s the incidence of sexually transmitted diseases in the UK has increased dramatically (Figure 1), with *C trachomatis* representing the most common infection. In 1986 an estimated 4.6 million cases of *C trachomatis* infections occurred in the United States and 120,000 cases in Scandinavia (Sweden, Norway and Finland) in 1986–87. Currently the promiscuous heterosexual population is at most risk although infection also occurs among homosexuals.

All members of the Chlamydia genus replicate in eukaryotic cells undergoing a complex developmental cycle [5]. They are antigenically diverse, possessing species, subspecies and type-specific antigens but share a common, genus-specific lipopolysaccharide antigen (LPS) [5]. Within the *C trachomatis* species, numerous serotypes are recognised and their association with different clinical conditions is shown in Table 1. Those serotypes causing trachoma (A–C) and lymphogranuloma venereum (serotypes 1, 2, 3) will not be considered further. *C trachomatis* (serotypes D–K) cause oculogenital infection and primarily infect

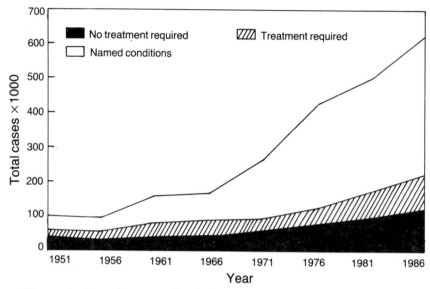

Figure 1 Sexually transmitted diseases, England and Wales 1951–85. Data from PHLS Communicable Disease Surveillance Centre (CDSC).

the epithelium of the male urethra and the female cervix and urethra. Symptomatic genital infection, presenting as an urethritis, is common in males [6,7] but a low incidence of asymptomatic infection (0–7%) is also

Table 1 Clinical spectrum of *Chlamydia trachomatis* infections

Serotypes	Disease	Complications
A–C	Trachoma	
1–3	Lymphogranuloma venerum	
A–K		
Male	Non-gonococcal urethritis (NGU)	Epididymitis
	Post gonococcal urethritis (PGU)	Reiter's syndrome
	Conjunctivitis	Sexually acquired
	Proctitis	arthritis (SARA)
Female	Mucopurulent cervicites	Salpingitis
	Urethritis	Perihepatitis
	Conjunctivitis	Endometritis
	Proctitis	Infertility
		Ectopic pregnancy
Neonates	Conjunctivitis	Pneumonia

recognised [7]. In contrast, asymptomatic infection in females is much higher, with a reported incidence of approximately 60% in some groups [8,9]. The consequences of an unrecognised and therefore untreated cervical or urethral infection in the female may be far-reaching.

Genital infection in men
Genital *C trachomatis* infection in men usually presents as an urethritis. This may occur following infection and treatment of *Neisseria gonorrhoeae* (post gonococcal urethritis) [6,7,9] or alternatively without a concominant gonococcal infection (non-gonococcal urethritis) [6,7].

NON-GONOCOCCAL URETHRITIS (NGU)
The recognition of a distinct disease (NGU) which was not caused by *N gonorrhoea* was first established in 1879. Some 20 years later, an association with *C trachomatis* was made. This condition currently accounts for more than 100,000 cases reported each year by genito-urinary medicine (GUM) clinics in England and Wales (Figure 2) and is the most common symptomatic infection seen at these clinics. It is estimated that 30%–58% of these NGU cases are attributable to *C trachomatis* infection [7], and as each case represents infection in at least two people, the pool of infection in the community is extremely large.

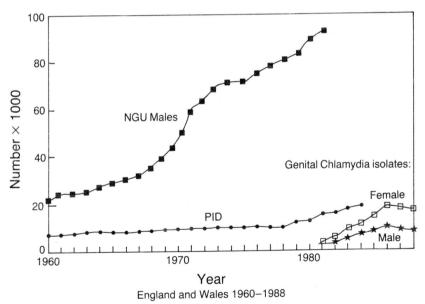

Figure 2 Pelvic inflammatory disease, non-specific genital infection and genital chlamydia. Data from CDSC.

207

This syndrome occurs in patients who have been infected with both *N gonorrhoeae* and *C trachomatis*. Current antimicrobial therapy for *N gonorrhoeae* infection does not eradicate *C trachomatis* and as a result a post-gonococcal urethritis resulting from the replication of *C trachomatis* in the urethra occurs.

Isolation of *C trachomatis* in patients with *N gonorrhoeae* infection before treatment ranges from 17.5%–32% [6,10–16] in different studies. In contrast the detection rate of *C trachomatis* in patients treated for *N gonorrhoeae* infection with urethritis (PGU) ranges from 38%–88% [6,10–16] and from 0–50% [6,10–16] in treated patients without urethritis. The differences in isolation rates of *C trachomatis* in these various studies may partly be accounted for by the composition of the study group, where the proportions of homosexuals vary. It is known that the prevalence of *C trachomatis* is lower in homosexual men when compared with heterosexual populations [17]. The observation that the isolation rate of *C trachomatis* is lower in untreated patients with *N gonorrhoeae* infection than in treated patients with urethritis (PGU) can be explained by the greater concentration of *C trachomatis* in the urethra in PGU as a result of several replication cycles. It follows that the efficiency of isolation is influenced by the number of infectious elementary bodies present in a clinical sample, being greater in patients with PGU as compared with patients with untreated *N gonorrhoeae* infections who are also infected with *C trachomatis*.

Complications of *C trachomatis* infection in men

A number of complications in men can arise following infection with *C trachomatis*. These include epididymitis [18], Reiter's syndrome [19], sexually acquired reactive arthritis (SARA) [19] and possibly endocarditis [20]. There is convincing evidence that epididymitis in younger men (<35 years) is associated with *C trachomatis* infection and the organism has been demonstrated in epididymal aspirates [21]. Urethral carriage of *C trachomatis* can also be demonstrated in these patients, although urethral carriage alone is not conclusive evidence of a causal role in epididymitis. Approximately 50% of the estimated 500,000 annual cases of acute epididymitis in the United States are caused by *C trachomatis* infection [22].

REITER'S SYNDROME

Urethritis, conjunctivitis and arthritis are the classical triad of clinical manifestations associated with this syndrome. In Asia, North Africa and Europe, it is related to gastrointestinal infections. In contrast, in North America and the UK it is recognised as a sexually-acquired syndrome following *C trachomatis* infection [7].

The syndrome is more common in males than females [19] and a link with *C trachomatis* was established more than 25 years ago [23]. This association is based on the isolation of *C trachomatis* from the male urethra in 6%–39% of patients [24–27] and on the isolation of *C trachomatis* directly from synovial fluid, membranes and conjunctivae of some affected patients [28]. It is likely that sexually-acquired reactive arthritis (SARA) is a manifestation of the same disease and clinical presentation involving conjunctivitis and either urethritis or arthritis occurs in 35% of cases [19]. Approximately 1%–2% of all NGU cases develop SARA [19] and *C trachomatis* has been isolated from biopsy material and the cellular fraction of synovial fluids [29–31]. It remains to be proved whether the arthritis has an immunological basis or whether degraded chlamydia antigens in association with HLA determinants are a critical step in the pathogenesis.

OTHER COMPLICATIONS

The role of *C trachomatis* in the pathogenesis of chronic abacterial prostatitis remains to be established, as convincing evidence is not currently available [32]. Similarly, although rare reports [20] incriminate *C trachomatis* in endocarditis, further studies are required to establish this link.

C trachomatis has been demonstrated in rectal specimens from patients, particularly homosexuals, with and without proctitis. There is little doubt that oculogenital serotypes of *C trachomatis* are associated with proctitis which has also been documented in the heterosexual population. It is recognized that the prevalence of urethral *C trachomatis* infections in the male heterosexual in the USA is three times as great as in the homosexual population [33,34] and 4%–8% of homosexual men attending GUM clinics have demonstrable rectal chlamydial infection [17]. It is apparent that further studies are necessary to establish the role of oculogenital serotypes of *C trachomatis* in the pathology of rectal infection.

Genital infection in women

In contrast to *C trachomatis* infections in males, infection in the female is commonly asymptomatic [8,9]. In spite of this, the consequences of female infection, within our present knowledge, are far more serious and can give rise to considerable morbidity in later years. The organism infects and replicates within the epithelium of the cervix and urethra. An ascending infection with involvement of the upper genital tract occurs and can result in clinical or subclinical pelvic inflammatory disease (PID), presenting as endometritis, salpingitis or a perihepatitis (Fitz-Hugh Curtis Syndrome) [35,36]. Tubal damage can occur and in a proportion of such patients ectopic pregnancy and infertility will be the outcome [35]. *C trachomatis* is the most important cause of PID in the developed world accounting for 25%–50% of the one million cases in the USA [17] and is increasing in the UK (Figure 2). Upper genital tract infection gives rise to significant morbidity but also results in considerable costs to health authorities in investigating and treating the acute disease. Each year in the United States an estimated one billion dollars are spent on the management of women with PID, including the treatment of infants with chlamydia pneumonia [17]. In addition there are considerable costs associated with the investigation of infertility in patients during their childbearing years as a result of previous tubal damage. The psychological effect in later life on women who find themselves infertile as a result of previous *C trachomatis* infections can be devastating, highlighting the importance of early recognition and diagnosis with appropriate treatment.

As in other sexually transmitted diseases, *C trachomatis* has an important role in the family setting. Infection in the male can result in transmission of the organism to his pregnant partner which is a source of neonatal infection during delivery. The consequences of acquiring infection during delivery may be ophthalmia neonatorum, which if inappropriately treated may result in severe pneumonitis.

CERVICAL INFECTION

C trachomatis is recovered from the cervix in 12%–31% of women attending GUM clinics in the UK [7]. The majority of infections are asymptomatic but *C trachomatis* has an important role in mucopurulent cervicitis [17]. Epidemiological studies have demonstrated a striking correlation between men with chlamydia-positive NGU and their consorts as compared with consorts of men with chlamydia-negative NGU. Thus the isolation rate of *C trachomatis* from the cervix of contacts of men with chlamydia positive NGU ranges between 45%–67% [7] in contrast to a cervical isolation rate of 4%–18% in contacts of men with chlamydia-negative NGU.

The recovery of *C trachomatis* from the cervix of pregnant women attending family planning clinics in the UK [37] is much lower (approx 3%), but is epidemiologically important in the light of the neonatal infections which occur at delivery. Higher recovery rates in pregnant women have been reported from Sweden and the USA (4%–21%) [38–42]. In other clinical settings *C trachomatis* has been recovered from the cervix in 10%–16% of women undergoing termination of pregnancy in the UK (personal observation) and Sweden [42]. This indicates the need to screen women prior to termination of pregnancy if PID following termination is to be avoided. This is particularly relevant in the young, sexually active woman to avoid infertility in later life.

Complications of genital infection in women

PELVIC INFLAMMATORY DISEASE

In general *C trachomatis* is the most important cause of PID in the Western world and the organism has been recovered from the cervix in 31% of cases [35]. However, it should be noted that in San Francisco *N gonorrhoeae* has been shown to be more important [43]. Canicular spread of *C trachomatis* from the lower genital tract gives rise to symptomatic or subclinical salpingitis [35]. It has been estimated that 83,000 cases of salpingitis, which resulted in more than 8,000 infertile women, occurred in England and Wales in 1978 [35]. It is unlikely that salpingitis has decreased in the intervening years. The end result of inflammation and tubal damage may well be infertility and an increased risk of ectopic pregnancy [35].

The aetiological role of *C trachomatis* in PID was originally considered when early workers recognised salpingitis in mothers of babies who had neonatal ophthalmia [44]. The recovery of *C trachomatis* from the cervix of women with salpingitis strengthens this association but it is not conclusive evidence of upper genital tract infection. In studies where clinical salpingitis was diagnosed but not confirmed by laparoscopy, *C trachomatis* had been isolated from the cervix in 20%–26% of cases [45,46]. More convincing evidence has been reported in women with laparoscopically-proven salpingitis where cervical carriage was shown to be between 29%–36% [47,48]. Definitive evidence for the role of *C trachomatis* in salpingitis has been demonstrated by the direct isolation of the organism from fallopian tubes in 5%–30% of women but not from the fallopian tubes of women without salpingitis [42,48]. In practice, diagnostic laboratories will rarely receive fallopian tubes from women with acute PID and the usual investigation would be the detection of the organism in endocervical swabs. Although the demonstration of the organisms in endocervical swabs does not constitute proof of a causal role in PID, it is essential to treat such patients and their partners appropriately. Clearly a negative laboratory result for *C trachomatis* from an endocervical swab does not exclude infection of the pelvic organs and this highlights the need for improved or different laboratory assays to diagnose upper genital tract infections.

Risk factors associated with PID include the use of an intrauterine device (IUD) as a method of contraception [17,35], as well as termination of pregnancy in sexually active young women. Both groups of women should be screened for *C trachomatis* and they and their partners treated intensively.

PERIHEPATITIS

N gonorrhoeae is well established as a cause of acute perihepatitis (Curtis–Fitz-Hugh Syndrome) [36]. Women present with an acute right-sided upper abdominal pain, and in the early stages a localised peritonitis may be present in association with the lower and adjacent abdominal wall. There is now good evidence that *C trachomatis* is also involved [36]. The precise incidence of this syndrome is unknown but a greater clinical awareness is currently apparent [36]. Recovery of *C trachomatis* from endocervical swabs in women presenting with compatible symptoms is again suggestive but not conclusive evidence of perihepatitis. Nevertheless, treatment should be instigated immediately. As in salpingitis, a clinical diagnosis of perihepatitis poses problems for the diagnostic laboratory to establish an aetiological role for *C trachomatis* and again emphasises the need for improved diagnostic procedures for upper genital tract *C trachomatis* infections.

INFERTILITY

The role of *C trachomatis* in the aetiology of upper genital tract infection is now well established. Animal models [49,50] have demonstrated patchy areas of epithelial damage similar to that seen in women with salpingitis [51]. The consequence of tubal damage in humans may be infertility and the role of *C trachomatis* in females who are infertile is now well recognised. Attempts to demonstrate *C trachomatis* in the endocervix of infertile women are largely unrewarding because tubal damage has often occurred years earlier and as a result serological techniques have been used, with success, in the management of this group of women [52,53].

OTHER SITES OF INFECTION IN WOMEN

In addition to cervical infection in women, *C trachomatis* has also been isolated simultaneously from the urethra [54]. However, some reports have documented the isolation of *C trachomatis* from the urethra only [55]. The 'acute urethral' syndrome is now recognised as a sequelae of urethral infection and presents as dysuria and frequency, most commonly in young, sexually active women [56]. This provides evidence that sampling of women for *C trachomatis* infection should include both the cervix and urethra for optimum identification of infected women. Studies have also shown that infection of the Bartholin's glands [57] and the rectum occurs, although it is questionable whether laboratory diagnosis is necessary from this latter site as significant rectal disease has not been documented [58].

Conjunctivitis

Ocular infections caused by *C trachomatis* are common in sexually active individuals [7,59,60]. Usually, infection presents, after an incubation period of 1–2 weeks, as a follicular conjunctivitis which may be clinically indistinguishable from viral infection. In these patients and their partners cervical or urethral infection can usually be demonstrated

[59] and ocular infection may result from mechanical transmission of the organism from the genitalia [7]. Laboratory diagnosis is essential to identify the causative agent, which in turn ensures that both partners have appropriate therapy.

Neonatal infection

Cervical infection in pregnant women is the source of neonatal *C trachomatis* infection as a result of the baby's intimate contact with the genital tract at delivery. The risk of neonatal infection at delivery may be more than 50% and colonisation of the conjunctiva, middle ear, nasopharynx, trachea, lung, rectum and vagina occurs [7,61]. The most obvious clinical sign of neonatal infection in the first instance is neonatal conjunctivitis, which usually presents between the third and thirteenth day of life in 18%–74% of babies at risk [61]. The severity of the conjunctivitis may range from mild, presenting as a 'sticky' eye, to severe inflammation and discharge with closure of the eye. Undiagnosed neonatal conjunctivitis often leads to inappropriate treatment (*eg* topical chloramphenicol) and persistence of symptoms. The consequences of an untreated or unrecognised neonatal infection may be a lower respiratory infection, presenting as a severe pneumonitis between the fourth and twelfth week of life. It has been shown that 3%–18% of babies born to infected mothers in the USA will present as a chlamydial pneumonitis [62] and this emphasises the need for rapid laboratory diagnosis and treatment. In the UK, precise figures for neonatal *C trachomatis* pneumonitis are unavailable but it appears to be a relatively rare complication when compared to the incidence in the USA [61]. In contrast, chlamydial ophthalmia neonatorum is considered to be five times as common as gonococcal conjunctivitis in London [63] and three times as common in Liverpool [64]. The recognition and diagnosis of chlamydia ophthalmia neonatorum is often the first indication of maternal infection, which is most common in young, sexually-active single girls. The laboratory diagnosis and confirmation of chlamydia ophthalmia neonatorum indicates the need to treat the parents as well as the infant.

Laboratory diagnosis of *C trachomatis*

Traditionally, cell culture has been used for the laboratory diagnosis of *C trachomatis* infections. Techniques have now become sufficiently sophisticated to attain maximum sensitivity, detecting 80%–90% of actual infections [65]. In spite of this, *in vitro* cultivation of *C trachomatis* requires considerable expertise to attain optimal isolation rates, as well as fundamental cell culture facilities. This has limited a diagnostic service to those UK laboratories with the necessary facilities and expertise. Furthermore, cell culture methods, although universally recognised as the 'gold standard' by which all other methods are assessed, are costly

and time consuming, so their use is limited to certain high risk populations. Methods used for the isolation of *C trachomatis* in cell culture
have been described [66] and will not be repeated here. The limitation
of cell culture diagnostic services in the UK (and elsewhere) has led to
consumer demand for alternative assays to fill the gap left by the lack of
services in many districts. The increasing awareness that *C trachomatis*
is an important sexually-transmitted infection has been addressed by
many commercial companies who have now developed a wide range of
products. Currently the majority of these new assays are directed
towards antigen detection and they have the considerable advantage
that viable organisms are not required. Thus the major problem of rapid
transport of specimens, on which cell culture was dependent, does not
arise. Most of the assays available to date are based on the application
of monoclonal or polyclonal antibodies as an antigen capture detector
system and are either incorporated into an immunofluorescence reagent
or utilised in an enzyme-linked immunosorbent assay (ELISA). As a
result of the rapid expansion of available commercial assays many
papers have been published, comparing them with the cell culture 'gold
standard' assay [67–77]. It is recommended that any laboratory under-
taking *C trachomatis* diagnosis should either systematically evaluate new
products or take advice from experienced laboratories before introduction into their routine laboratory. Considerable differences in both
sensitivity and specificity among some of the commercial products
compared with cell culture have been documented. Quality control,
both nationally and in-house, will be an essential feature of these assays
in future.

MONOCLONAL ANTIBODIES

Studies have demonstrated that monoclonal antibodies directed at specific *C trachomatis* epitopes have an extremely high specificity [67,76]
and in certain clinical situations an associated high sensitivity [67,76]
when compared with cell culture. *C trachomatis* monoclonal antibodies
have been produced with a specificity for the genus-specific lipopoly-
saccaride (LPS) or the species-specific major outer membrane protein
(MOMP). It follows that the former monoclonal antibodies (LPS) will
react with *C psittaci*, *C trachomatis* and TWAR antigens, which can be
extremely useful in certain clinical settings [78], while the latter (MOMP)
react with all known *C trachomatis* serotypes.

IMMUNOFLUORESCENT TECHNIQUES

Both genus-specific (Imagen™, Celltech Diagnostics) and species-
specific (Syva, Microtrak™) fluorescein labelled monoclonal antibodies
are available for the rapid diagnosis of *C trachomatis* infection. Both
products have been universally used and most information is available
on these two reagents, although gradually other products are becoming

more widely available and are being evaluated. Considerable experience is required by the microscopist if these reagents are to be used routinely. Immunofluorescence is a subjective procedure and the images observed using the genus-specific reagent are quite different to those with the species-specific reagent. For the relatively inexperienced worker the results obtained with the species-specific reagent are considerably easier to interpret as only elementary bodies (EBs) are stained. The uniformity in shape and size of EBs ensures that experience and expertise in visualising them is rapidly acquired and differentiation from other stained material is relatively easy. Cross-reactions with other bacteria, if and when they occur, are not a serious problem as morphological criteria can be used in their differentiation.

In contrast genus-specific monoclonal antibodies react with both elementary bodies and genus specific LPS. Experience with these monoclonal antibodies has shown that cell-free LPS in addition to EBs is common in clinical samples [77]. The LPS antigen is considerably smaller and more variable in size than EBs and visualisation and recognition of a positive clinical sample requires more experience than that required with the species-specific monoclonal antibody. The specificity and characteristics of the different antibodies used in commercial assays are shown in Table 2.

APPLICATION OF MONOCLONAL REAGENTS (IF)

All these commercial reagents can be usefully applied to the identification of *C trachomatis* in cell culture. However, their main use is in the rapid diagnosis of *C trachomatis* in clinical samples, where impressive results can be obtained by experienced workers.

In general genus-specific monoclonal antibodies are suitable for the diagnosis of *C trachomatis* in patients with conjunctivitis [76,79], where intracellular inclusions are readily visible as well as cell free EBs and LPS antigen. Use of these antibodies requires considerably more experience when dealing with clinical samples received from GUM clinics. The reasons are clear to those laboratory workers familiar with the examination of urethral or endocervical smears from this population. The interpretation of brightly fluorescing EBs and particulate LPS antigen can be difficult because of the higher background of stainable material in these specimens. Because of this, species-specific monoclonal antibodies are more suitable to the examination of urethral/endocervical smears from GUM clinic patients, as these antibodies stain only EBs. Cell-free genus specific LPS is not stained and 'background' noise is therefore not such a problem. It is recommended that inexperienced laboratory workers should familiarise themselves with IF techniques on clinical samples from GUM clinics before their routine introduction into the laboratory. For the examination of smears from patients with conjunctivitis, there is little to choose between the two products and

Table 2 Characteristics of antibodies used in commercial assays for detection of oculogenital *C trachomatis*

Antibody	Specificity	Sensitivity	Advantage	Disadvantage
Polyclonal	LPS*	Generally excellent	Genus specific. Enhanced sensitivity where clinical sample contain 'free' LPS.	Variable affinity both in the antiserum and between animals. Specificity affected by purity of immunogen which may result in cross reactions. Removal of non-reactive protein (<99%) necessary to improve sensitivity. High degree of quality control and farm or animal facilities required.
Monoclonal	LPS†	Excellent	Genus specific. Reproducibility in production. Antibody is sub-class specific.	High initial investment costs in production and characterisation of clones.
	MOMP‡	Excellent in certain situations.	Defined affinity for each antibody. Purity of immunogen not critical. Recognises one epitope. Consistent supply of antibody. Low production costs.	May be too specific when used in diagnosing viral or bacterial infections due to strains of differing antigenic determinants.

*Polyclonal antibodies to *C trachomatis* lipoplysaccaride (LPS) antigen are used in most commercial ELISA assays
†*C trachomatis* lipooysaccaride (LPS) monoclonal antibodies have an excellent sensitivity and specificity and are used in some commercial ELISA assays. Variability of antigen staining by immunoflourescent techniques limits their use in the laboratory diagnosis of *C trachomatis* infections where a high degree of expertise is required
‡*C trachomatis* major outer membrane (MOMP) monoclonal antibodies lack sensitivity in ELISA assays. Ideal for use in immunofluorescent techniques when appropriate samples are collected

personal preference dictates their use. It should be noted that in the examination of smears from patients with conjunctivitis there is a possibility that staphylococci may be present in the sample. These organisms may be stained with these reagents if the Fc portion of the antibody molecule has not been cleaved or the product does not incorporate a blocking IgG antibody. In practice, this should not present as a diagnostic problem because of the much larger size of the staphylococcus in comparison with *C trachomatis* EBs. Problems may arise, however, in terms of interpretation, if a clinical sample contains both staphylococci and *C trachomatis*!

LIMITATIONS OF IF TECHNIQUES
Although immunofluorescence techniques using monoclonal antibodies have a high specificity, their use in routine laboratory practice is limited by the workload carried out by an individual laboratory [77]. The assay is subjective and tiresome, which is a major limiting factor if large numbers of clinical samples are processed daily (>30). They are, however, ideal for the 'one off' sample or for the examination of smears from patients with conjunctivitis, including neonates, where a definitive answer can be given to the clinician within one hour of receipt of the sample. In addition, IF techniques using species-specific reagents are the method of choice for the diagnosis of rectal infections where cell culture may be difficult due to contamination with antibiotic-resistant organisms.

A prerequisite of the IF technique is that smears, either genital or conjunctival, contain representative cellular material for reliable diagnosis. Recent reports suggest that for optimal staining of *C trachomatis* EBs, monoclonal antibodies directed against MOMP are consistently better than those raised against LPS [80]. In addition, it appears that methanol fixation of EBs may be preferable to acetone fixation prior to staining with some of the species-specific monoclonal antibodies (MOMP) commercially available, and this may prove a further important point to consider when inexperienced laboratory workers embark on these techniques. A major advantage of the IF technique is that it is the only method where the quality of the sample can be directly assessed.

ENZYME IMMUNOASSAYS
These assays are gaining favour on a worldwide basis for the large scale screening of populations for *C trachomatis* infection. This is largely dictated by the lack of local cell culture facilities but also by the improvement commercial companies are achieving with their assays. An important distinction is that enzyme immunoassays rely on a colour change as an indicator of positivity, rather than direct visualisation of inclusions (cell culture) or EBs (immunofluorescence). For these reasons, the specificity of any enzyme immunoassay must be reliably

assessed in order to avoid reporting false positives. Furthermore, in the light of possible medico-legal implications, particularly in cases of sexual assault, cell culture remains the only legally acceptable assay for the demonstration of *C trachomatis* infection.

Numerous assays are now available to laboratories in the UK and elsewhere in the world. At the time of writing two immunoassays (IDEIA™, Celltech Diagnostics; Chlamydiazyme™, Abbott) have been studied and evaluated in depth by many workers throughout the world [68–75]. Inevitably these studies do not always agree as to which assay currently has the best predictive values and sensitivity/specificity. One of the major reasons for these discrepancies is whether a representative clinical sample was applied to the assays under evaluation. Other reasons include the sensitivity of the cell culture used as the 'gold standard', and as a result it has been suggested that certain enzyme immunoassays are not yet suitable for widespread use [74].

Specifically, most studies have evaluated these two immunoassays using cell culture as the 'gold standard'. This entails multiple swabs, compounded when swabs are also collected for *N gonorrhoeae*, which cannot be reliably expected to yield equal amounts of antigen on each swab. In spite of these problems, both the IDEIA and Chlamydiazyme assays have documented high sensitivities and specificities. Although improvements have been achieved by commercial companies, it should be noted that the sensitivity and specificity of these assays are calculated from comparisons with cell culture which itself may only detect 80%–90% of actual infections.

These assays differ fundamentally in their principles: the IDEIA assay is based on a genus-specific monoclonal antibody with a double enzyme amplification step [72]. In practice this allows a greater sensitivity because of the presence of cell-free LPS available as a further amplification to the reaction reagents in the assay. The Chlamydiazyme assay uses a polyclonal antibody, which also reacts with LPS, but an amplification step is not incorporated. Studies have suggested that Chlamydiazyme has a suspect specificity problem, in that cross-reactions with some other bacteria have been documented [74,81]. The recent introduction of a neutralisation step will go some way towards correcting this problem although the confirmation of all positive samples will have cost implications. Laboratory workers should be familiar with problems arising from the use of the washing equipment used in ELISA assays. We have recently noted that if contamination of the equipment occurs, then false positives will appear due to organisms possessing protein 'A' binding-specific antibody. All equipment should be thoroughly cleansed and decontaminated before use in ELISA assays.

Other commercial assays are gradually appearing and the slightly inferior sensitivity of some of these assays is difficult to explain. Some include the use of a genus-specific monoclonal antibody as well as an amplification step, and the inferior sensitivity may be due to differences in the avidity of the monoclonal antibody used. This question requires further investigation as there is little data currently available on the avidity of the monoclonal antibodies used. A recent publication documented large numbers of elementary bodies by immuno-fluorescence in an enzyme immunoassay buffer which provided a nega-tive result in the ELISA under evaluation [74]. In this example the avidity of the antibody used in the assay may be a problem and is unlikely to be an isolated observation.

An interesting observation made with some of these enzyme-immunoassays is the lack of correlation between optical density readings and the number of *C trachomatis* inclusions detected in cell culture when comparative studies are carried out. In particular those immunoassays which incorporate a genus-specific monoclonal antibody (LPS) show this lack of correlation. It can be readily explained by the presence of cell-free LPS in clinical samples (non-infectious) amplifying the signal detected in the immunoassay, whereas cell culture only detects infec-tious elementary bodies. A similar lack of correlation is noted when the number of EBs in a clinical sample are enumerated by IF techniques and compared with enzyme immunoassay optical densities (Figure 3). This merely confirms the observations and experiences obtained with the genus-specific enzyme immunoassays and emphasises the need to cap-ture the cell-free LPS present in clinical samples for maximum sensi-

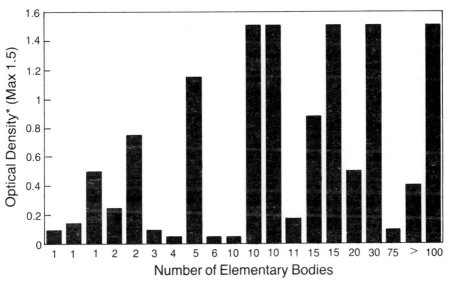

Figure 3 Comparison of optical densities and number of elementary bodies.

tivity. Those immunoassays which do not use a genus-specific monoclonal antibody will inevitably have an inferior sensitivity.

In the final analysis, an immunoassay needs thorough evaluation before its introduction into routine laboratory practice. Its performance is markedly affected by the quality of the clinical sample, which is particularly relevant to GUM male clinic patients where 'routine' urethral swabs may be less than optimum. It is well recognised that the efficiency of swabbing in these settings can be extremely variable. The choice of test (IF or ELISA) will have to take all the above aspects into account as well as reproducibility, workloads and costs, with their associated budgetary implications. It should be noted that ELISAs should not be used in the detection of *C trachomatis* from rectal swabs as a high false positive incidence has been documented in this clinical setting. The exact reason for this lack of specificity is currently unknown. Finally, ELISAs are not yet suitable for the large-scale screening of low risk populations as the predictive values documented are unacceptable.

It is not within the scope of this chapter to include details of specimen collection, but a note of caution is relevant. Clinical staff are well versed in the method of collecting endocervical/urethral swabs [7]. Recent studies have established, however, that the type of swab used may have a detrimental effect on the signal of some enzyme immunoassays. It has been noted that a reduction of the signal in positive samples can occur on storage of the specimen prior to testing. In some instances, a previously positive swab can produce a negative signal after a few days of storage. The implications of these recent observations are that laboratory users of commercial kits should contact the manufacturers for precise instructions as to what make of swab is appropriate to their assay. It is also preferable to carry out the ELISA within 24 hours of receipt of sample, if swabs are a problem. This is reasonable in laboratories with a large workload. In smaller laboratories where only a few samples are received daily, immunofluorescence techniques may be more applicable. Certainly, if rapid diagnostic methods are applied to the identification of *C trachomatis* infections for clinical reasons, then batch testing on a weekly basis is inappropriate.

EVALUATION OF ENZYME-IMMUNOASSAYS

The problems associated with producing accurate data on the evaluation of enzyme-immunoassays have been discussed above. Recently the Bristol laboratory has addressed the particular problem of sampling errors associated with multiple swabbing. Preliminary studies have demonstrated that an early morning 'first catch' urine sample is an acceptable sample, free of the sampling errors associated with multiple swabs [82], and a very suitable specimen for the uniform evaluation of enzyme-immunoassays. Briefly, equal aliquots of urine were centrifuged and the deposit used to evaluate three enzyme-immunoassays. The

Table 3 Comparison of three EIAs with direct IF on urine samples

EIA		Direct IF	
		+	**−**
Assay 1	+	19	2*
	−	0	70
Assay 2	+	16	10
	−	3	62
Assay 3	+	8	0
	−	11	72

*Urethral swabs from both patients were positive for *C trachomatis*

results of these studies (Table 3) have demonstrated significant differences in sensitivity and specificity, although it should be stated that two of the manufacturers do not claim that their assays will detect antigen in urine. Confirmation of positive urine samples was achieved by the demonstration of EBs in the urine deposits using a species-specific fluorescein labelled monoclonal antibody (Microtrak^{TM}, Syva).

These preliminary studies indicate that an early morning first catch urine sample represents the optimal sample for evaluation purposes and will play an important part in studies where sampling error is a problem. Further studies may help to resolve whether such early morning urine samples will replace the collection of urethral swabs in male populations. It is well accepted that male urethral swabbing is an invasive technique resented by the majority of patients and is unpopular with the staff who carry it out. Furthermore, recognition and identification of asymptomatic male carriers will be greatly facilitated by testing early morning urine samples [82]. It is already clear that an early morning urine sample is not an appropriate sample for the maximum detection of *C trachomatis* infections in female populations [82], because endocervical carriage is more common than urethral.

Who should be tested?

This aspect of *C trachomatis* infection has always been a difficult issue. In the past, cell culture techniques have been available to a restricted number of laboratories and a number of workers have discussed the feasibility of population screening [7,83]. Most favoured a restricted service, often because cell culture is costly, time consuming and dependent on highly trained personnel with considerable expertise, as the method demands attention to fine detail. As a result, most clinics in this country selectively test all females attending GUM clinics, as asymptomatic infection in females is particularly high. Using this policy, all positive female patients and their consorts would be treated. In addition, all symptomatic male patients attending these clinics would be treated and others would be treated on epidemiological grounds. Unfortunately it is clear that this policy has not reduced the incidence of *C trachomatis* infections, as laboratory reports of *C trachomatis* and NGU in the UK continue to rise (see Figure 2).

A number of reasons might account for the increasing incidence of NGU: the restricted diagnostic services available in the country in the past is a major reason; the low but significant asymptomatic carriage in males (approx 7%) who are currently not tested and remain undetected and untreated is a further factor; in addition, re-infections may be important in maintaining the pool of *C trachomatis* carriers. To reduce new *C trachomatis* infections significantly and to reduce the pool of infection in the community it may be necessary for radical changes in sexual relationships to occur. If the human immunodeficiency virus pandemic spreads significantly into the heterosexual population, changes in the number of an individual's sexual partners may result which in turn would reduce the pool of *C trachomatis* infections.

It is clear that only significant changes in the prevalence of *C trachomatis* in the community will have a long-term effect on the complications of *C trachomatis* infection such as PID and associated infertility. Testing must become far more widespread, although it is improving at district level. A suggested scheme for screening at-risk patients is shown in Table 4.

The role of serology

This is the most controversial issue associated with *C trachomatis* infections. Within the UK and elsewhere in the world, immuno-fluorescent antibody techniques, utilising either whole cell inclusions (WIF) [84] (species- and genus-specific) or EBs (MIF) [85], which are both species- and type-specific, are used in specialised laboratories. Recently it has been shown that the WIF assay is more sensitive than the MIF assay for the detection of species-specific *C trachomatis* antibody [86]. Commercial assays have been developed but they have not yet been comprehensively evaluated in the UK. All workers in the field agree that immunofluorescent antibody techniques are useful epidemiological tools for assessing the prevalence of chlamydial antibody in at-risk populations, but have little value in the routine diagnosis of current *uncomplicated* genital infections. The basis for these conclusions is numerous studies which have shown that 78%–100% of women with active cervical *C trachomatis* infection have species-specific antibody, while 31%–87% of women who have no demonstrable infection have *C trachomatis* antibody. It is clear that the presence of species-specific antibody does not correlate with the isolation of *C trachomatis* from the genital tract [7]. It follows that antibody estimations on a single sample

Table 4 Suggested screening for *C trachomatis* infection

Clinical setting	Population	Specimen	Assay	Comment
GUM clinics*	All females	Endocervical/urethral swabs	ELISA	
	Males with proctitis	Rectal swab	IF	False positives with EIA
	Asymptomatic males	First catch EMU	ELISA	Urine non-invasive
Antenatal clinics*	1 Adolescents (<20 years) 2 Unmarried women (sexually active) 3 Married women with multiple partners or with a history of STD 4 Termination of pregnancy 5 Patients with suspected PID	Endocervical/urethral swabs	ELISA	For large scale screening ELISA preferable
Eye hospitals*	Babies or sexually active adults with conjunctivitis	Conjuctival swabs/smears	IF	IF ideal for rapid 'one-off' diagnosis
General practice*	Patients with suspected PID	Endocervical/urethral swab Blood sample	ELISA serology	

*All partners of positive consorts would be treated. All patients with gonorrhoea, NGU, PGU, urethral syndrome, PID, epididymitis (men <35 years old) and contacts of such patients should be treated on epidemiological grounds

of serum may reflect either current or past lower genital tract infection and therefore have no diagnostic value. As such, antibody tests should not be carried out when uncomplicated genital tract infection is suspected. In this clinical situation, culture or antigen detection is always the diagnostic method of choice. Similarly in ophthalmia neonatorum or neonatal pneumonia, culture or antigen detection is the method of choice. However, in this clinical setting serology does have a useful role, since chlamydia specific IgM can be demonstrated by immunofluorescent techniques [61], which allows differentiation from passively-acquired maternal antibody.

INFERTILITY

In contrast, it has now been conclusively demonstrated that IF antibody techniques are particularly useful in the management of infertile women [52,53,87]. Studies have demonstrated that high antibody levels (>1,024) correlate with but are not diagnostic of tubal damage due to *C trachomatis* infection, either recently or in the distant past [53]. Thus the single estimation of *C trachomatis* antibody in serum samples by the WIF or MIF tests is a valuable investigation to carry out in infertile women and indicates the need for early laparoscopy, which also reduces the overall costs of other expensive investigations. It must be emphasised that high IF antibody levels correlate with previous tubal damage and are not necessarily an indicator for the instigation of antimicrobial therapy.

SEROLOGY IN ACUTE OR CHRONIC UPPER GENITAL TRACT INFECTION ON WOMEN

Upper genital tract infection (PID): Our own experience and that of other workers suggests that serology does have a role in complications of *C trachomatis* infection, particularly in cases of PID where antigen detection fails. This is not uncommon as it is widely recognised that *C trachomatis* is often not detected in endocervical swabs when upper genital tract infection exists. The combination of complement fixation tests (genus specific) and whole inclusion immunofluorescent (WIF) antibody tests (genus- and species-specific) often gives additional information on the aetiology in patients presenting with PID and is a clinically useful assay. High titres or seroconversions/rising antibody titres by both CF and IF have been documented in acute PID.

The salient point of comparing the ratio of genus-reactive antibody (CF) with genus/species-specific antibody (WIF) in upper genital tract infections is that the species specific IF antibody is commonly more than eight times higher than the genus-specific CF antibody. Raised levels of CF and significantly raised IF antibody titres (>1,024) are clinically useful indicators of acute or chronic upper genital tract infection and indicate the need to instigate appropriate antichlamydial therapy in the patient and partner. The CF and IF serological responses in patients with clinical PID compared with patients with uncomplicated lower genital tract infections are shown in Figures 4 and 5. An important consideration when carrying out serological assays in these groups of patients is the need to establish accurately the clinical history and any recent antimicrobial therapy. This is highlighted by the observation that CF and IF antibody may decline slowly over a period of months. The use of the CF antibody alone for the investigation of the complications which arise from simple *C trachomatis* infections should be discouraged, as previous infections with *C psittaci* or the TWAR agent may confuse the serological interpretation. Referral to laboratories specialising in species-specific serology is appropriate in these situations. Chlamydia specific IgM, although occasionally useful in primary *C trachomatis* infections associated with PID, is not generally applicable and other micro-

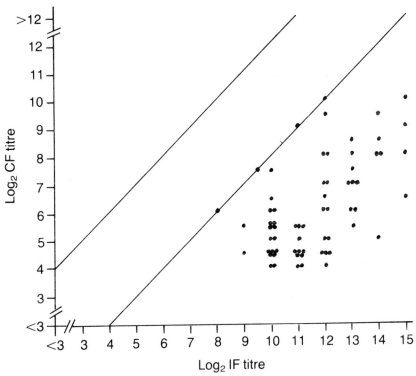

Figure 4 Comparison of CF antibody (genus-specific) with IF antibody (WIF) (species-specific) in patients with clinical PID. Note the exceptionally high levels of IF antibody.

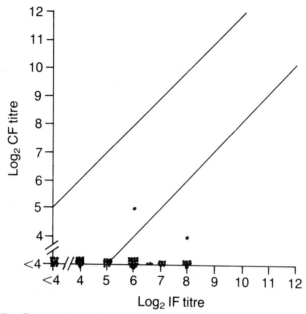

Figure 5 Comparison of CF antibody with IF antibody (WIF) in sexually active patients with uncomplicated lower genital tract infection. Note the lower titres (IF) in this group in comparison to upper genital tract infections (Figure 4) and the absence of CF antibody.

biological methods to supplement serology are needed. It is apparent that there is a need for the development and production of highly purified or synthetically-prepared specific chlamydial epitopes, which can be applied to more precise serological assays for the diagnosis of *C trachomatis* in upper genital tract infections.

Summary
Over the last decade, the spectrum of diseases associated with *C trachomatis* infections has increased dramatically, with a concomitant increase in the prevalence of infection. Present screening policies have not successfully contained or controlled the epidemic and as a result the morbidity and infertility associated with the complications arising from *C trachomatis* infections have also increased. We are now at the stage where *C trachomatis* is not only the most common STD infection but also, with the exception of human immunodeficiency virus, the most costly when all aspects of infection, including complications, are considered in terms of health care. With the development of increasingly sensitive antigen detection methods [88] and the involvement of com-

mercial institutions, more widespread testing is now possible. This will hopefully have an impact in reducing the morbidity associated with these infections. The recent exciting development of amplifying appropriate microbial genomes by the polymerase chain reaction offers a potentially highly sensitive technique for the diagnosis of *C trachomatis* in both high- and low-risk populations.

References

1   Grayston JT, Kuo C-C, Wang S-P, Altman J. A new *Chlamydia psittaci* strain TWAR, isolated in acute respiratory tract infections. *N Engl J Med* 1986, **315**, 161–8.

2   Campbell LA, Kuo C-C, Grayston JT. Characterization of the new *Chlamydia* agent, TWAR, as a unique organism by restriction endonuclease analysis and DNA-DNA hybridization. *J Clin Microbiol* 1987, **25**, 1911–6.

3   Schacter J. Chlamydial infections (first of three parts). *N Engl J Med* 1978, **298**, 428–35.

4   Schacter J. Chlamydial infections (second of three parts). *N Engl J Med* 1978, **298**, 490–5.

5   Ward ME. Chlamydial classification, development and structure. *Br Med Bull* 1983, **39**, 109–15.

6   Richmond SJ, Hilton AL, Clarke SKR. Chlamydial infection: role of Chlamydia sub-Group A in non-gonococcal and post gonococcal urethritis. *Br J Vener Dis* 1972, **48**, 437–44.

7   Oriel JD, Ridgway GL. Genital infection by *Chlamydia trachomatis. Current topics in Infection* 1982, Series 2, 41–52.

8   Burns DC MacD, Darougar S, Thin RN, Lothian L, Nicol CS. Isolation of Chlamydia from women attending a clinic for sexually transmitted diseases. *Br J Vener Dis* 1975, **51**, 314–18.

9   Oriel JD, Johnson AL, Barlow D, Thomas BJ, Nayyar K, Reeve P. Infection of the uterine cervix with *C. trachomatis. J Infect Dis* 1978, **137**, 443–51.

10   Vaughan-Jackson JD, Dunlop EMC, Darougar S, Treharne JD, Taylor-Robinson D. Urethritis due to *C. trachomatis. Br J Vener Dis* 1977, **53**, 180–3.

11   Oriel JD, Reeve P, Thomas BJ, Nicol CS. Infection with Chlamydia Group A in men with urethritis due to Neisseria gonorrhoeae. *J Infect Dis* 1975, **131**, 376–82.

12   Oriel JD, Ridgway GL, Reeve P, Beckingham DC, Owen J. The lack of effect of ampicillin plus probenecid given for genital infections with Neisseria gonorrhoea on associated infections with *C. trachomatis. J Infect Dis* 1976, **133**, 568–71.

13   Oriel JD, Ridgway GL, Tchamouroff S, Owen J. Spectinomycin hydrochloride in the treatment of gonorrhoea: its effect on associated *C. trachomatis* infections. *Br J Vener Dis* 1977, **53**, 226–9.

14 Holmes KK, Handsfield HH, Wang S-P, Wentworth BB, Turck M, Anderson JB, Alexander ER. Etiology of non-gonococcal urethritis. *N Engl J Med* 1975, **292**, 1199–205.

15 Bowie WR, Lee CK, Alexander ER. Prediction of efficacy of antimicrobial agents in treatment of infections due to *C. trachomatis*. *J Infect Dis* 1978, **138**, 655–9.

16 Perroud HM, Miedzybrodzka K. Chlamydial infection of the urethra in men. *Br J Vener Dis* 1978, **54**, 45–9.

17 *Chlamydia trachomatis infections*: Policy guidelines for prevention and control. MMWR 1985; 34: Suppl 3S 1985, **34**, 53s–74s.

18 Heap G. Acute epididymitis attributable to Chlamydial infection – preliminary report. *Med J Aust* 1975, **i**, 718–19.

19 Keat A, Thomas BJ, Taylor-Robinson D. Chlamydial infection in the aetiology of arthritis. *Br Med Bull* 1983, **39**, 168–74.

20 Van der Bel-Khan JM, Watanakunakorn C, Menefee MG, Long HD, Dicter R. *C. trachomatis* endocarditis. Am Heart J 1978, **95**, 627–36.

21 Berger RE, Alexander ER, Monda GD, Ansell J, McCormick G, Holmes KK. *Chlamydia trachomatis* as a cause of acute 'idiopathic' epididymitis. *N Engl J Med* 1978, **298**, 301–4.

22 Summaries and panel recommendations. In: United States National Institute of Allergy and Infectious Diseases Study Group on Sexually Transmitted Diseases. 1980 Status Report: National Institutes of Health, 1981: 215–64.

23 Siboulet A, Galistin P. Arguments in favour of a virus aetiology of non-gonococcal urethritis illustrated by three cases of Reiter's disease. *Br J Vener Dis* 1962, **38**, 209–11.

24 Vaughan-Jackson JD, Dunlop EMC, Daroughar S, Dwyer RStC, Jones BR. Chlamydial infection: Results of tests for Chlamydia in patients suffering from acute Reiter's disease compared with results of tests of the genital tract and rectum in patients with ocular infection due to TRIC agent. *Br J Vener Dis* 1972, **48**, 445–51.

25 Gordon FB, Quan AL, Steinman TI, Philip RN. Chlamydial isolates from Reiter's syndrome. *Br J Vener Dis* 1973, **49**, 376–80.

26 Schacter J. Can chlamydial infections cause rheumatic disease? In: Dumonde DC (ed.) Infection and immunology in the rheumatic diseases. Oxford: Blackwell Scientific, 1976, 151–7.

27 Kousa M, Saikku P, Richmond S, Lassus A. Frequent association of Chlamydial infection with Reiter's syndrome. *Sex Trans Dis* 1978, **5**, 57–61.

28 Schacter J, Barnes MG, Jones JP, Engelman EP, Meyer KF. Isolation of bedsoniae from the joints of patients with Reiter's syndrome. *Proc Soc Exp Biol Med* 1966, **122**, 283–5.

29 Schacter J, Dawson CR. Human chlamydial infections. Littleton, MA: PSG Publishing, 1978, 144–6.

30 Dunlop EMC, Freedman A, Garland JA, Harper IA, Jones BR, Race JW, duToit MS, Treharne JD. Infection by bedsoniae and the possibility of spurious isolation. 2. Genital infection, disease of the eye, Reiter's disease. *Am J Ophthalmol* 1967, **63**, Suppl 1073–81.

31 Ostler H, Dawson CR, Schacter J, Engleman EP. Reiter's syndrome. *Am J Ophthalmol* 1971, **71**, 986–91.

32 Mardh P-A, Ripa KT, Colleen S, Treharne JD, Darougar S. Role of *C. trachomatis* in non-acute prostatitis. *Br J Vener Dis* 1978, **54**, 330–4.

33 Stamm WE, Koutsky LA, Benedetti JK, Jourden JL, Brunham RC, Holmes KK. *Chlamydia trachomatis* urethral infections in men: prevalence. *Ann Intern Med* 1984, **100**, 47–51.

34 Judson FN. Epidemiology and control of non gonococcal urethritis and genital Chlamydial infections: a review. *Sex Trans Dis* 1981, **8**, 117–26.

35 Westrom L, Mardh P-A. Chlamydial salpingitis. *Br Med Bull* 1983, **39**, 145–50.

36 Bolton JP, Darougar S. Perihepatitis. *Br Med Bull* 1983, **39**, 159–62.

37 Hilton AL, Richmond SJ, Milne JD, Hindley F, Clarke SKR. Chlamydia A in the female genital tract. *Br J Vener Dis* 1974, **50**, 1–10.

38 Alexander ER, Chandler J. Pheifer TA, Wang S-P, English M, Holmes KK. Prospective study of perinatal *Chlamydia trachomatis* infection. In: Holmes KK, Hobson D (eds) Non gonococcal urethritis and related infections. Washington DC: Am Soc Microbiol, 1977, 148–52.

39 Schacter J, Grossman M, Holt J, Sweet R, Goodner E, Mills J. Prospective study of Chlamydial infection in neonates. *Lancet* 1979, **ii**, 377–80.

40 Frommell GT, Rothenberg R, Wang S-P, McIntosh K. Chlamydial infection of mothers and their infants. *J Pediatr* 1979, **95**, 28–32.

41 Hammerschlag MR, Anderka M, Semine DZ, McComb D, McCormack WM. Prospective study of maternal and infantile infections with *C. trachomatis*. *Pediatrics* 1979, **64**, 142–8.

42 Mardh P-A, Helin I, Bobeck S, Laurin J, Nilsson T. Colonisation of pregnant and puerperal women and neonates with *C. trachomatis*. *Br J Vener Dis* 1980, **56**, 96–100.

43 Sweet RL, Mills J, Hadley KW, Blumenstock E, Schacter J, Robbie MO, Draper DL. Use of laparoscopy to determine the microbiologic etiology of acute salpingitis. *Am J Obstet Gynecol* 1979, **134**, 68–74.

44 Dunlop EMC, Jones BR, Al-Hussaini MK. Genital infection in association with TRIC virus infection of the eye, III. Clinical and other findings. Preliminary report. *Br J Vener Dis* 1964, **40**, 33–42.

45   Eschenbach DA, Buchanan TM, Pollock HM, Forsyth PS, Alexander ER, Lin J-S, Wang S-P, Wentworth BB, McCormack WM, Holmes KK. Polymicrobial etiology of acute pelvic inflammatory disease. *N Engl J Med* 1975, **293**, 166–71.

46   Paavonen JA, Saikku P, Vesterinen E, Aho K. *C. trachomatis* in acute salpingitis. *Br J Vener Dis* 1979, **55**, 203–6.

47   Hamark B, Brorsson J-E, Tonnes E, Forssman L. Salpingitis and Chlamydia subgroup A. *Acta Obstet Gynecol Scand* 1976, **55**, 377–8.

48   Mardh P-A, Ripa T, Svensson L, Westrom L. *C. trachomatis* infection in patients with acute salpingitis. *N Engl J Med* 1977, **296**, 1377–9.

49   Ripa KT, Moller BR, Mardh P-A, Freundt EA, Melsen F. Experimental acute salpingitis in grivet monkeys provoked by *Chlamydia trachomatis*. *Acta Pathol Microbiol Scand* B 1979, **87**, 65–70.

50   Moller BR, Mardh P-H. Experimental salpingitis in grivet monkeys by *C. trachomatis*. *Acta Pathol Microbiol Scand B* 1980, **88**, 107–14.

51   Moller BR, Westrom L, Ahrons S, Ripa KT, Svensson L, von Mecklenburg C, Henrikson H, Mardh P-A. *C. trachomatis* infection of the fallopian tubes: histological findings in two patients. *Br J Vener Dis* 1979, **55**, 422–8.

52   Moore DE, Spadoni LR, Foy HM *et al.* Increased frequency of serum antibodies to *Chlamydia trachomatis* in infertility due to distal tube disease. *Lancet* 1982, **ii**, 574–7.

53   Conway D, Glazener CMA, Caul EO *et al.* Chlamydia serology in fertile and infertile women. *Lancet* 1984, **i**, 191–3.

54   Dunlop EMC, Darougar S, Hare MJ, Treharne JD, Dwyer RstC. Isolation of Chlamydia from the urethra of a woman. *Br Med J* 1972, **ii**, 386.

55   Woolfitt JMG, Watt L. Chlamydial infection of the urogenital tract in promiscuous and non-promiscuous women. *Br J Vener Dis* 1977, **53**, 93–5.

56   Paavonen J. *C. trachomatis* induced urethritis in female partners of men with non-gonococcal urethritis. *Sex Trans Dis* 1979, **6**, 69–71.

57   Davies JA, Rees E, Hobson D, Karayiannis P. Isolation of *C. trachomatis* from Bartholin's ducts. *Br J Vener Dis* 1978, **54**, 409–13.

58   Munday PE, Taylor-Robinson D. Chlamydial infection in proctitis and Crohn's disease. *Br Med Bull* 1983, **39**, 155–8.

59   Viswalingam ND, Wishart MS, Woodland RM. Adult chlamydial opthalmia (paratrachoma). *Br Med Bull* 1983, **39**, 123–7.

60   Tullo AB, Richmond SJ, Easty DL. The presentation and incidence of paratrachoma in adults. *J Hyg (Lond)* 1981, **87**, 63–9.

61   Hobson D, Rees E, Viswalingam ND. Chlamydial infections in

neonates and older children. *Br Med Bull* 1983, **39**, 128–32.

62 Alexander ER, Harrison HR. Role of *Chlamydial trachomatis* in perinatal infection. *Rev Infect Dis* 1983, **5**, 713–19.

63 Dunlop EMC. Non specific genital infections: laboratory aspects. In: Morton RS, Harris JRW (eds) Recent advances in sexually transmitted diseases. Vol. 1. Edinburgh: Churchill Livingstone, 1975, 267–95.

64 Rees E, Tait IA, Hobson D, Johnson FWA. Perinatal chlamydial infection. In: Hobson D, Holmes KK (eds) Non-gonococcal urethritis and related infections. Washington DC: American Society for Microbiology, 1977, 140–7

65 Schacter J. Biology of *C. trachomatis*. In: Holmes KK, Mardh P-A, Sparling PF, Wiesner PF (eds) Sexually transmitted diseases. New York: McGraw Hill, 1984, 243–57.

66 Richmond SJ. *Chlamydia trachomatis* infection. In: Jephcott AE (ed.) Sexually transmitted diseases: a rational approach to their diagnosis. London: Public Health Laboratory Service, 1988, 42–52.

67 Thomas BJ, Evans RT, Hawkins DA, Taylor-Robinson D. Sensitivity of detecting *Chlamydia trachomatis* elementary bodies in smears by use of fluorescein-labelled monoclonal antibody: comparison with conventional chlamydial isolation. *J Clin Pathol* 1984, **37**, 812–16.

68 Jones MF, Smith TF, Houglum AJ, Hermann JE. Detection of *Chlamydia trachomatis* in genital specimens by Chlamydiazyme test. *J Clin Microbiol* 1984, **20**, 465–7

69 Caul EO, Paul ID. Monoclonal antibody based ELISA for detecting *Chlamydia trachomatis*. *Lancet* 1985, **i**, 279.

70 Hambling MH, Kurtz JB. Preliminary evaluation of an enzyme immunoassay test for the detection of *Chlamydia trachomatis*. *Lancet* 1985, **i**, 53.

71 Mumtaz G, Mellars BJ, Ridgway GL, Oriel JD. Enzyme immunoassay for the detection of *Chlamydia trachomatis* antigen in urethral and endocervical swabs. *J Clin Pathol* 1985, **38**, 740–2.

72 Pugh SF, Slack RCB, Caul EO, Paul ID, Appleton PN, Gatley S. Enzyme amplified immunoassay: a novel technique applied to direct detection of *Chlamydia trachomatis* in clinical specimens. *J Clin Pathol* 1985, **38**, 1139–41.

73 Chernesky MA, Mahony JB, Castriciano S, *et al.* Detection of *Chlamydia trachomatis* antigens by enzyme immunoassay and immunofluorescence in genital specimens from symptomatic and asymptomatic men and women. *J Infect Dis* 1986, **154**, 141–8.

74 Taylor-Robinson D, Thomas BJ, Osborn MF. Evaluation of enzyme immunoassay (Chlamydiazyme) for detecting *Chlamydia trachomatis* in genital tract specimens. *J Clin Pathol* 1987, **40**, 194–9.

75 Wiesmeier E, Bruckner DA, Malotte CK, Manduke L. Enzyme-linked immunosorbent assays in the detection of *Chlamydia trachomatis*: How valid are they? *Diagn Microbiol Infect Dis* 1988, **9**, 219–23.

76 Potts MJ, Paul ID, Roome APCH, Caul EO. Rapid diagnosis of *Chlamydia trachomatis* infection in patients attending an ophthalmic casualty department. *Br J Ophthalmol* 1986, **70**, 677–80.

77 Alexander I, Paul ID, Caul EO. Evaluation of a genus reactive monoclonal antibody in rapid identification of *Chlamydia trachomatis* by direct immunofluorescence. *Genitourin Med* 1985, **61**, 252–4.

78 McGivern D, White R, Paul ID, Caul EO, Roome APCH, Westmoreland D. Concomitant zoonotic infections with ovine Chlamydia and 'Q' fever in pregnancy: Clinical features, diagnosis, management and public health implications. Case report. *Br J Obstet Gynecol* 1988, **95**, 294–8.

79 Hawkins DA, Wilson RS, Thomas BJ, Evans RT. Rapid, reliable diagnosis of Chlamydial ophthalmia by means of monoclonal antibodies. *Br J Ophthal* 1985, **69**, 640–4.

80 Cles LD, Bruch K, Stamm WE. Staining characteristics of six commercially available monoclonal immunofluorescence reagents for direct diagnosis of *Chlamydia trachomatis* infections. *J Clin Microbiol* 1988, **26**, 1735–7.

81 Saikku P, Puoloakkainen M, Leinonen M, Nurminen M, Nissinen A. Cross-reactivity between Chlamydiazyme and Acinetobacter strains. *N Engl J Med* 1986, **314**, 922–3.

82 Caul EO, Paul ID, Milne JD, Crowley T. Non-invasive sampling method for detecting *Chlamydia trachomatis*. *Lancet* 1988, **ii**, 1246–7.

83 Richmond SJ, Paul ID, Taylor PK. Value and feasibility of screening women attending STD clinics for cervical chlamydial infections. *Br J Vener Dis* 1980, **56**, 92–5.

84 Richmond SJ, Caul EO. Single-antigen indirect immunofluorescence test for screening venereal disease clinic populations for chlamydial antibodies. In: Hobson D, Holmes KK (eds) Nongonococcal urethritis and related infections. Washington DC: American Society for Microbiology, 1977, 259–65.

85 Wang S-P, Grayston JT, Alexander ER, Holmes KK. Simplified microimmunofluorescence test with Trachoma-lymphogranuloma venereum (*C. trachomatis*), antigens for use as a screening test for antibody. *J Clin Microbiol* 1975, **1**, 250–5.

86 Peterson EM, Oda R, Tse P, Gastaldi C, Stone SC, De La Maza LM. Comparison of a single-antigen microimmunofluorescence assay and inclusion fluorescent – antibody assay for detecting

Chlamydial antibodies and correlation of the results with neutralizing ability. *J Clin Microbiol* 1989, **27**, 350–2.

87 Robertson JN, Ward ME, Conway D, Caul EO. Chlamydial and gonococcal antibodies in sera of infertile women with tubal obstruction. *J Clin Pathol* 1987, **40**, 377–83.

88 Mearns G, Richmond SJ, Storey C. Sensitive immune dot blot test for diagnosis of *Chlamydia trachomatis* infection. *J Clin Microbiol* 1988, **26**, 1810–3.

Chapter 11

Chronic Q fever

RJC Hart

Q fever was discovered little more than 50 years ago in Brisbane, where an outbreak occurred among the workers in a meat processing factory. Derrick [1] described the disease, which he named Q (or Query) fever, and the organism which he isolated was identified as a rickettsia by Burnet and Freeman [2]. At almost the same time a similar organism was isolated from ticks in the United States of America by Davis and Cox [3]. The two organisms were subsequently shown to be identical and are now known as *Coxiella burnetii*.

Q fever has a worldwide distribution and is the manifestation of overt infection by *C burnetii*; subclinical infections are common judging from the number of people in whom phase 2 antibodies are discovered coincidentally during routine serological examinations. The majority of cases occur in men of working age, but may occur in either sex and at all ages. Children occasionally suffer from the disease: Richardus *et al* [4] reviewed 18 acute cases in children aged three years or less and referred to a seven-year-old boy who had Q fever endocarditis.

Q fever was first observed outside Australia in 1944 when it was reported in Italy among American and British troops [5]. The local inhabitants had a high incidence of antibodies and the organism appears to have been widespread in the Mediterranean littoral. Subsequent investigations have shown that it has worldwide distribution, though infections are uncommon in northern Europe. *C burnetii* is probably spread into new areas by the importation of infected cattle. Infection spreads rapidly from infected to uninfected cattle, as shown by Huebner and Bell [6], who reported that 40% of uninfected cattle imported into an area where the infection was enzootic became infected within six months. *C burnetii* is a zoonotic infection affecting chiefly cattle, sheep and goats, but has been found in a large number of species of animals and birds. It has also been isolated from various species of tick, but the role of these insects in the spread of disease is not clear. It is probable that they are associated with infection in animals, but play no part in infecting man. Human infections generally result from the inhalation of infected aerosols, arising particularly from the products of conception of cattle or sheep, but it is possible that raw milk from infected cows may also be responsible for human infections. A milk-borne outbreak was reported in a detention centre [7] and antibodies to *C burnetii* can be demonstrated in a higher proportion of people who drink raw milk than those who use only milk which has been heat-treated [8]. The possibility of infection by human milk and of infection *in utero* is discussed by Richardus *et al* [9] and by Prasad *et al* [10] who isolated the organism from a human placenta.

During the period when Q fever was spreading in various parts of the world, outbreaks were not uncommon and there was often a high incidence of severe disease among those exposed to infection, for example workers in abattoirs, sheep shearers or those exposed to aerosols from infected straw [11]. Serological surveys have shown that once the organism has become established in an area, antibodies can be detected in a proportion of the population, including young children. This suggests that non-apparent infections must be widespread. The majority of cases of Q fever in recent years has been single, sporadic and often not readily linked to a recognisable source of the organism. However, outbreaks do still occur and they are probably the result of the production of a heavily-infected aerosol in a place where many of the population are susceptible to infection by the organism, for example in a laboratory [12] or a Swiss Alpine valley [13].

The causative organism

C burnetii resembles rickettsiae in being an intracellular or extracellular parasite or pathogen of vertebrates and arthropods, but differs from them in growing in vacuoles (phagolysosomes) of host cells rather than in the cytoplasm or nucleus. Its cycle of development includes the formation of an endospore-like body in the yolk sac of the chick embryo [14]. It is more resistant than rickettsiae to heat and chemical agents. It can be cultured in the yolk sac or monolayers of chick embryo cells, and by intraperitoneal inoculation of many species of laboratory animals, whence it can be demonstrated in the spleen or by the development of an antibody response. It has been isolated in the laboratory from the sputum of patients suffering from pneumonia, from infected heart valves from cases of endocarditis and from the milk of infected animals and women. The inoculum, after appropriate treatment if necessary, is injected into the yolk sac of the developing hen's egg or intra-peritoneally into guinea-pigs. The organism can readily be demon-strated in infected yolk sacs by making impression smears of the washed yolk sac and staining with a Romanovsky or other suitable stain. After intraperitoneal inoculation of a guinea-pig, antibodies develop which can be detected 2–4 weeks after infection by complement fixation or other appropriate tests. *C burnetii* is, however, an organism that has been associated with many laboratory infections and stringent safety precautions, only available in high security laboratories, must be observed if attempts are to be made to isolate the organism.

C burnetii undergoes a phase variation somewhat akin to the smooth–rough variation observed in many bacteria. The smooth hydrophilic organisms in phase 1 are stable in suspension and not phagocytosed in the absence of specific antibodies. After repeated passage, eg in yolk sacs, they become rough, hydrophobic and autoagglutinable. These readily phagocytosed organisms are in phase 2. When phase 2 organisms are inoculated into laboratory animals they rapidly revert to phase 1. Organisms in phase 1 are virulent for guinea-pigs, while those in phase 2 are avirulent or of very low virulence [15]. The antigens used for serological testing are made from whole organisms, but lipopolysaccharide antigens have been extracted and recent work [16] suggests that the extraction of protein antigens may be important in the future. The most effective method for diagnosis of Q fever is the examination of paired sera for rising titres which are demonstrated in the complement fixation, ELISA or immunofluorescence test. ELISA tests have also been used for the early diagnosis of Q fever by the demonstration of high-titred IgM responses [17]. In the great majority of acute cases of Q fever antibodies are produced only to the phase 2 antigen, but in a few cases there may be a transient phase 1 antibody response. In chronic Q fever antibodies are produced to both antigens, usually at high titres (see p. 241).

Acute Q fever

Infection by *C burnetii* is frequently not apparent, but often gives rise to Q fever. The acute disease is more common in men than women and varies from a comparatively trivial febrile infection to a full-blown pneumonia which may be fatal. After an incubation period of 2–3 weeks, usually 17–19 days, an influenza-like illness commences with severe headache, pain in the limbs, shivering and anorexia. In some cases the temperature falls after two or three days and the patient recovers. In other cases the illness progresses and a cough develops, which is usually dry at first, but later may be associated with blood-streaked sputum. Physical signs are often unremarkable, but the chest X-ray shows one or more areas of opacity. Laboratory investigations are unrewarding. The leucocyte count is unchanged, but there may be derangement of liver function tests. Severe cases of Q fever are often ill for several weeks despite effective antibiotic therapy and the illness may be complicated by myocarditis [18], hepatitis [19] or encephalitis [20]. Thrombocytopaenia and haemolysis [21] have also been described. The diagnosis is made serologically by the demonstration of a rising titre of antibody in paired sera against the phase 2 antigen of *C burnetii*.

Treatment is symptomatic in the milder cases, but in more severe cases tetracylines are the antibiotics of choice.

Clinical features of chronic Q fever

Non-specific febrile illness may be the result of chronic Q fever. A series of seven patients was reported [22] in whom endocarditis was not diagnosed, although they had pre-existing cardiovascular disease. One died suddenly without antibiotic treatment. Material from the aortic aneurysm of another caused seroconversion in guinea-pigs. All survivors were followed for several years or until they died and none developed signs of cardiac or hepatic infection. However, all but one received prolonged courses of antibiotics which may have prevented the development of endocarditis.

ENDOCARDITIS

This is the most common manifestation of chronic Q fever, amounting to 11% of 839 cases of Q fever in one series in England and Wales [23,24]. In this series, Palmer and Young found that two thirds of the patients were men and that almost all cases occurred in people between the ages of 15 and 65. Although Palmer and Young reported no cases in children, two were reported by Laufer *et al* [25], and Sawyer *et al* [15] collected reports of five children with Q fever endocarditis. They analysed a further 28 cases and reported that the median age was 50.5 years and that 68% were in men. Fourteen patients were reported from Spain [26]; seven were men and one a boy of 11 years. Sixteen cases were reported by Turck *et al* [27]; 11 were in men aged between 36 and 63 years. Nine of the 16 patients with chronic Q fever reported from Scotland had endocarditis and eight were men aged between 23 and 66 years [28]. Five of eight patients with endocarditis reported from Belfast were men between 24 and 67 [29], but only five of 10 from Dublin were men (ages 29–51) [30]. The classical presentation is thus in a man of working age, which contrasts with the incidence of other types of infective endocarditis in which there is only a small preponderance of males and the patients are usually older. *C burnetii* causes 2%–3% of all cases of infective endocarditis in England and Wales [24]. In the majority of cases of Q fever endocarditis, it is impossible to elicit a history of acute Q fever, but this may be due to the fact that the period between infection and the development of endocarditis appears to be very variable and may be as long as 20 years. There is usually a history of previous valvular damage or abnormality such as a bicuspid aortic value, or the infection may attack an artificial valve, but in some cases there is no evidence of previous endocardial damage. Underlying cardiac lesions were noted in 33% of cases reported to the Communicable Disease Surveillance Centre and 23 out of 30 had prosthetic heart valves [23]. The incidence of infection in heart valve prostheses is approximately 2% and two of a series of 44 such infections were due to *C burnetii* [31]. The importance of heart valve prostheses in the aetiology of *C burnetii* endocarditis was emphasised by Young [24] who stated that about half the patients with *C burnetii* endocarditis reported to the CDSC had heart valve prostheses.

The onset is insidious and the patient presents with the symptoms and signs associated with chronic or subacute infective endocarditis, usually involving the aortic valve, though the mitral valve may also be attacked either alone or together with the aortic. The time between the onset of symptoms and reporting may be as long as 39 months [15]. There are no particular symptoms or signs associated with Q fever endocarditis, and in fact it may go unrecognised as 'blood culture negative endocarditis'. All such cases should, therefore, be serologically tested for antibodies to phase 1 and phase 2 antigens of *C burnetii*.

Unusual presentations of Q fever endocarditis have been observed. A 37-year-old man who helped his neighbour with his sheep in the Ardeche was admitted to hospital in acute cardiac failure [32] and a myocardial infarct in a 39-year-old woman in Wales probably resulted from large vegetations on the right coronary cusp of her aortic valve [33]. Endocarditis arising on a valve prosthesis was first described by Kristinsson and Bentall [34] and has also affected porcine xenograft valves [35,36]. Ellis, Smith and Moffat [28] reported infection of dacron arterial grafts in two patients and pointed out that chronic Q fever is not invariably associated with endocarditis.

The possibility that Q fever endocarditis may be activated by steroid therapy is raised by the report of a patient who had an aortic valve replacement [37]. The valve showed no evidence of infection by *C burnetii*. On the day after operation she became febrile and was treated with prednisone for a week for post-cardiotomy syndrome after amino-glycoside and beta-lactam antibiotics had produced no response. She subsequently became febrile again and Q fever endocarditis was diagnosed about six weeks after the operation. The administration of steroids after surgery may have provoked the reactivation of *C burnetii* which was lying dormant.

If the disease is untreated, the valvular destruction is severe, leading to increasing cardiac failure. Embolic phenomena are common in Q fever endocarditis. Turck *et al* [27] reported that pulmonary embolism occurred in three of their patients and arterial emboli in 10, resulting in amputations of the legs of two.

HEPATITIS

Many patients suffering from Q fever endocarditis have abnormal liver function tests without overt evidence of hepatitis, though some develop hepatomegaly, often with splenomegaly. Turck *et al* [27] found abnormal hepatic histology in all eight of their patients in whom it was studied, two of whom presented with febrile illnesses and enlarged livers and spleens and no evidence of endocarditis. One patient died as a result of cirrhosis. Four of the Scottish cases of endocarditis had hepatomegaly, and one without any endocardial involvement had hepatosplenomegaly [28]. A fatal case of endocarditis and hepatitis was reported by Saginur *et al* [38]. Palmer and Young's series [23] included 10 cases of hepatitis without endocarditis. Two patients with chronic Q fever hepatitis were followed for up to two years [39,40]. Granulomata and increasing fibrosis were observed but neither patient developed cirrhosis. Chronic Q fever endocarditis and hepatosplenomegaly was seen in two children [24].

GLOMERULAR NEPHROPATHY

Haematuria has been described in many cases of chronic Q fever, but three cases of endocarditis complicated by glomerular nephropathy were reported from Spain [41]. Two had focal segmental proliferative glomerulonephritis and the other developed diffuse intracapillary proliferative glomerulonephritis. One of the former cases went into renal failure and died, the others responded to treatment.

OSTEMYELITIS

This was observed in four of Ellis, Smith and Moffat's cases [28]. Two patients presented with spinal lesions, the other two had endocarditis with lesions in the bones of the lower limbs, presumably resulting from emboli.

PLACENTITIS

This has been mentioned earlier as a possible cause of intrauterine infection. Ellis, Smith and Moffat [28] reported a mother, suffering from chronic Q fever, who gave birth to an infant who died on the sixth day after birth from *Escherichia coli* septicaemia.

THROMBOCYTOPAENIA

This was reported in 12 out of 16 patients in one series [27], six of whom had purpuric rashes. A pregnant woman with a purpuric rash was found to have chronic Q fever [42]. Her general condition improved after treatment with tetracycline, but her thrombocytopaenia persisted. Labour was induced after three weeks' treatment, at 28 weeks, and the placenta was macroscopically and microscopically necrotic. *C burnetii* was isolated from it. The mother's thrombocytopaenia resolved after treatment and she was treated with tetracycline for nine months. The infant, though premature, did well but developed yellow teeth as a result of the antenatal tetracyline.

ENCEPHALITIS

A case of endocarditis complicated by encephalitis was described by Brooks, Licitra and Peacock [20], who reviewed the reports in the literature of neurological complications of Q fever.

Diagnosis

Because there are no clinical criteria for the diagnosis of chronic Q fever, recourse to laboratory methods is necessary. The isolation of the organism is not practicable in the majority of laboratories and is less certain and more time-consuming than serology. The complement fixation test has been most widely used, but immunofluorescence and ELISA techniques have established a place for themselves in the diagnosis of acute and chronic Q fever. In acute Q fever, antibody to phase 2 is always present; phase 1 antibodies, if they can be found at all, are usually transient and of low titre, but persistence of low titre phase 1 antibody has been observed some months after acute Q fever [17,43]. In chronic Q fever, phase 1 antibodies are always present, and the titres are often very high, but phase 2 antibodies are usually present also, often at considerably higher titre than the phase 1 antibodies. A single specimen of serum is sufficient to produce the diagnosis; rising titres are seldom sought and rarely found in chronic infections.

The differences in antibody responses between acute and chronic infections may be because the phase 2 antigen is more superficial, or perhaps because in chronic infections organisms persist in phase 1. There has been much discussion about diagnostic levels of phase 1 antibody in chronic Q fever. Serological tests differ in sensitivity from laboratory to laboratory, but most authors accept a phase 1 complement fixation titre of 200 or more as diagnostic of chronic infection [27,28]. The indirect immunofluorescent (IF) antibody test has been shown to be valuable because the demonstration of specific IgA to phase 1 antigen appears to be diagnostic of chronic infection [44–46]. ELISA tests were said to be simpler and more sensitive than IF, though they failed to detect IgA in four out of 13 cases [45].

Management of Q fever endocarditis

The prognosis once Q fever endocarditis has been diagnosed should be very guarded. There are many reports of fatal cases in the literature. Some patients die within a few months, despite appropriate antibiotic treatment. It is difficult to assess the mortality from this condition because the length of time for which patients have been followed up varies and the outcome of the illness is not recorded in some cases, but useful information may be gained from some series. Five of the patients reported by Turck *et al* [26] died after varying periods of up to 28 months after diagnosis, and two of the seven Scottish patients with endocarditis died [28]. Five of ten patients in a series from Dublin died

[30], while a fatal outcome was reported in only one of eight from Belfast. Although two patients out of five reported from Nova Scotia [48] died, in neither case was death due to Q fever. It can be said that there is a high risk of fatal outcome which is increased by unstable cardiac conditions and by embolic or other complications. The role of surgical treatment has been discussed for many years [27,34]. It is indicated when the valves are very severely damaged or when the patient's cardiac condition continues to deteriorate despite antibiotic treatment. Prosthetic valves, which may be specially manufactured or porcine xenografts, are used to replace the aortic or mitral valve or both; however, numerous cases of infection of prosthetic valves have been recorded.

Most patients respond satisfactorily to antibiotic therapy but treatment needs to be kept up for a long period or even for life. *C burnetii* is sensitive to many antibiotic agents, including tetracyclines, clindamycin, trimethoprim-sulphamethoxazole, rifampicin and the quinolones. Many different treatment regimens have been described using antibiotics singly or in combination, often starting by the intravenous route and changing to oral administration later. Tetracyclines have been used most extensively and these drugs form the mainstay of treatment, either used alone [25,30,49] or in combination with other antibiotics. They are however, rickettsiostatic agents and this apparent disadvantage has led to the suggestion by many authors that rickettsiocidal agents should be combined with them. However, combinations of tetracycline with lincomycin and clindamycin, also rickettsiostatic agents, have been used extensively [27,28]. Chloramphenicol has been used in short courses only, because of the risks of haematological complications [27,48]. Of the rickettsiocidal agents, cotrimoxazole has been the most widely used, usually with tetracycline [29,30,36,48] but sometimes alone [30,49]. Rifampicin has been combined with tetracycline [37,50,52] or cotrimoxazole [53]. Erythromycin is used in acute Q fever, but has not established itself in the treatment of the chronic disease. No consensus as to the best choice of antibiotics has been reached, and no comparative trials have been carried out. Successful outcomes and failures have been reported from the use of tetracyclines alone, and for the various combinations. Doxycycline, because of its activity and because the once or twice daily dosage makes for better patient compliance, is clearly established as the tetracycline of choice for long-term administration. The combination of a tetracycline with lincomycin or clindamycin is recommended by some authors [27,28] but others prefer a tetracycline and cotrimoxazole [29,30,36,48] and yet others doxycline and rifampicin [31,32], while the possible value of the quinolones has been discussed by Dupuis and Peter [54].

Tetracycline has been used successfully to treat a pregnant woman [42] and children [25,55], its choice as the most effective drug for treatment of a potentially fatal condition overriding the undesirable side effects.

In the present state of knowledge, doxycycline given orally is clearly the antibiotic of choice, unless there are obvious and serious contra-indications. If combined therapy is desired, add either cotrimoxazole because of its rickettsiocidal activity or rifampicin because it is especially active against intracellular organisms. The latter combination does not cause side effects when used for long periods [56].

There has long been controversy over the duration of antibiotic therapy; some authors suggest that treatment should be continued indefinitely [30,50], while others consider that treatment should be for periods of at least 12 months [27], or until there is clinical evidence of resolution of the endocarditis and the phase 1 complement fixing (CF) antibody titres have fallen below 200 [29]. The variable response of patients and the possibility of relapse during or after treatment suggests that it is better to treat each patient individually rather than lay down hard and fast rules, and to continue antibiotic treatment until there is clinical and serological evidence of the cessation of activity of the infection. Brecker and Eykyn [56] described the clinical course over 25 years of a patient with Q fever endocarditis. His aortic and mitral valves were both affected and required surgical repair or replacement on three occasions. He was treated with tetracycline for many years. For five years after his last valve replacements he had combined treatment with doxycycline and rifampicin, but his antibody titres, especially to phase 1, remained very high. In most cases there is a steady fall followed by a stable but unchanging titre. Cessation of antibiotic treatment should not be a reason for abandoning surveillance, which should continue indefinitely.

Recent reviews [57–59] suggest that arguments over the duration of antibiotic therapy will continue and the report of a case apparently cured by six months' doxycycline [59] will add to the controversy. Many patients are eager to stop the antibiotic treatment and in fact some do so against medical advice. It is probably wise to warn the patient that antibiotic treatment will need to continue for at least two years. The patient is usually in hospital for diagnosis, assessment and the commencement of treatment, but once oral treatment has been stabilised the patient can be discharged to resume such activities as he is capable of carrying out without over-tiring himself. Careful clinical and serological follow-up is essential and is usually carried out every three months for at least the first year [55]. Provided that the clinical and cardiological findings are satisfactory and the phase 1 antibody titres are falling steadily, it may then be desirable to lengthen the intervals to six months and eventually to one year. Many patients appear to make a complete recovery, but others, having responded initially to treatment, deteriorate and may die despite medical and surgical treatment.

Most patients with Q fever endocarditis respond to antibiotic therapy but some do not, and the question of surgery with replacement of the aortic valve by an artificial prosthesis has to be considered. This may be a difficult decision because of the clinical condition of the patient. The operation should be performed under antibiotic cover and antibiotic treatment should continue for some weeks after the operation. The question of stopping antibiotic therapy is one that needs very careful consideration of the clinical and serological progress of the patient.

If it is decided to take a patient off antibiotics, very careful surveillance is necessary for the first six months so that antibiotic treatment can be reintroduced at once if there is evidence of relapse. One patient stopped his treatment after about ten months and did not attend for follow-up. He was killed in a road traffic accident five years later, and necropsy revealed no evidence of active infection [49]. This case and the one reported cured after six months' treatment suggest that there may be an element of over-cautiousness in the length of antibiotic treatment prescribed. It is wise to consider each case on its merits and to decide the duration of treatment as a result of cardiac and laboratory surveillance.

Follow-up of acute Q fever

The majority of cases do not require follow-up. They make a rapid clinical recovery with or often without tetracycline treatment. Some of the more severe cases require more careful consideration and the complications occurring in acute Q fever such as myocarditis, hepatitis, encephalitis or haemolysis are indications that the disease is more serious. In these cases, as well as serological testing for phase 2 antibodies, tests should also be carried out for phase 1 antibodies. If they are present, the titre is usually significantly less than the diagnostic level of 200, but careful clinical and serological surveillance is indicated for some weeks until the patient is well and the antibody titre to phase 1 has fallen considerably or disappeared. Rarely the antibody persists for months without any sign of chronic infection [43]. It must be pointed out that in the now quite extensive literature on the subject, little has been written about the development of chronic Q fever following the acute attack, and it is not possible at present to postulate any predisposing factors which may lead to development of the chronic disease. However, if a patient suffering from a valvular abnormality of the heart develops Q fever, this is clearly an indication for thorough and probably prolonged antibiotic treatment and careful follow-up. The same will also apply to patients who have valve prostheses or arterial grafts.

Conclusion

Q fever is not an uncommon infection, particularly in rural Britain. The incidence of chronic Q fever is sufficiently high for it to be in the minds of all doctors dealing with patients suspected of suffering from endo-carditis, and all bacteriologically-negative cases should be screened serologically for Q fever, using both phase 1 and phase 2 antigens of *C burnetii*. Those patients with chronic Q fever, either endocarditis or hepatitis, who respond to antibiotic therapy need careful follow-up and prolonged treatment, which may include cardiac surgery.

References

1   Derrick EH. Q fever, a new fever entity: clinical features, diagnosis and laboratory investigation. *Med J Aust* 1937, **ii**, 281–99.

2   Burnet FM, Freeman M. Experimental studies on the virus of Q fever. *Med J Aust* 1937, **ii**, 299–305.

3   Davis GE, Cox HR. A filter-passing infectious agent isolated from ticks. Isolation from *Dermacentor andersoni*, reactions in animals and filtration experiments. *Public Health Rep* 1938, **53**, 2259–67.

4   Richardus JH, Dumas AM, Huisman J, Schaap GJP. Q fever in infancy: a review of 18 cases. *Pediatr Infect Dis* 1985, **4**, 369–73.

5   Robbins FC, Gault RL, Warner FB. Q fever in the Mediterranean area: report of its occurrence in Allied troops. II Epidemiology. *Am J Hyg* 1946, **44**, 23–50.

6   Huebner RJ, Bell JA. Q fever studies in southern California. *JAMA* 1951, **145**, 301–5.

7   Brown GL, Colwell DC, Hooper WL. An outbreak of Q fever in Staffordshire. *J Hyg (Camb)* 1968, **66**, 649–55.

8   Marmion BP, Stoker MGP, McCoy JH, Malloch RA, Moore B. Q fever in Great Britain: An analysis of 69 sporadic cases, with a study of the prevalence of infection in humans and cows. *Lancet* 1953, **i**, 503–10.

9   Richardus JH, Donkers A, Dumas AM, Schaap GJP, Akkermans JPWM, Huisman J, Valkenburg HA. Q fever in the Netherlands: a sero-epidemiological survey among human population groups from 1968 to 1983. *Epidemiol Infect* 1987, **98**, 211–9.

10 Prasad BN, Chandiramani HK, Wagle A. Isolation of *Coxiella burnetii* from human sources. *Int J Zoonoses* 1986, **13**, 112–7.

11 Hart RJC. The epidemiology of Q fever. *Postgrad Med J* 1973, **49**, 535–8.

12 Hall CJ, Richmond SJ, Caul EO, Pearce NH, Silver IA. Labora-tory outbreak of Q fever acquired from sheep. *Lancet* 1982, **i**, 1004–6.

13 Dupuis G, Petite J, Péter O, Vouilloz M. An important outbreak of human Q fever in a Swiss Alpine valley. *Int J Epidemiol* 1987, **16**, 282–7.

14 Kreig NR, Holt JG (eds). *Bergy's manual of systematic bacteri-ology*. Williams & Wilkins 1984, 701.

15 Williams JC, Thomas LA, Peacock MG. Humoral immune re-

sponse to Q fever: Enzyme-linked immunosorbent assay antibody response to *Coxiella burnetii* in experimentally infected guinea pigs. *J Clin Microbiol* 1986, **24**, 935–9.

16 Muller H-P, Schmeer N, Rantamaki L, Semler B, Krauss H. Isolation of a protein antigen from *Coxiella burnetii*. *Zentralbl Bakteriol Mikrobiol Hyg [A]* 1987, **265**, 277–89.

17 Schmeer N, Krauss H, Werth D, Schiefer HG. Serodiagnosis of Q fever by enzyme linked immunosorbent assay (ELISA). *Zentralbl Bakeriol Mikrobiol Hyg [A]* 1987, **267**, 57–63.

18 Sheridan P, MacGaig JN, Hart RJC. Myocarditis complicating Q fever. *BMJ* 1974, **ii**, 155–6.

19 Sawyer LA, Fishbein DB, McDade JE. Q fever: current concepts. *Rev Infect Dis* 1987, **9**, 935–46.

20 Brooks RG, Licitra CM, Peacock MG. Encephalitis caused by *Coxiella burnetii*. *Ann Neurol* 1986, **20**, 91–3.

21 Spellman DW. Q fever: A study of 111 consecutive cases. *Med J Aust* 1982, **i**, 547–53.

22 Fergusson RJ, Shaw TRD, Kitchen AH, Matthews MB, Inglis JM, Peutherer JF. Subclinical chronic Q fever. *Q J Med* 1985, **57**, 669–76.

23 Palmer SR, Young SEJ. Q fever endocarditis in England and Wales, 1975–81. *Lancet* 1982, **ii**, 1448–9.

24 Young SEJ. Aetiology and epidemiology of infective endocarditis in England and Wales. *J Antimicrob Chemother* 1987, **20 Suppl A**, 7–14.

25 Laufer D, Lew PD, Oberhansli I, Cox JN, Longson M. Chronic Q fever endocarditis with massive splenomegaly in childhood. *J Pediatr* 1986, **108**, 535–9.

26 Tellez A, Sainz C, Echevarria C, de Carlos S, Fernandez MV, Leon P, Brezina R. Q fever in Spain: acute and chronic cases 1981–1985. *Rev Infect Dis* 1988, **10**, 198–202.

27 Turck WPG, Howitt G, Turnberg LA, Fox H, Longson M, Matthews MB, Das Gupta R. Chronic Q fever. *Q J Med* 1976, **45**, 193–217.

28 Ellis ME, Smith CC, Moffat MAJ. Chronic or fatal Q fever infection: a review of 16 patients seen in North-East Scotland (1967–80). *Q J Med* 1983, **52**, 54–66.

29 Varma MPS, Adgey AAJ, Conolly JH. Chronic Q fever endocarditis. *Br Heart J* 1980, **43**, 695–9.

30 Tobin MJ, Cahill N, Gearty G, Maurer B, Blake S, Daly K, Hone R. Q fever endocarditis. *Am J Med* 1982, **72**, 396–400.

31 Braimbridge MV, Eykyn SJ. Prosthetic valve endocarditis. *J Antimicrob Chemother* 1987, **20 Suppl A**, 173–80.

32 Etienne J, Delahaye F, Raoult D, Frieh J-P, Loire R, Delaye J.

Acute heart failure due to Q fever endocarditis. *Eur Heart J* 1988, **9**, 923–6.

33 Watt AH, Fraser AG, Stephens MR. Q fever endocarditis presenting as myocardial infarction. *Am Heart J* 1986, **112**, 1333–5.

34 Kristinsson A, Bentall HH. Medical and surgical treatment of Q fever endocarditis. *Lancet* 1967, **ii**, 693–7.

35 Ross PJ, Jacobson J, Muir JR. Q fever endocarditis of porcine xenograft valves. *Am Heart J* 1983, **105**, 151–3.

36 Fernandez-Guerrero ML, Muelas JM, Aguado JM, Renedo G, Fraile J, Soriani F, de Villalobos E. Q fever endocarditis on porcine bioprosthetic valves: Clinicopathologic features and microbiologic findings in three patients treated with doxycycline, cotrimoxazole and valve replacements. *Ann Intern Med* 1988, **108**, 209–13.

37 Lev BI, Shachar A, Segev S, Weiss P, Rubinstein E. Quiescent Q fever endocarditis exacerbated by cardiac surgery and coticosteroid therapy. *Arch Intern Med* 1988, **148**, 1531–2.

38 Saginur R, Silver SS, Bonin R, Carlier M, Orizaga M. Q fever endocarditis. *Can Med Assoc J* 1985, **133**, 1228–30.

39 Atienza P, Ramond M-J, Degott C, Lebrec D, Rueff B, Benhamou J-P. Chronic Q fever hepatitis complicated by extensive fibrosis. *Gastroenterology* 1988, **95**, 478–81.

40 Yebra M, Marazuela M, Albarrán F, Moreno A. Chronic Q fever hepatitis. *Rev Infect Dis* 1988, **10**, 1229–30.

41 Perez-Fontan M, Huarte E, Tellez A, Rodriguez-Carmona A, Picazo ML, Martinez-Ara J. Glomerular nephropathy associated with chronic Q fever. *Am J Kidney Dis* 1988, **11**, 298–306.

42 Riechman N, Raz R, Keysary A, Goldwasser R, Flatau E. Chronic Q fever and severe thrombocytopenia in a pregnant woman. *Am J Med* 1988, **85**, 253–4.

43 Winner SJ, Eglin RP, Moore VIM, Mayon-White RT. An outbreak of Q fever affecting postal workers in Oxfordshire. *J Infect* 1987, **14**, 255–61.

44 Peacock MG, Philip RN, Williams JC, Faulkner RS. Serological evaluation of Q fever in humans: enhanced phase 1 titres of immunoglobulins G and A are diagnostic for Q fever endocarditis. *Infect Immun* 1983, **41**, 1089–98.

45 Dupuis G, Péter O, Luthy R, Nicolet J, Peacock M, Burgdorfer W. Serological diagnosis of Q fever endocarditis. *Eur Heart J* 1986, **7**, 1062–6.

46 Raoult D, Urvologye J, Etienne J, Roturier M, Puel J, Chaudet H. Diagnosis of endocarditis in acute Q fever by immunofluorescence serology. *Acta Virol (Praha)* 1988, **32**, 70–4.

47 Péter O, Dupuis G, Bee D, Lüthy R, Nicolet J, Burgdorfer W. Enzyme linked immunosorbent assay for diagnosis of chronic Q fever. *J Clin Microbiol* 1988, **26**, 1978–82.

48 Haldane EV, Marrie TJ, Faulkner RS, Lee SHS, Cooper JH, MacPherson DD, Montague TJ. Endocarditis due to Q fever in Nova Scotia: Experience with five patients in 1981–1982. *J Infect Dis* 1983, **148**, 978–85.

49 Hart RJC, Shaw DB. Q fever endocarditis. *BMJ* 1973, **iii**, 233.

50 Kimbrough RC, Ormsbee RA, Peacock M, Rogers WR, Bennetts RW, Raaf J, Krause A, Gardner C. Q fever endocarditis in the United States. *Ann Intern Med* 1979, **91**, 400–2.

51 Freeman R, Hodson ME. Q fever endocarditis treated with trimethoprim and sulphamethoxazole. *BMJ* 1972, **i**, 419–20.

52 Raoult D, Piquet P, Gallais H, de Micco C, Drancourt M, Casanova P. *Coxiella burnetii* infection of a vascular prosthesis. *N Engl J Med* 1986, **315**, 1358–9.

53 Subramanya NI, Wright JS, Khan MAR. Failure of rifampicin and cotrimoxazole in Q fever endocarditis. *BMJ* 1982, **285**, 343–4.

54 Dupuis G, Péter O. Infections aigues et chroniques a *Coxiella burnetii* (fievre Q): du diagnostic au traitement. *Rev Med Suisse Romande* 1988, **108**, 677–82.

55 Beaufort-Krol GCM, Storm CJ. Chronic Q fever endocarditis. *J Pediatr* 1987, **110**, 330–1.

56 Brecker SJD, Eykyn SJ. Q fever endocarditis twenty-five years on. *Lancet* 1989, **ii**, 684–5.

57 Geddes AM. Q fever. *BMJ* 1983, **287**, 927–8.

58 Pierce MA, Saag MS, Dismukes WE, Cobbs CG. Q fever endocarditis. *Am J Med Sci* 1986, **292**, 104–6.

59 Edlinger EA. Chronic Q fever. *Zentralbl Bakteriol Mikrobiol Hyg [A]* 1987, **267**, 51–6.

Chapter 12

Mycoplasma pneumoniae

TG Wreghitt

History

The first respiratory tract mycoplasma to be isolated (*Mycoplasma mycoides* subsp *mycoides*) was from cattle with contagious bovine pleuropneumonia [1]. Although 10 different species of mycoplasma have been isolated from the human respiratory tract (Table 1), most are commensal organisms and only *M pneumoniae* is commonly isolated from the lower respiratory tract and has been unequivocally linked to human respiratory tract disease.

The disease associated with the most severe symptoms produced by *M pneumoniae*, pneumonia, has been known by a variety of names: in the 1920s and 1930s it was known as viral pneumonia; in 1938 Reimann

Table 1 Mycoplasma species isolated from the human respiratory tract

Species	Respiratory disease
M pneumoniae	Yes
M hominis	?*
U urealyticum	?*
M buccale	No
M faucium	No
M fermentans	No
M lipophilum	No
M orale	No
M primatum	No
M salivarium	No

*Ureaplasma urealyticum and *M hominis* have been isolated from the lungs of neonates with pneumonia in the absence of other pathogens [2,3]. However, there is no conclusive evidence of disease association, as yet, for these two mycoplasmas.

described the syndrome as primary atypical pneumonia [4]. It is now recognised that primary atypical pneumonia may be caused by a number of organisms, but that *M pneumoniae* is the most important.

Studies on human volunteers conducted during World War II by the Commission on Acute Respiratory Diseases demonstrated that primary atypical pneumonia was caused by a filterable agent [5–8]. In 1944, Eaton and colleagues reported the isolation of an organism (later called the Eaton agent) in the lungs of cotton rats, hamsters and in chick embryos [9]. Definitive proof of the association of the Eaton agent with primary atypical pneumonia was provided by Liu *et al* [10], who employed indirect immunofluorescence to show rising titres in paired serum samples from patients with the disease. Marmion and Goodburn [11] provided evidence that the Eaton agent was not a virus, since they showed it was sensitive to several antibiotics and morphologically resembled mycoplasmas. The Eaton agent was successfully grown in cell-free medium and identified as a mycoplasma by Chanock *et al* [12]. Chanock *et al* [13] named it *M pneumoniae*.

Clinical features

M pneumoniae is primarily a pathogen of the human respiratory tract, where it causes a range of illness from mild upper respiratory tract symptoms to pneumonia [14–17]. The most frequent symptom associated with *M pneumoniae* infection is tracheobronchitis, which is seldom recognised by clinicians as an indicator of *M pneumoniae* disease [18].

In 1967, Watson reported symptoms associated with *M pneumoniae* infection during an outbreak in south-east England [19]. Patients characteristically presented with headache (which was worse in adults), sore throat and fever. A non-productive cough of increasing severity was noted on the second or third day of illness in 75% patients. The absence of a cough makes the diagnosis of *M pneumoniae* infection most unlikely. *M pneumoniae* pneumonia usually has an insidious onset, and is clinically indistinguishable from that caused by a range of other organisms such as *Chlamydia psittaci*, *Chlamydia pneumoniae* and *Legionella pneumophila*. Pneumonia develops in 3%–10% of patients infected with *M pneumoniae* and in epidemic years is responsible for approximately 20% of community-acquired pneumonias [20]. Clinical signs of consolidation are unusual, in contrast to the X-ray findings which frequently show extensive shadowing [20]. Pneumonia is usually unilateral, affecting one of the lower lobes, but bilateral and multilobe involvement is also seen in approximately 20% of patients [21]. The course of the disease is variable but is often protracted, with a persistent cough a common feature, along with relapses.

Twenty-five per cent of *M pneumoniae* infections are asymptomatic [21]. Extrapulmonary manifestations of *M pneumoniae* infection include central and peripheral nervous system disease [22–27], myalgia [28,29],

Table 2 Symptoms associated with *M pneumoniae* infection

Cough	99%
Fever >37.8°C	94%
Malaise	89%
Fever >38.9°C	77%
Headache	66%
Chills	58%
Sore throat	54%

From Foy *et al* [36]

erythema multiforme and Stevens-Johnson syndrome [29–31]. Arthritis [33,34], haemolytic anaemia [16,29,32], hepatitis [16], disseminated intravascular coagulation [16] and myringitis [35] may also be seen in association with *M pneumoniae* infection. These symptoms arise with varying frequency with haemolytic anaemia, Stevens-Johnson syndrome and neurological complications being found most frequently [29]. Symptoms associated with *M pneumoniae* infection in patients with pneumonia in an urban area are shown in Table 2.

Epidemiology
M pneumoniae infections are found worldwide and arise in epidemics which last one to two years and occur every four to six years [36,37]. During epidemics, the frequency of infection may be three to 10 times that seen in non-epidemic periods. The incidence of *M pneumoniae* infection in England and Wales between 1977 and 1987 is shown in Figure 1. Infections arise throughout the year and since the incubation period ranges from two to three weeks [19], spread usually occurs slowly. *M pneumoniae* spreads most effectively where there is repeated

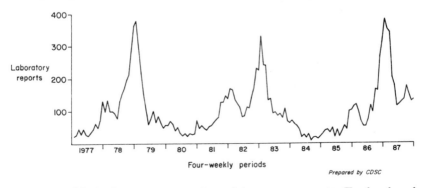

Figure 1 *Mycoplasma pneumoniae* – laboratory reports, England and Wales, 1977–87

251

and close contact, for example in the family [38,39] or in closed communities such as army barracks and other institutions [40,41]. *M pneumoniae* infections arise most frequently in children aged five to 10 years, less often in older children and young adults, and are found with decreasing frequency in people aged over 40 years [36]. Asymptomatic infections are found most frequently in children under five years old and pneumonia is found most frequently in children aged between five and 15 years old [42]. Foy *et al* [43] showed that *M pneumoniae* reinfections are relatively common, being noted between two and 10 years after previously-documented infection in a long-term surveillance study in Washington. Naturally-acquired immunity to infection with *M pneumoniae* was of limited duration.

Pathogenicity

Since mycoplasmas are not intracellular pathogens, their attachment to host cells is vital to their survival within the host respiratory tract [44]. This is particularly difficult because of the mucous gel covering the respiratory epithelium. *M pneumoniae* can attach to cell surfaces, even when non-viable [45], and is thought to do so by means of a specialised tip structure [46]. However, it has been noted that *M pneumoniae* organisms lie parallel to the plasma membrane in non-ciliated cells but position themselves vertically in ciliated cells [47,48]. This would suggest that the specialised tip structure is not essential for attachment. Adsorption to cells has been linked with two surface proteins, P1 (190 kiladalton) and P2 (78 kiladalton), which are removed when *M pneumoniae* preparations are treated with trypsin [49]. Adsorption and attachment to cells is unlikely to be the only factor associated with pathogenicity, since some non-pathogenic strains of *M pneumoniae* are capable of attachment to cells.

Sialic acid sites on membranes have been implicated as receptors for *M pneumoniae* attachment [50,51] which is inhibited by pretreatment of cells with neuraminidase. Attachment of *M pneumoniae* to cell surfaces leads to various biochemical changes within the cell, eventually results in loss of ciliary motion [52] and may also lead to the sloughing of affected cells [53]. Ciliary function can be impaired for many months after *M pneumoniae* infection [54].

Another factor which may affect the pathogenicity of *M pneumoniae* strains is their ability to produce peroxide. Strains with a reduced ability to produce peroxide have been shown to be less virulent [55].

M pneumoniae infection produces both a cell-mediated and an antibody response in man [56]. Antibodies to both P1 and P2 antigens are usually found in human serum samples from patients with *M pneumoniae* infection, although individuals vary in their ability to produce antibodies to other *M pneumoniae* proteins [57].

Laboratory diagnosis

This can be achieved in three ways: culture of *M pneumoniae*, detection of *M pneumoniae* antigen or DNA, or detection of specific antibody.

CULTURE

M pneumoniae isolation is routinely performed in only a few laboratories with appropriate facilities. The most common method is inoculation of throat swabs (in virus transport medium) or sputum (mixed with an equal volume of N-acetyl cysteine to promote mucolysis) on to solid and liquid media. The method employed by Dr Margaret Sillis in the Norwich Public Health Laboratory is the inoculation of 2ml fluid from throat swab or sputum onto solid medium and into 2ml glucose broth; 0.2ml of this broth is then inoculated into a 2ml glucose broth to give a 1 in 10 dilution. The glucose broth consists of Oxoid PPLO broth (60% v/v), horse serum (10% v/v), glucose, phenol red, Oxoid yeast extract, penicillin and thallous acetate [58]. The solid medium employed in Norwich comprises PPLO agar base (70% v/v), porcine serum (10% v/v), boiled blood extract (10% v/v), Oxoid yeast extract, DNA, thallous acetate and penicillin [12].

Plates are incubated at 37°C in 5% carbon dioxide in nitrogen and examined at 3–4 day intervals for three weeks. Colonies are subcultured from primary plates and confirmed as *M pneumoniae* by means of growth inhibition tests employing filter paper discs impregnated with *M pneumoniae* antiserum (DMRQC, Central Public Health Laboratory, Colindale). Broth cultures are incubated at 37°C for three weeks. If the broth turns yellow in the absence of turbidity, it is subcultured on to solid medium and colonies identified as above.

Although the isolation of *M pneumoniae* from sputum and other respiratory tract secretions is a valuable indication that the organism is contributing to a particular episode of respiratory tract symptoms, culture and identification may take 10–14 days. Therefore this is not an ideal means of diagnosing infection, but is a very valuable asset when evaluating the worthiness of other techniques such as antigen or DNA detection or antibody assays. Also, isolation alone is not a wholly reliable diagnostic procedure, since *M pneumoniae* can be isolated from symptom-free individuals [59–62] and is known sometimes to exist as a commensal organism in man.

ANTIGEN DETECTION

There are several techniques which have been employed to demonstrate *M pneumoniae* antigens in tissues or body fluids. Immuno-electro-osmophoresis has been used by some workers [63,64]. The detection of *M pneumoniae* by immunofluorescence has also been reported [65,66], although this technically-demanding method has not been adopted widely. ELISA has been used to detect *M pneumoniae* antigen. Helbig and Witzleb [67] employed the technique, but not on clinical specimens, while Weiner *et al* [68] used ELISA for the detection of *M pneumoniae* in clinical specimens after a short incubation in liquid culture. More recently, Kok *et al* [69] described the use of direct an indirect ELISAs for the detection of *M pneumoniae* antigen in nasopharyngeal aspirates and sputum from adults and children. The indirect method was found to be the most convenient but both ELISA systems had similar sensitivity. There were no cross-reactions found with 10 other species of mycoplasma and a wide range of bacteria and chlamydiae found in the human respiratory tract, but low level cross-reaction was noted with *M genitalium* which is not found in the human respiratory tract. These ELISAs detected *M pneumoniae* in 90% of the samples from which the organism was isolated and *M pneumoniae* antigen was found in 43% samples from patients with serological evidence of recent infection, but from which *M pneumoniae* was not cultured. With this ELISA, *M pneumoniae* can be detected on the next working day.

An immunoblot assay was developed by Madsen *et al* [70], employing a monoclonal antibody to a 43-kiladalton, *M pneumoniae* membrane-associated protein. The sensitivity of this method was found to be similar to the ELISA antigen detection technique of Kok *et al* [69].

MOLECULAR TECHNIQUES

Several reports have described the use of DNA probes for the detection of *M pneumoniae* in clinical specimens [71–73]. Dular *et al* [71] compared the Gen-Probe® kit with culture and found identical results with 89% specimens. The probe was 89% sensitive and specific when compared with culture. Although the probe method is sensitive, specific and rapid, it is labour-intensive, expensive and employs radio-labelled reagents with a short shelf life.

ANTIBODY DETECTION

Many methods have been employed for detecting *M pneumoniae* antibodies in human serum. The methods which detect *M pneumoniae* IgG are suitable for the diagnosis of *M pneumoniae* infection in patients of all ages, since they are capable of detecting rising antibody titres if appropriate paired serum samples are available. However, since *M pneumoniae* IgM is found less frequently in patients experiencing re-infection (and hence in older patients), tests for the detection of *M pneumoniae* IgM are most useful in younger patients (particularly those less than 40 years of age).

The complement fixation test (CFT) [74] has been the mainstay of routine laboratory diagnosis of *M pneumoniae* infections for many years. However, antibody rises may not be detectable by this method until 7–10 days after the onset of symptoms and not all culture-positive patients develop CFT antibody or a significant rise in titre [28]. It is also very difficult to determine the significance of CFT titres obtained with single samples of serum, unless they are very high (≥ 256). Such high titres can be found many months after infection and so particularly in the months following periods of high incidence of *M pneumoniae* infection, interpretation of CFT titres is particularly difficult in the absence of other confirmatory tests, particularly those detecting specific IgM. Serum samples are most often available from patients at a time when the CFT titre has already reached its maximum. Demonstration of a fourfold or greater rise in antibody titre is required to be reasonably sure of the diagnosis, and even then, some workers have questioned the specificity of the lipid antigen commonly used in the CFT [75,76]. *M pneumoniae* glycolipids may cross-react with similar antigens on Streptococcus MG [77] and *Streptococcus pneumoniae* [78] and the I antigen on human erythrocytes [79].

Cold agglutinins showing specificity within the I/i antigen system are found in the plasma of approximately 50% patients with acute *M pneumoniae* infection [80–82]. Occasionally the titre and thermal range of these IgM autoantibodies may be sufficient to give rise to acute haemolysis and anaemia approximately 10–20 days after the onset of respiratory symptoms. The cold agglutinin titre usually remains high for about a week after the onset of anaemia, but falls gradually to normal 2–3 months later. The advantage of this technique is that while being fairly non-specific, it is rapid and simple to perform and a useful indicator when positive in epidemic years [83].

Several investigators have used the indirect immunofluorescence test to detect *M pneumoniae* antibody response [82,84–89]. This technique has the advantages of speed (taking only two hours to perform) and the fact that specific IgM, IgA and IgG can be detected individually. However, while the test is relatively simple to perform, reading is subjective and the kind of antigen used in the test is most important. A washed suspension of *M pneumoniae* antigen grown on glass is best, but it needs to be in a suitably clumpy form [84]. The test can be as sensitive and specific as ELISA if properly performed, particularly for the detection of *M pneumoniae*-specific IgM [84], for which it is a very useful assay.

The growth-inhibiting property of antibody can be demonstrated by means of the metabolism inhibition test [90], which involves the inhibition (by specific antibody) of a colour change in liquid medium that indicates pH change, due to glucose, arginine or urea utilisation. For *M pneumoniae*, the test is based on the inhibition of acid production from glucose. Antibody titres reflect the dilution of serum that inhibits the colour change produced by a standard inoculum of *M pneumoniae*. This test is used only in specialised laboratories, but has a similar sensitivity to CFT [91–93], is less sensitive than ELISA, radio immunoprecipitation and the mycoplasmacidal test, and much less convenient than ELISA.

The mycoplasmacidal test [94] is based on the principle that in the presence of antibody and complement, some mycoplasma are rapidly killed by lysis. Standard amounts of antigen and complement are incubated with serum dilutions, and after incubation culture medium is added. The test has similar sensitivity to ELISA [91–93] but is more complicated to perform and thus is suitable only as a research tool.

Radioimmunoassay and radioimmunoprecipitation tests for the detection of *M pneumoniae* antibody have been described by a number of workers [95–97]. These tests are very sensitive and specific and can distinguish between *M pneumoniae*-specific IgM, IgG and IgA. However they are in very limited use, principally because of the need to use radioactive reagents, which may involve practical difficulties and have a short shelf life as compared with ELISA conjugates.

The indirect haemagglutination test has been used for many years to study the serological response to *M pneumoniae* infections. A test employing tanned erythrocytes was first described by Dowdle and Robinson [98]. The test was later modified by Lind [99] and Ghyka *et al* [100]. The test is rapid, taking less than an hour to perform, and is relatively sensitive. However, although the test detects total antibody produced early in infection (including IgM), this antibody cannot be specifically recognised as IgM and thus low titres are as difficult to interpret as low level CFT titres in the early stages of the disease. Diamed Diagnostics have introduced a test kit (Fujirebio Serodia-Myco) which utilises freeze-dried sensitised erythrocytes. This test compares favourably with CFT for the serological diagnosis of *M pneumoniae* infections, although giving non-specific agglutination with some sera [101]. Rousseau and Tettmar [102] compared CFT, Serodia-Myco haemagglutination test and indirect immunofluorescence (to assess IgA, IgM and IgG) for the diagnosis of *M pneumoniae* infection. They found the Serodia-Myco haemagglutination test the most reliable single test to use, although the most useful combination of tests was the haemagglutination test with the indirect immunofluorescence test for detecting specific IgM and IgG. The specific IgA status for the patients was not found helpful.

Fujirebio have recently introduced their Serodia Myco-II gel particle agglutination test for the diagnosis of *M pneumoniae* infection. This test is based on the same principle as the Serodia-Myco haemagglutination test, but instead of sensitised chicken red cells, sensitised gel particles are employed. We have examined the Serodia-Myco II test [103], employing serum samples from our previous study [104] of patients with respiratory symptoms typical of *M pneumoniae* infection. We found good comparison with the Serodia-Myco II test, the indirect immuno-fluorescence IgM test and μ-capture ELISA for the detection of *M pneumoniae*-specific IgM, but there were several serum samples which were positive in the Serodia-Myco II test but negative in other two assays. The Serodia-Myco II test is a good screening assay for the diagnosis of recent *M pneumoniae* infection, but is probably less specific in younger patients than indirect immunofluorescence and μ-capture ELISA. However, in older patients who frequently fail to produce a good specific IgM response, the Serodia-Myco II test may be of more value. Certainly, the test is less demanding than the other two tests for a routine microbiology laboratory.

Various ELISA techniques have been used to examine the serological response to *M pneumoniae* infection [76,84,91–93,104–109]. All of these reports have involved the use of indirect ELISAs for detecting *M pneumoniae*-specific IgM or IgG or both. Busolo, Tonin and Conventi [91] used an indirect ELISA with the *M pneumoniae* either grown in the microtitre wells and fixed with formalin, or grown in Roux bottles, washed and then used as antigen. They found good correlation between ELISA and metabolism inhibition titres and showed that their ELISA was less sensitive than radioimmunoprecipitation and had similar sensitivity to the mycoplasmacidal test, but was more sensitive than CFT and the metabolism inhibition test [91–93].

With an indirect ELISA technique modified from that of Busolo *et al* [91–93], Miyaji [108] found that *M pneumoniae* antigen prepared by centrifugation of broth culture and resuspension in PBS gave reproducible results but that additional sonication of the antigen gave optimum performance. *M pneumoniae* grown in the ELISA plates [91] gave more variable results in the ELISA, probably because of variable growth in different wells.

Raisanen *et al* [76] developed an indirect ELISA using Tween/ether-treated protein antigen, because Brunner *et al* [95] had reported that *M pneumoniae* lipid antigen gave non-specific results in some assays.

Dussaix, Slim and Tournier [107] also performed an indirect ELISA for detecting *M pneumoniae* IgM and IgG, using a protein antigen. They found general agreement between ELISA IgG and CFT titres but there was a wide scatter of results and some sera with undetectable *M pneumoniae* CFT antibody had high ELISA IgG titres. *M pneumoniae* ELISA IgM antibody was found in sera from all patients with a four-fold rise in CFT antibody or a stable high CFT titre; 2% of sera gave false positive results thought to be due to the presence of rheumatoid factor.

Van Griethuysen *et al* [105] also used an indirect ELISA to detect *M pneumoniae*-specific IgM and IgG. They used sonicated *M pneumoniae* antigen similar to that of Miyaji [108] and employed a similar preparation of *M pulmonis* as a control antigen. ELISA IgM antibody was found to correlate better than IgG with CFT and IHA antibody. Diagnostic levels of ELISA IgM antibody were detected in the second week of disease, although they persisted for up to a year, which made careful evaluation of medical history a necessity for accurate interpretation of results.

Wreghitt and Sillis [84] were the first to report the development of a

μ-capture ELISA for detecting *M pneumoniae*-specific IgM, although a similar μ-capture radioimmunoassay was reported by Price [96]. They compared the test with both IFA and indirect ELISA and concluded that the μ-capture ELISA and IFA were both better than indirect ELISA for detecting *M pneumoniae*-specific IgM, having similar sensitivity and specificity. Complement fixation test antigen was not found suitable for use in the μ-capture ELISA. The indirect ELISA, using sonicated whole antigen, was both less sensitive and less specific; it gives high assay values with several sera having undetectable or low titre *M pneumoniae* CFT antibody but low or negative μ-capture ELISA and IFA values. Furthermore, some sera with high influenza A virus and Paul-Bunnell antibody titres gave high levels of *M pneumoniae* IgM in the indirect ELISA, but not in the μ-capture ELISA or IFA tests. Using μ-capture ELISA and IFA in another study, Wreghitt and Sillis [104] showed that *M pneumoniae*-specific IgA and IgM were most often found together in sera from patients with recent *M pneumoniae* infection and that, in general, the level of *M pneumoniae*-specific IgM correlated with the CFT titre of sera. In a modification of the μ-capture technique, Coombs *et al* [110] developed a μ-capture haemagglutination test which had similar sensitivity and specificity as μ-capture ELISA, but which took just over one hour to perform.

In establishing the significance of an assay value, certain details relating to the patient have to be considered, in particular day of disease and age. This is less important when considering IgG and IgA antibody, but essential when the significance of IgM values as indicators of recent infection is being evaluated. High concentrations of *M pneumoniae* IgM can be found by indirect ELISA for a year or more after infection [84,107], and this is not influenced by the type of antigen used in the assay [84,105,107]. It is not clear whether this long-lasting IgM is specific or not, but a test which can detect IgM more than a year after infection is not ideal. Thus, even when high levels of *M pneumoniae* IgM are found in serum taken from patients with pneumonia, interpretation of indirect ELISA results is difficult. In contrast, maximum levels of *M pneumoniae* IgM are found in the first month by μ-capture ELISA. Thereafter, the level of *M pneumoniae* IgM detected by μ-capture ELISA gradually falls with time, so that most patients have lost detectable *M pneumoniae* IgM by six months after their *M pneumoniae* infection (Figure 2). Thus, the level of *M pneumoniae* IgM is directly related to the length of time from the onset of *M pneumoniae* infection.

The age of the patient is also important, particularly in deciding the significance of concentrations of *M pneumoniae* IgM. Older patients (>40 years) are much less likely than younger ones to produce *M pneumoniae* IgM after *M pneumoniae* infection (Figure 3) [85,86,111–113]. This seems likely to be because more older patients experience a re-infection with *M pneumoniae* without detectable IgM response, where-

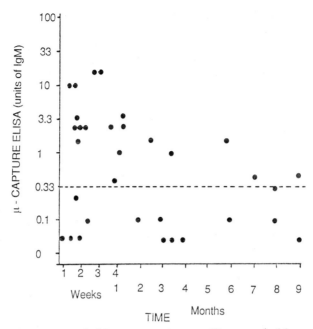

Figure 2 Amounts of *M pneumoniae*-specific IgM (arbitrary units) detected by indirect and μ-capture ELISA in sera taken at various times after proven *M pneumoniae* infection. Dotted lines indicate positive cut-off values.

as the younger patients are more likely to be suffering from a primary attack, where an IgM response is usually found. Moule *et al* [113] showed that younger patients, less than 20 years of age, were more likely to produce high levels of *M pneumoniae*-specific IgM. The number of patients who produced low or undetectable amounts of IgM increased with age.

Therefore, in older patients, tests which reliably detect *M pneumoniae*-specific IgG or IgA or both are most useful. The relative merits of these tests have been discussed, but probably the best is the Fujirebio Serodia Myco II gel particle agglutination test.

Clearly, further work needs to be done to improve the diagnosis of *M pneumoniae* infections. While the recent availability of DNA probes and antigen detection techniques have been important steps towards that goal, there is still a need for a more rapid, simple, cheap and durable test for the detection of *M pneumoniae* antigen.

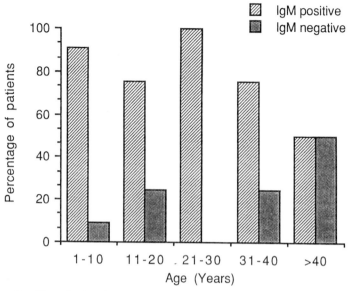

Figure 3 Correlation between the presence of *M pneumoniae*-specific IgM and age of patient, for sera with *M pneumoniae* CFT titres >256

Acknowledgements

Figure 1 is reproduced by kind permission of the PHLS Communicable Disease Surveillance Centre.

Figure 2 is reproduced by kind permission of the editor of the *Journal of Hygiene*.

References

1   Nocard E, Roux E. Le microbe de la peripneumonie. *Ann Inst Pasteur* 1898, **12**, 240–62.

2   Cassell GH, Davis RO, Waites KB, *et al*. Isolation of *Mycoplasma hominis* and *Ureaplasma urealyticum* from amniotic fluid at 16–20 weeks of gestation: potential effect on outcome of pregnancy. *Sex Trans Dis* 1983, **10**, 294–302.

3   Quinn PA, Butany J, Chipman M, Taylor J, Hannah W. A prospective study of microbial infection in stillbirths and early neonatal death. *Am J Obstet Gynecol* 1985, **151**, 238–49.

4   Reimann H. An acute infection of the respiratory tract with atypical pneumonia. *JAMA* 1938, **111**, 2377–84.

5   Commission on Acute Respiratory Diseases: Fort Bragg, North Carolina. Transmission of primary atypical pneumonia to human

volunteers. 1. Experimental methods. *Bull Hopkins Hosp* 1946, **79**, 97–108.

6   Commission on Acute Respiratory Diseases: Fort Bragg, North Carolina. Transmission of primary atypical pneumonia to human volunteers. II. Results of inoculation. *Bull Hopkins Hosp* 1946, **79**, 109–24.

7   Commission on Acute Respiratory Diseases: Fort Bragg, North Carolina. Transmission of primary atypical pneumonia to human volunteers. III. Clinical features. *Bull Hopkins Hosp* 1946, **79**, 125–52.

8   Commission on Acute Respiratory Diseases: Fort Bragg, North Carolina. Transmission of primary atypical pneumonia to human volunteers. IV. Laboratory studies. *Bull Hopkins Hosp* 1946, **79**, 153–67.

9   Eaton MD, Meiklejohn G, van Herick W. Studies on the etiology of primary atypical pneumonia. A filterable agent transmissible to cotton rats, hamsters, and chick embryos. *J Exp Med* 1944, **79**, 649–68.

10   Liu C, Eaton MD, Heyl JT. Studies on primary atypical pneumonia. II. Observations concerning development and immunological characteristics of antibody in patients. *J Exp Med* 1959, **109**, 545–56.

11   Marmion BP, Goodburn GM. Effect of an organic gold salt on Eaton's primary atypical pneumonia agent and other observations. *Nature* 1961, **189**, 247–8.

12   Chanock RM, Hayflick L, Barile MF. Growth on artificial medium of an agent associated with atypical pneumonia and its identification as a PPLO. *Proc Nat Acad Sci USA* 1962, **48**, 41–9.

13   Chanock RM, Dienes L, Eaton MD *et al*. *Mycoplasma pneumoniae*: proposed nomenclature for atypical pneumonia organism (Eaton agent). *Science* 1963, **140**, 662.

14   Clyde WA. *Mycoplasma pneumoniae* infections of man. In: Tully JG, Whitcomb RF, eds. The mycoplasmas. Vol II. Human and animal mycoplasmas. New York: Academic Press, 1979: 275–306.

15   Miller AC, Hanson GC. Respiratory failure due to *M pneumoniae* in young adults. *BMJ* 1983, **287**, 1028.

16   Murray HW, Masur M, Senterfit LB, Roberts RB. The protean manifestations of *Mycoplasma pneumoniae* infections in adults. *Am J Med* 1975, **58**, 229–42.

17   Koletsky RJ, Weinstein AJ. Fulminant *Mycoplasma pneumoniae* infection: Report of a fatal case and review of the literature. *Am Rev Resp Dis* 1980, **122**, 491–6.

18   Komaroff AL, Aronson MD, Pass TM, Ervin CT, Branch WT. Serologic evidence of chlamydial and mycoplasmal pharyngitis in adults. *Science* 1984, **222**, 927–9.

19   Watson GI. *Mycoplasma pneumoniae* in general practice. *J Coll Gen Prac* 1967, **13**, 174–96.
20   Dhillon P, Collins J. Recent developments in pneumonia. *Practitioner* 1983, **227**, 1695–8, 1701, 1704–5.
21   Cassell GH, Clyde WA, Davis JK. Mycoplasmal respiratory infections. In: Razin S, Barile MF, eds. The mycoplasmas IV. Mycoplasma pathogenicity. London: Academic Press, 1985, 65–106.
22   Mårdh PA, Ursing B, Lind K. Persistent cerebellar symptoms after infection with *Mycoplasma pneumoniae*. *Scand J Inf Dis* 1975, **7**, 157–60.
23   Twomey AJ, Espir MLE. Neurological manifestations of *Mycoplasma pneumoniae* infection. *BMJ* 1979, **2**, 832–3.
24   Urquhart GED. *Mycoplasma pneumoniae* infection and neurological complications. *BMJ* 1979, **2**, 1512.
25   Rothstein TL, Kenny GE. Cranial neuropathy, myeloradiculopathy, and myositis: complications of *Mycoplasma pneumoniae* infection. *Arch Neurol* 1980, **36**, 476–7.
26   Cotter FE, Bainbridge D, Newland AC. Neurological deficit associated with *Mycoplasma pneumoniae* infection reversed by plasma exchange. *BMJ* 1983, **286**, 22.
27   Koskiniemi M, Vaheri A. Effect of measles, mumps, rubella vaccination on pattern of encephalitis in children. *Lancet* 1989, **i**, 31–4.
28   Grayston JT, Foy HM, Kenny GE. The epidemiology of mycoplasma infections of the human respiratory tract. In: Hayflick L, ed. The Mycoplasmatales and the L-phase of bacteria. Amsterdam: North-Holland Publishing Co., 1969: 651–82.
29   Ali NJ, Sillis M, Andrews BE, Jenkins PF, Harrison BDW. The clinical spectrum and diagnosis of *Mycoplasma pneumoniae* infection. *Quart J Med* 1986, **58**, 241–51.
30   Finland M, Peterson OL, Allen HE, Samper BA, Barnes MW. Cold agglutinins. II. Cold isohemagglutinins in primary atypical pneumonia of unknown etiology with a note on the occurrence of hemolytic anemia in three cases. *J Clin Invest* 1945, **24**, 458–73.
31   Lyell A, Gordon AM, Dick HM, Sommerville RG. The role of *Mycoplasma pneumoniae* infection in erythema multiforme. *Proc R Soc Med* 1968, **61**, 1330–1.
32   McCormack JG. *Mycoplasma pneumoniae* and erythema multiforme – Stevens Johnson syndrome. *J Infect* 1981, **3**, 32–6.
33   Hernandez LA, Urquhart GED, Carson-Dick W. *Mycoplasma pneumoniae* infection and arthritis in man. *BMJ* 1977, **2**, 14–16.
34   Taylor-Robinson D, Gumpel JM, Hill A, Swannell AJ. Isolation of *Mycoplasma pneumoniae* from synovial fluid of a hypogammaglobulinaemic patient in a survey of patients with inflammatory polyarthritis. *Ann Rheum Dis* 1978, **37**, 180–2.

35   Rifkind D, Chanock RM, Kravetz H, Johnson K, Knight V. Ear involvement (myringitis) and primary atypical pneumonia following inoculation of volunteers with Eaton agent. *Am Rev Resp Dis* 1962, **85**, 479–89.

36   Foy HM, Kenny GE, McMahan R, Mansy AM, Grayston JT. *Mycoplasma pneumoniae* pneumonia in an urban area. *JAMA* 1970, **214**, 1666–72.

37   Noah ND. *Mycoplasma pneumoniae* infection in the United Kingdom 1967–73. *BMJ* 1974, **2**, 544–6.

38   Balassanian N, Robbins FC. *Mycoplasma pneumoniae* infection in families. *N Eng J Med* 1967, **277**, 719–25.

39   Foy HM, Grayston J, Kenny GE, Alexander E, McMahan R. Epidemiology of *Mycoplasma pneumoniae* infection in families. *JAMA* 1966, **197**, 859–66.

40   Mogabgab WJ. *Mycoplasma pneumoniae* and adenovirus respiratory illnesses in military and university personnel, 1959–1966. *Am Rev Resp Dis* 1968, **97**, 345–58.

41   Mufson MA, Chang V, Gill V, Wood SC, Romansky MJ, Chanock RM. The role of viruses, mycoplasmas and bacteria in acute pneumonia in civilian adults. *Am J Epidemiol* 1967, **86**, 526–44.

42   Taylor-Robinson D. Mycoplasmas and chlamydiae. In: Scadding JG, Cumming G, Thurlbeck WM, eds. Scientific foundations of respiratory medicine. London: Heinemann, 1981: 413–29.

43   Foy HM, Kenny GE, Sefi R, Ochs HD, Allan ID. Second attacks of pneumonia due to *Mycoplasma pneumoniae*. *J Infect Dis* 1977, **135**, 673–7.

44   Razin S. The mycoplasmas. *Microbiol Rev* 1978, **42**, 414–70.

45   Gabridge MG, Barden-Stahl YD, Polisky RB, Engelhardt JA. Differences in the attachment of *Mycoplasma pneumoniae* cells and membranes to tracheal epithelium. *Infect Immun* 1977, **16**, 766–72.

46   Collier AM. Mycoplasmas in organ culture. In: Tully JG, Whitcomb RE, eds. The mycoplasmas. II. Human and animal mycoplasmas. New York: Academic Press, 1979: 475–93.

47   Gabridge MG, Taylor-Robinson D, Davies HA, Dourmashkin RR. Interaction of *Mycoplasma pneumoniae* with human lung fibroblasts: characterization of the in vitro model. *Infect Immun* 1979, **25**, 446–54.

48   Gabridge MG. A review of the morphological and biochemical features of the attachment process in infections with *Mycoplasma pneumoniae*. *Rev Infect Dis* 1982, **4**, S179–84.

49   Hu PC, Collier AM, Baseman JB. Surface parasitism by *Mycoplasma pneumoniae* of respiratory epithelium. *J Exp Med* 1977, **145**, 1328–43.

50   Sobeslavsky O, Prescott B, Chanock RM. Adsorption of *Myco-*

plasma pneumoniae to neuraminic acid receptors of various cells and possible role in virulence. *J Bacteriol* 1968, **96**, 695–705.

51 Gabridge MG, Taylor-Robinson D. Interaction of *Mycoplasma pneumoniae* with human lung fibroblasts: role of receptor sites. *Infect Immun* 1979, **25**, 455–9.

52 Collier AM, Clyde WA, Denny FW. Biologic effects of *Mycoplasma pneumoniae* and other mycoplasmas from man on hamster tracheal organ culture. *Proc Exp Biol Med* 1969, **132**, 1153–8.

53 Collier AM, Clyde WA. Appearance of *Mycoplasma pneumoniae* in lungs of experimentally infected hamsters and sputum from patients with natural disease. *Am Rev Resp Dis* 1974, **110**, 765–73.

54 Camner P, Jarstrand C, Philipson K. Tracheobronchial clearance 5–15 months after infection with *Mycoplasma pneumoniae*. *Scand J Infect Dis* 1978, **10**, 33–5.

55 Yayoshi M. Association between *M. pneumoniae* hemolysis, attachment and pulmonary pathogenicity. *Yale J Biol Med* 1983, **56**, 685–9.

56 Brunner H, Bayer AG, Beck J, Kirchner H. Effect of *Mycoplasma pneumoniae* and some of its fractions on host defense mechanisms. *Yale J Biol Med* 1983, **56**, 863–4.

57 Biberfeld G, Biberfeld P, Sterner G. Cell-mediated immune response following *Mycoplasma pneumoniae* infection in Man. I. Lymphocyte stimulation. *Clin Exper Immunol* 1974, **17**, 29–41.

58 Hu PC, Huang C-H, Collier AM, Clyde WA Jr. Demonstration of antibodies to *Mycoplasma pneumoniae* attachment protein in human sera and respiratory secretions. *Infect Immun* 1983, **41**, 437–9.

59 Tully JG, Rose DL, Whitcomb RF, Wenzel RP. Enhanced isolation of *Mycoplasma pneumoniae* from throat washings with a newly modified culture medium. *J Infect Dis* 1979, **139**, 478–82.

60 Lind K. Isolation of *Mycoplasma pneumoniae* (Eaton agent) from patients with primary atypical pneumonia. *Acta Pathol Microbiol Scand* 1966, **66**, 124–34.

61 Smith CB, Friedewald WT, Chanock RM. Shedding of *Mycoplasma pneumoniae* after tetracycline and erthromycin therapy. *N Eng J Med* 1967, **276**, 1172–5.

62 Foy HM, Grayston JT, Kenny GE, Alexander ER, McMahan R. Epidemiology of *Mycoplasma pneumoniae* infection in families. *JAMA* 1966, **197**, 859–66.

63 Biberfeld G, Sterner G. Study of *Mycoplasma pneumoniae* infections in families. *Scand J Infect Dis* 1969, **1**, 39–46.

64 Menonna J, Chmel H, Menegus M, Dowling P, Cook S. Precipitating antibodies in mycoplasma infection. *J Clin Microbiol* 1977, **5**, 610–12.

65 Wiernik A, Jarstrand C, Tunevall G. The value of immunoelec-

tro-osmophoresis (IEOP) for the etiological diagnosis of acute respiratory tract infections due to pneumococci and *Mycoplasma pneumoniae*. *Scand J Infect Dis* 1978, **10**, 173–6.

66   Hers JFPh. Fluorescent antibody technique in respiratory viral diseases. *Am Rev Resp Dis* 1963, **88**, 3ii, 316–38.

67   Hers JFPh, Masurel N. Infection with *Mycoplasma pneumoniae* in civilians in the Netherlands. *Ann NY Acad Sci* 1967, **143**, 477–60.

68   Helbig JH, Witzleb W. Enzyme-linked immunosorbent assay (ELISA) zum Antigennachweis von *Mycoplasma pneumoniae*. *Z Gesamte Hyg* 1984, **30**, 106–17.

69   Weiner L, McMillan J, Poe L, Gidding S, Lamparella V, Patti A. *Mycoplasma pneumoniae* (Mpn) etiology of pharyngitis. Use of enzyme immunoassay (EIA) for rapid and direct detection of infection. Abstract 1177, *Ped Res* 1985, **19**, 307a.

70   Kok TW, Varkanis G, Marmion BP, Martin J, Esterman A. Laboratory diagnosis of *Mycoplasma pneumoniae* infection: 1. Direct detection of antigen in respiratory exudates by enzyme immunoassay. *Epidemiol Inf* 1988, **101**, 669–84.

71   Madsen RD, Weiner LB, McMillan JA, Saeed FA, North JA, Coates SR. Direct detection of *Mycoplasma pneumoniae* antigen in clinical specimens by a monoclonal antibody immunoblot assay. *Am J Clin Pathol* 1988, **89**, 95–9.

72   Dular R, Kajioka R, Kasatiya S. Comparison of Gen-Probe commercial kit and culture technique for the diagnosis of *Mycoplasma pneumoniae* infection. *J Clin Microbiol* 1988, **26**, 1068–9.

73   Saglie R, Cheng L, Sadighi R. Detection of *Mycoplasma pneumoniae*-DNA within diseased gingiva by in-situ hybridization using a biotin-labeled probe. *J Periodontol* 1988, **59**, 121–3.

74   Hyman HC, Yogev D, Razin S. DNA probes for detection and identification of *Mycoplasma pneumoniae* and *Mycoplasma genitalium*. *J Clin Microbiol* 1987, **25**, 726–8.

75   Bradstreet CMP, Taylor CED. Technique of complement-fixation test applicable to the diagnosis of virus diseases. *Monthly Bull Min Health Public Health Labo Service* 1962, **21**, 96–104.

76   Ponka A, Ponka T, Sarna S, Penttinen K. Questionable specificity of lipid antigen in *Mycoplasma pneumoniae* complement fixation test in patients with extrapulmonary manifestations. *J Infect* 1981, **3**, 332–8.

77   Räisänen SM, Suni JI, Leinikki P. Serological diagnosis of *Mycoplasma pneumoniae* infection by enzyme immunoassay. *J Clin Pathol* 1980, **33**, 836–40.

78   Marmion BP, Plackett P, Lemcke RM. Immunochemical analysis of *Mycoplasma pneumoniae*. I. Methods of extraction and reaction of fractions from *M pneumoniae* and from *M mycoides* with homologous antisera and with antisera against Streptococcus MG.

Aust J Exper Biology Med Sci 1967, **45**, 163–87.

79 Allen PZ, Prescott B. Immunochemical studies on a *Mycoplasma pneumoniae* polysaccharide fraction: cross-reactions with type 23 and 32 antipneumococcal rabbit sera. *Infect Immun* 1978, **20**, 421–9.

80 Costea N, Yakulis VJ, Heller P. The mechanism of induction of cold agglutinins by *Mycoplasma pneumoniae*. *J Immunol* 1971, **106**, 598–604.

81 Peterson OL, Ham TH, Finland M. Cold agglutinins (auto-hemagglutinins) in primary atypical pneumonias. *Science* 1943, **97**, 167–8.

82 Finland M, Peterson OL, Allen HE, Samper BA, Barnes MW. Cold agglutinins. II. Cold isohemagglutinins in primary atypical pneumonia of unknown etiology with a note of the occurrence of hemolytic anemia in these cases. *J Clin Invest* 1945, **24**, 458–73.

83 Biberfeld G. Antibody responses in *Mycoplasma pneumoniae* infection in relation to serum immunoglobulins, especially IgM. *Acta Pathol Microbiol Scand* Section B, 1971, **79**, 620–34.

84 Garrow DH. A rapid test for the presence of increased cold agglutinins. *BMJ* 1958, **ii**, 206–8.

85 Wreghitt TG, Sillis M. A *μ*-capture ELISA for detecting *Mycoplasma pneumoniae* IgM: comparison with indirect immunofluorescence and indirect ELISA. *J Hyg* Camb, 1985, **94**, 217–27.

86 Biberfeld G, Sterner G. Antibodies in bronchial secretions following natural infection with *Mycoplasma pneumoniae*. *Acta Pathol Microbiol Scand* Section B, 1971, **79**, 599–605.

87 Sillis M, Andrews BE. A simple test for *Mycoplasma pneumoniae* IgM. *Zentralbl Bakteriol [Orig A]* 1978, **241**, 239–40.

88 Skaug K, Eng J, Örstavik I, Haug KW. The diagnostic value of determination of IgM antibodies against *Mycoplasma pneumoniae* by the indirect immunofluorescent antibody test. *Acta Pathol Microbiol Scand* Section B, 1976, **84**, 170–6.

89 Umetsu M, Ogawa S, Chiba S, Nakao T. Immune responses in *Mycoplasma pneumoniae* infections. *Tohoku J Exp Med* 1975, **116**, 213–18.

90 Lind K. Preparation of antigen for the indirect fluorescent antibody test in diagnosis of *Mycoplasma pneumoniae* infection. *Acta Pathol Microbiol Scand* Section B, 1970, **78**, 149–52.

91 Purcell RH, Wong D, Chanock RM, Taylor-Ribonson D, Canchola J, Valdesuso J. Significance of antibody to mycoplasma as measured by metabolic-inhibition techniques. *Ann NY Acad Sci* 1967, **143**, 664–75.

92 Busolo F, Tonin E, Conventi L. Enzyme-linked immunosorbent assay for detection of *Mycoplasma pneumoniae* antibodies. *J Clin Microbiol* 1980, **12**, 69–73.

93   Busolo F, Tonin E, Meloni GA. Enzyme-linked immunosorbent assay for serodiagnosis of *Mycoplasma pneumoniae* infections. *J Clinical Microbiology* 1983, **18**, 432–5.

94   Busolo F, Meloni GA. Serodiagnosis of *M. pneumoniae* infections by enzyme-linked immunosorbent assay (ELISA). *Yale J Biol Med* 1983, **56**, 517–21.

95   Brunner H, James WD, Horswood RL, Chanock RM. Measurement of *Mycoplasma pneumoniae* mycoplasmacidal antibody in human serum. *J Immunol* 1972, **108**, 1491–8.

96   Brunner H, Schaeg W, Brück, U, Schummer U, Sziegoleit D, Schiefer HG. Determination of IgG, IgM and IgA antibodies to *Mycoplasma pneumoniae* by an indirect staphylococcal radioimmunoassay. *Med Microbiol Immunol* 1978, **165**, 29–41.

97   Price PC. Direct radioimmunoassay for the detection of IgM antibodies against *Mycoplasma pneumoniae*. *J Immunol Met* 1980, **32**, 261–73.

98   Brunner H, Schaeg W, Brück U, Schummer U, Schiefer HG. A staphylococcal radioimmunoassay for detection of antibodies to *Mycoplasma pneumoniae*. *Med Microbiol Immunol* 1977, **163**, 25–35.

99   Dowdle WR, Robinson RQ. An indirect hemagglutination test for diagnosis of *Mycoplasma pneumoniae* infections. *Proc Soc Exper Biol Med* 1964, **116**, 947–50.

100   Lind K. An indirect haemagglutination test for serum antibodies against *Mycoplasma pneumoniae* using formalinized, tanned sheep erythrocytes. *Acta Pathol Microbiol Scand* 1968, **73**, 459–72.

101   Ghyka GR, Stoian N, Sorodac G, Peiulescu P, Osman J, Puca D. An antigen for the diagnosis of *M. pneumoniae* infections by passive hemagglutination. *Virol* 1977, **28**, 111–15.

102   Taylor P. Evaluation of an indirect haemagglutination kit for the rapid serological diagnosis of *Mycoplasma pneumoniae* infections. *J Clin Pathol* 1979, **32**, 280–3.

103   Rousseau SA, Tettmar RE. The serological diagnosis of *Mycoplasma pneumoniae* infection: a comparison of complement fixation, haemagglutination and immunofluorescence. *J Hyg* Camb, 1985, **95**, 345–52.

104   Barker CEH, Wreghitt TG, Sillis M. Evaluation of the Serodia-Myco II gel particle agglutination test for selecting *Mycoplasma pneumoniae* antibody: comparison with μ-capture ELISA and indirect immunofluorescence. *J Clin Pathol* 1990, **43**, 163–5.

105   Wreghitt TG, Sillis M. An investigation of the *Mycoplasma pneumoniae* infections in Cambridge in 1983 using μ-capture enzyme-linked immunosorbent assay (ELISA), indirect immunofluorescence (IF) and complement fixation (CF) tests. *Isr J Med Sci* 1987, **23**, 704–8.

106 van Griethuysen AJA, de Graaf R, van Druten JAM, Heessen FWA, van der Logt JTM, van Loon AM. Use of the enzyme-linked immunosorbent assay for the early diagnosis of *Mycoplasma pneumoniae* infection. *Eur J Clin Microbiol* 1984, **3**, 116–21.

107 Mizutani H, Mizutani H. Immunologic responses in patients with *Mycoplasma pneumoniae* infections. *Am Rev Resp Dis* 1983, **127**, 175–9.

108 Dussaix E, Slim A, Tournier P. Comparison of enzyme-linked immunosorbent assay (ELISA) and complement fixation test for detection of *Mycoplasma pneumoniae* antibodies. *J Clin Pathol* 1983, **36**, 228–32.

109 Miyaji T. Determination of antibody to *Mycoplasma pneumoniae* by enzyme-linked immunosorbent assay. I. Factors in the assay. *Sci Rep Res Inst* Tohoku Univ (Med) 1984, **31**, 1–12.

110 Miyaji T. Determination of antibody to *Mycoplasma pneumoniae* by enzyme-linked immunosorbent assay. II. Serum antibody in patients with *Mycoplasma pneumoniae* infections. *Sci Rep Res Inst* Tohoku Univ (Med) 1984, **31**, 13–26.

111 Coombs RRA, Easter G, Matejtschuk P, Wreghitt TG. Red-cell IgM-antibody capture assay for the detection of *Mycoplasma pneumoniae*-specific IgM. *Epidemiol Inf* 1988, **100**, 101–9.

112 Grayston JT, Alexander ER, Kenny GE, Clarke ER, Fremont JC, MacColl WA. *Mycoplasma pneumoniae* infections. Clinical and epidemiological studies. *JAMA* 1965, **191**, 369–74.

113 Chamberlain P. Saeed AA. A study of the specific IgM response in *Mycoplasma pneumoniae* infection in man. *J Hyg* Camb, 1983, **90**, 207–11.

114 Moule JH, Caul EO, Wreghitt TG. The specific IgM response to *Mycoplasma pneumoniae* infection: interpretation and application to early diagnosis. *Epidemiol Infect* 1987, **99**, 685–92.

Chapter 13

Microbiology in the National Blood Transfusion Service

J Barbara

In this chapter I do not intend to describe in detail the series of infections which can be transmitted by transfusion. These have been covered in various reviews and were the subject of two recent articles [1,2]. Instead I hope to provide an overall view of the kinds of problems encountered in transfusion microbiology and some approaches to solving them. Transfusion microbiology has only recently emerged as a comprehensive discipline in its own right [3,4]; it was previously regarded more as a sometimes troublesome side effect of transfusion, to be dealt with in a corner of the blood-grouping laboratory. However, the increasing range and understanding of transfusion-transmitted infections (some, like HIV infection, with very dramatic consequences) and the growing complexity of laboratory testing methods have led to the increased prominence of transfusion microbiology. In addition, the Transfusion Service is now subject to product liability, so that as a supplier of blood to hospitals it is directly responsible for the safety of the blood which must be 'fit for the purpose' for which it was supplied.

The main concerns of transfusion microbiology are as follows.

- Maintaining a safe blood supply in relation to infectious agents by: donor education, so that those at high risk of contracting infections such as HIV will exclude themselves from donating; serological screening of blood donations to exclude potentially infectious microbial agents and the quality control of this screening; and investigating reports of infections in donors which have manifested themselves shortly after donation since transmission to recipients is possible if the donor was infectious at the time of donating; follow-up of reports of likely transfusion-transmitted infection and attempting to identify infectious donors with supplementary, non-routine tests to prevent further transmissions; bacteriological quality control and environmental monitoring; inactivation of viruses in products manufactured from plasma pools.

- Selection of donors with high levels of antibody against HBV, herpes varicella-zoster, CMV and tetanus so that specific immuno-globulins can be prepared at the Bio-Products Laboratory, Elstree.
- Research into: epidemiological aspects of infections transmitted by transfusion; methods for inactivating viruses in pooled-plasma products; and the development, modification and assessment of methods for screening for microbial agents important in transfusion.

Maintaining a safe blood supply

DONOR EDUCATION

The realisation that certain 'high-risk' activities such as intravenous drug use and male homosexuality increase the risk of contracting HIV infection allowed identification of minority groups of potential donors who could be asked to exclude themselves from giving blood. Where heterosexual transmission of HIV is common (eg Central Africa) this strategy is obviously not applicable. In the UK, the risk groups are specified in leaflets prepared by the Department of Health which are given to each donor. These people must not give blood:

- Men and women who know they are infected with the AIDS virus or who have AIDS.
- Men who have had sex with another man at any time since 1977.
- Men and women who have injected themselves with drugs at any time since 1977.
- Men and women who have had sex at any time since 1977 with men or women living in African countries, except those on the Mediterranean.
- Men and women who have had sex with anyone in these groups. Sexual partners of haemophiliacs.
- Men and women who are prostitutes.

In addition, at the North London Blood Transfusion Centre (NLBTC) which serves the North West Thames Region (where by the end of 1988 41% of all UK AIDS cases had been reported), a confidential exclusion option is provided. Donors are given a form which specifies the risk groups, on which they can indicate that their blood should not be used for transfusion purposes. This is completed in a private booth and allows the donor to exclude his or her donation from clinical use, saving the embarrassment of walking away without donating; sometimes family or peer pressure makes it impossible for high-risk donors to avoid attending the donor clinic. Overall, donor education can be calculated to be more than 95% successful when the potential number of donors positive for anti-HIV in the absence of 'self-exclusion' is compared with the actual statistics for donors found anti-HIV positive. These statistics are collated nationally at the Manchester Transfusion Centre in conjunction with the PHLS Communicable Disease Surveillance Centre. Between the start of routine screening in October 1985 and early 1989 more than eight million donations had been tested (the majority from donors who give blood repeatedly), and the overall confirmed anti-HIV positivity rate in donors was 1 in 73,000. For first-time or previously-untested donors (constituting roughly one fifth of the donor population), the rate was 1 in 27,000. Altogether 113 donors (83% male) had been confirmed anti-HIV positive and most were in recognised risk groups, mainly male homosexuals and intravenous drug users. The numbers of donors found anti-HIV positive declined steadily, probably because of the continued effects of donor education, self-exclusion by at-risk donors and the ready availability of alternative testing sites. In the UK, the overall prevalence was 1 in 50,000 in 1986, 1 in 108,000 in 1987 and 1 in 120,000 in 1988. For first-time donors, the corresponding rates were 1 in 17,000, 1 in 29,000 and 1 in 53,000. However, in 1990 the rate in first-time donors had risen to 1 in 20,000.

Another important function performed at the Manchester Transfusion Centre, in conjunction with the Central Public Health Laboratory, Colindale, is to compile data on the performance and quality control aspects of anti-HIV screening performed by the transfusion service. The Division of Microbiological Quality Control and Reagents at Colindale provides low-level anti-HIV sera at the limits of detectability for frequent daily quality control of assays at each centre. This has allowed integrated and extensive quality monitoring of anti-HIV screening in the Blood Transfusion Service (BTS) at a national level and is an encouraging example of the value of cooperation between the BTS and the PHLS. Trends in false positivity can also be assessed. The comparatively few repeatably-reactive sera detected in screening are sent to reference laboratories for further investigation. If there is any doubt about the HIV status of the donor, samples from the next donation are tested at the reference laboratory. If considered negative, that donation would not be used but the subsequent donation would be released for transfusion, provided it is found to be anti-HIV negative and a suitable time interval has elapsed since the first 'suspicious' result.

SEROLOGICAL SCREENING

The large range of microbial agents with potential for transmission by transfusion is shown in Table 1.

Many such agents are also transmissible sexually or by intravenous drug use and a donor at risk of being infected and transmitting one infection is often also at risk of infection by some of the others as well. It has been observed at NLBTC that with the exclusion of male homosexual blood donors (because of HIV risk), there has been a concomitant decrease in the number of donors found HBsAg positive. From an average of approximately 50 HBsAg positive donors in 170,000

Table 1   Agents transmissible by blood transfusion

Bacteria
  Syphilis
  Brucellosis: donors with a history are not accepted in the UK.
Parasites
  Malaria
  Trypanasoma cruzi: Chagas' disease, endemic in Latin America. The parasite is present in 75% of seropositive individuals. Between 1% and 22% of donors in Latin America are seropositive
  Toxoplasma gondii: only a risk with granulocytes transfused to immunosuppressed recipients
  Babesia microti: Nantucket fever; potential transfusion risk in North America.
Plasma-borne viruses
  Hepatitis B (HBV and ?HBV2) and delta agent
  Hepatitis A (HAV): rarely transmitted
  Hepatitis non-A, non-B (NANBH) – hepatitis C virus (and others?)
  Serum parvovirus B19
  HIV 1 and HIV 2 (causative agents of AIDS), the human immunodeficiency viruses (also cell-associated).
Cell-associated viruses
  Cytomegalovirus (CMV)
  Epstein-Barr virus (EBV): >95% of adults are immune
  HTLV-I, causing human T-cell leukaemia or tropical spastic paraparesis.
  HTLV-II.

On rare occasions, Colorado tick fever virus and *Coxiella burnetii* (Q fever) can cause transfusion complications.

donations per annum prior to AIDS publicity, only 31 donors were found to be HBsAg-positive in 1987, 40 in 1989 and 23 in 1990. Even more striking has been the marked reduction of acutely-infected HBsAg positive donors, where previously 16% of HBsAg-positive donors were undergoing acute infections [5]. Another example of this phenomenon relates to the decrease in non-A, non-B hepatitis (NANBH) after transfusion in the USA. This decrease followed the introduction of anti-HIV screening (with the associated publicity for blood donors) and, interestingly, preceded the introduction of non-specific (anti-HBc and alanine aminotransferase) screening of donations as surrogate tests for NANBH infectivity [6]. The decrease is likely to have been due in part at least to the change in the blood donor base following exclusion of donors at risk of transmitting HIV.

  In the UK there have been no recent extensive studies to assess the extent of post-transfusion NANBH prospectively. A large study in the early 1970s and a later small survey suggest a figure of approximately 2% NANBH after transfusion. A prospective study at the North London Blood Transfusion Centre has so far confirmed that the rate is currently even lower (0.5%) [7].

Microbial agents transmitted by transfusion usually have several of the following characteristics:

- long incubation periods
- often cause asymptomatic infections
- persistent and/or high level viraemia
- expressed carrier states (eg HBsAg)
- latent states (eg the herpes viruses)
- stability in stored blood, components or products.

The persistent nature of the infections often means that antibody to the agent may be detectable, concomitant with infectivity in either plasma or cellular components. Thus many of the screening tests for transfusion-transmitted infections utilise detection of antibodies to the agent, rather than for the agent itself. The notable exception is HBsAg produced during HBV infection. In this case, the virus produces a vast excess of surface antigen allowing detectability over a large range of titre, which may be up to six logarithms. In contrast, the search for a test for the NANBH viral antigen had never been successful. Testing for HIV p24 antigen, even at picogram sensitivity, can cover a range of up to three logarithms at most, resulting in our current reliance on anti-HIV assays for the screening of blood donations prior to transfusion. The p24 antigen assay scores negative throughout most of the course of HIV infection, although anti-HIV-positive donations are almost invariably infectious. However, HIV antigen assays do seem useful when assessing the prognosis of infection (especially in association with β-2 microglobulin assays) and in monitoring drug treatment. In the context of screening blood donations, testing for anti-HIV remains the single most reliable assay for maintaining the safety of the blood supply with regard to HIV.

The strategy for screening for agents is as follows.

Hepatitis B virus: BsAg assay. Anti-HBc has also been suggested for screening, in a similar fashion to antibody tests for other agents. However, anti-HBc is absent early in acute HBV infection [8] when infectious virus is sufficiently abundant to contaminate large pools of plasma and the clotting factor concentrates made from them.

Hepatitis A virus: Rarely transmits [3]; there is no carrier state and donated blood is not screened for this agent.

Non A, non B hepatitis (hepatitis C): Although the virus has not been isolated, recent genetic cloning experiments have allowed the development of a screening test for antibody to one of the NANBH agents

(HCV); this test looks very promising but so far detects only IgG and not IgM classes of anti-HCV [9,10]. 'First-generation' assays employed HIV antigen cloned from the non-structural region of the virus. 'Second-generation' assays incorporating additional structural antigen are about to be released. These antigens also feature in recombinant-immunoblot strips or blocking arrays for supplementary testing. Non-specific tests for NANBH (ALT and anti-HBc) are not required prior to transfusion in the UK but a national study of these markers in blood donors is in progress. These studies, together with an assessment of the HCV assay, will be taken into account when formulating policies. The polymerase chain reaction has also proven to be a valuable research tool in HCV studies.

Human immunodeficiency viruses: As mentioned previously, screening is for antibody. The infection can be transmitted by transfusion of cellular components or plasma. Current screening tests are based on the detection of anti-HIV 1, with a 20% to 80% chance of detecting anti-HIV 2, depending on which test is used. In the UK, all donations from donors with West African associations are also tested for anti-HIV 2 at the Central Public Health Laboratory, Colindale. Because of this, the small number of UK identifications of HIV 2 infections in patients (five to date) and the very low rate of anti-HIV positivity in UK blood donors, routine screening for anti-HIV 2 is not undertaken in the UK. The recent availability of combined tests for anti-HIV 1 and 2 is one factor currently being taken into account in the constant monitoring of the situation. If such tests prove as sensitive, specific and reliable as the existing assays for anti-HIV, the options for screening may widen. At the moment, however, combined assays for anti-HIV 1 and 2 are still under evaluation.*

Human T-cell leukaemia viruses: Assays for antibody to HTLV-I (which causes human T-cell leukaemia or tropical spastic paraparesis) have been licensed in the USA and screening of blood donations began there in early 1989. Again, antibody is taken as an index of continuing infectivity. Infection by transfusion is restricted to components containing white blood cells. Worldwide, no cases of leukaemia have so far been reported after transfusion-transmitted infection, although tropical spastic paraparesis has been seen after transfusion. Although under constant review, anti-HTLV-I screening of blood donations is not thought to be justified currently in the UK. A six-month trial in which 100,000 blood donors will be screened for anti-HTLV started in January 1991 at the NLBTC. It is noteworthy that evidence is accumulating that

*Combined testing for anti-HIV 1 and 2 was introduced in the UK in mid 1990, following the detection (by cross-reaction) of a donor positive for anti-HIV 2. West African-associated anti-HIV 2 screening of donors at the Central Public Health Laboratory has therefore ceased.

much of what has been identified as anti-HTLV-I may well be anti-HTLV-II.

Herpes viruses: Epstein-Barr virus infection is associated with transfusion but most adult recipients have evidence of infection prior to any transfusion. Cytomegalovirus infection after transfusion is more common, but is only potentially serious (and sometimes even fatal) in immunosuppressed recipients. These include low birth weight neonates and transplant recipients, especially bone-marrow transplant cases. Approximately 50% of the UK adult population have antibody to CMV and are therefore considered capable of transmitting the infection through reactivation of virus latent in white cells. Reactivation of CMV latent in the *recipient's* own white cells can also occur after transfusion of white cell components, even if the donor is seronegative. Any seropositive donor is considered to have the potential to transmit CMV, although those most recently infected appear to be more likely to do so.

The role in transfusion of the recently described human herpes virus 6, if any, is not yet clear.

Serum parvovirus: The human serum parvovirus (B19) has been shown to be transmitted by uninactivated products prepared from pooled plasma [11] despite the short viraemia and lack of a carrier state in immunocompetent persons. When the virus is present in the blood, it is at extremely high titres and pooling plasma from several thousand donations enhances the likelihood of contamination of the starting material for Factor VIII preparation. British Factor VIII (heated at 80°C for 72h in the freeze-dried state) appears not to transmit the virus [12]. Since most infections are very mild, screening of blood donations is not indicated; this is fortunate since antibody assays would be of no value for this non-persistent infection and assays for the virus appropriate for transfusion screening may have insufficient sensitivity.

Parasites: Screening tests for antibody to malaria parasites (*Plasmodium* species), *Trypanosoma cruzi* (Chagas' disease), *Toxoplasma gondii* (toxoplasmosis) and *Babesia microti* (Nantucket fever) are available and are of potential value because of the persistent nature of these infections. However, donors likely to be infected by these agents are usually screened by appropriate history taking. *T cruzi* is restricted to Latin America and *B microti* to North America. ELISA is available for detecting antibody to *Plasmodium falciparum* (the most serious of the malarial infections) and cross-reacts quite well with *P vivax* [13]. The use of such malarial screening for selected blood donors is currently being considered in the UK. Toxoplasma-antibody negative donations would only be indicated for the provision of white-cell components to immunosuppressed seronegative recipients. Transmission has been documented in such patients.

Syphilis: Blood donations are screened for antibody to syphilis, but virtually no cases of transfusion-transmitted syphilis have been reported for several years [3]. Presence of antibody does, however, act as a marker for potential exposure to other sexually-transmitted diseases and in several UK transfusion centres the more specific *Treponema pallidum* haemagglutination assay [3] has replaced cardiolipin-based methods of screening. Donors found to have evidence of past or recent syphilis infection are informed after suitable confirmation of the findings and advised to consult their general practitioners to ensure that adequate treatment has been undertaken.

OTHER BACTERIA TRANSMITTED BY TRANSFUSION

Bacteria such as *Pseudomonas* and *Yersinia* species are occasionally transmitted by transfusion, causing endotoxaemia which has a high mortality. Such psychrophilic bacteria can multiply even at refrigerator temperatures. Furthermore, new techniques for prolonging the life of platelet preparations which are stored at 22°C have been associated in the USA with an increase in bacterial complications after platelet therapy. Transmission of *Yersina enterocolitica* after transfusion has also been reported in the UK [14]. Laboratory screening tests are impractical but care in cleansing the donor's arm and in meticulous quality control of blood collection, processing and storage is vital to minimise such occurrences. Environmental quality control and some product testing (especially of blood components prepared in 'part-open' systems) may assume increasing importance. Prior to transfusion, all units of blood should be checked for evidence of damage to the bag or signs of clotting or haemolysis, indicative of bacterial contamination.

Problems associated with pooling large numbers of plasma donations

Plasma pooling enhances the potential of several agents (HBV, NANBH, HIV, parvovirus) for dissemination by transfusion. Indeed, the viral aetiology of AIDS was corroborated by the reports of its transmission by Factor VIII. Haemophiliacs show high rates of infection by these agents when treated with non-inactivated Factor VIII. In future, cloned Factor VIII should solve this problem of infection; in the meantime, improved purification and viral inactivation techniques (either for the raw material or the freeze-dried product) are being developed. Heating seems effective for some agents [15] and considerable research is being devoted to the use of detergents and/or ultraviolet or gamma irradiation.

Follow-up of transfusion-transmitted infections

The most convincing proof of the effectiveness of policies to prevent transmission of infections by transfusion is the elimination of such cases. It is therefore vital that hospitals report any infections in recipients likely to have been caused by transfusion. Transfusion centres can then investigate the donors involved by checking stored blood samples or by testing fresh ones. Standard screening tests together with specialised assays for the agents or associated markers can then be undertaken. This action could well prevent further infections, provides a check on current methodologies and might also reveal shortcomings in the quality control of blood processing, so that steps can be taken to improve the system. The advent of product-liability has given added weight to the development of the use of bar-coded labels, computerised control of sample identification and direct interfacing with the output from the machines used to read the ELISA, RIA and even haemagglutination assays. Together with a recent surge in the development of automated sampling equipment featuring minimal sample carry-over, these approaches should minimise the chances of laboratory and clerical errors. When 1,000 samples are tested daily for a range of agents, the chances for error are obvious, especially if the blood or components are required for issue on the day of donation. HBsAg, anti-HIV and syphilis screening is mandatory. Several centres also screen a proportion of their donors for anti-CMV and some test donors at risk of malarial exposure for antibodies to *Plasmodium falciparum*. If anti-HBc, alanine amino-transferase, anti-HIV 2, anti-HTLV-I and anti-HCV is added to the list, the problems are obviously magnified. In the USA, there have been several cases of erroneous release of blood donations which had been found 'positive' for various markers, due to overloading of the system [16]. No HBsAg or anti-HIV positive units were involved, but the reports highlight the importance of developing logistical and quality control techniques in step with the screening methods to cope with the increasing complexity.

Any reports of onset of infection in blood donors immediately sub-sequent to donation must also be carefully followed up because of the persistent nature of transfusion-transmitted infections. For example, hepatitis B virus infection may render a donor infectious even though HBsAg might not be detectable several weeks prior to the onset of symptoms [17]. In such cases, infections in recipients are a distinct possibility, and clinical staff in the hospital would find prompt notification advantageous.

Collection of immune plasma

Normal immunoglobulin prepared from pools of random donor plasma contains sufficient anti-HAV to protect against hepatitis A infection. However, donors who have been shown by screening to have high levels of antibody to specific infections such as HBV, tetanus, CMV and herpes varicella-zoster can be selected for plasmapheresis and their plasma sent to the Bio Products Laboratory to prepare specific immune globulin. A series of standard preparations is available from the Bio Products Laboratory containing the minimum levels of antibody to the different specificities which are intended to define suitable 'high' levels of antibody when screening for immune plasma. The suitability of the different assays for high-titre antibodies is also being reviewed. The extent of screening for immune plasma varies between transfusion centres but if simple and economical assays suitable for detecting high-titre immune plasma become available it is hoped that more centres will be able to contribute to the provision of specific immune plasma.

Many questions remain unanswered in the field of specific immuno-globulins. Which antibodies to a given virus are the most protective? What is the relative importance of IgG versus IgM antibodies? Which are the best tests for this kind of screening? Unfortunately, the Transfusion Service receives very little feedback from hospitals on the efficacy of the products, so that some queries have been impossible to resolve.

Research and development

There is an enormous potential for research and development in the field of transfusion microbiology. When new transfusion-transmitted infections are identified, stored samples from high-risk (multiply-transfused) recipients are particularly useful to gain an understanding of the characteristics of transmission. Interesting aspects can sometimes be revealed. One example was the detection of transient, passively-acquired antibody to HIV in a patient transfused with anti-HIV positive platelets (before donation screening was initiated), followed subsequently by persistent seropositivity [18]. Long-term storage of donor sera, where possible, can also be valuable epidemiologically. General epidemiological characteristics of infections can also be gleaned from the mass testing that donor screening represents; but the advent of self-exclusion for HIV infection has meant that blood donor populations no longer mirror the general population as closely as they used to. Thus HBsAg carrier rates in blood donors used to be more closely related to overall population statistics than are anti-HIV positivity rates now. Nevertheless, the donor population remains an excellent sentinel for the spread of HIV into lower-risk groups such as heterosexuals. Again, transfusion-transmission can provide data about the character of a disease (such as AIDS) and accurate information about incubation periods because the time of infection can be defined exactly. Where there are facilities for continual follow-up of positive donors (such as HBsAg carriers), useful epidemiological information and a supply of reagents for assay development, quality control and even vaccine production can be made available.

Another huge area of research involves the development and assessment of screening tests for the various agents so that sensitivity and specificity can be improved.

Conclusion

The field of transfusion microbiology is a busy and expanding one. As techniques of medical intervention become more sophisticated, the requests for blood and blood products increase and the requirement for safe blood assumes greater importance. The Transfusion Service must face new challenges in terms of the speed and precision of the assays employed. The association with reference laboratories in confirming positive results has become more important as sensitivity of assays increase, with the concomitant demand for specificity, especially in low-prevalence populations. Unwarranted blood donor anxiety must be kept to a minimum, in the face of an ever-growing list of agents to be excluded from transmission by transfusion, if adequate blood supplies are to be maintained. In future, reference laboratories may well increase their use of more sophisticated technologies, such as polymerase chain reaction, as 'definitive' ultra-sensitive confirmatory tests. This will lead to further problems with specificity and interpretation and will require maximum cooperation between screening and reference laboratories.

Acknowledgements

I am grateful to Miss Marina Mobed for preparation of the manuscript. Dr H Gunson and Miss V Rawlinson provided data on anti-HIV positivity rates.

This chapter is based on an article which appeared in the *PHLS Microbiology Digest* 1990, **7**, 4–7.

References

1   Barbara JAJ, Contreras M. Infectious complications of blood transfusion: bacteria and parasites. *BMJ* 1990, **300**, 386–9.

2   Barbara JAJ, Contreras M. Infectious complications of blood transfusion: viruses. *BMJ* 1990, **300**, 450–3.

3   Barbara JAJ. Microbiology in blood transfusion. Bristol: Wright, 1983.

4   Tabor E. Infectious complications of blood transfusion. New York: Academic Press, 1982.

5   Barbara JAJ, Briggs M. Follow up of HBsAg positive donors to determine the proportion undergoing acute infections. *Transfusion* 1981, **21**, 605–6.

6   Alter M. In transcript of FDA Workshop on surrogate testing on non A, non B hepatitis (DHSS, DHS, FDA). Bethesda, Md. 1987, vol 1, part 1, 30–1.

7   Contreras M, Barbara JAJ, Anderson CC *et al.* Low incidence of non-A, non-B post transfusion hepatitis in London confirmed by hepatitis C virus serology. *Lancet* 1991, **337**, 753–57.

8   Barbara JAJ, Tedder RS, Briggs M. Anti-HBc testing alone not a reliable blood donor screen. *Lancet* 1984, **i**, 346.

9   Choo Q-L, Kuo G, Weiner AJ, Overby LR, Bradley DW, Houghton M. Isolation of a cDNA clone derived from a blood-borne non-A, non-B viral hepatitis genome. *Science* 1989, **244**, 359–62.

10 Kuo G, Choo Q-L, Alter HJ *et al*. An assay for circulating antibodies to a major etiologic virus of human non-A, non-B hepatitis. *Science* 1989, **244**, 362–4.

11 Mortimer PP, Luban NLC, Kelleher JF, Cohen BJ. Transmission of serum parvovirus-like virus by clotting-factor concentrates. *Lancet* 1983, **ii**, 482–4.

12 Mortimer PP, Cohen BJ, Williams MD, Hill FGH, Pasi J. Transmission of parvovirus B19 by factor concentrate. Abstr 0-F-2-5, 272. In: Educational Book 1988: the xx congress of the International Society of Blood Transfusion in association with the British Blood Transfusion Society. British Blood Transfusion Society, 1988, 27.

13 Wells L, Ala FA. Malaria and blood transfusion. *Lancet* 1985, **i**, 1317–9.

14 Mitchell R, Barr A. Transfusion reaction due to *Yersina enterocolitica*. *Comm Dis* Scotland, 1988, **88**, 4.

15 Study group: UK haemophilia Centre Directors. Effect of dry-heating of coagulation factor concentrates at 80°C for 72 hours on transmission on non-A, non-B hepatitis. *Lancet* 1988, **ii**, 814–6.

16 Blood recalled; FDA announces annual inspections. American Association of Blood Banks: *News Briefs* May 1988, **11**, (5), 1.

17 Barbara JAJ, Briggs M. A chain of infection with hepatitis B virus. *Comm Dis Rep* 1980, **43**, 40.

18 Apperley JF, Rice SJ, Hewitt P *et al*. HIV infection due to platelet transfusion after allogeneic bone-marrow transplantation. *Eur J Haem* 1987, **39**, 185–9.

Chapter 14

Quality assessment in virology

JM Hawkins

External quality assessment (EQA) has during recent years been recognised as an essential component of quality assurance, by which a diagnostic laboratory ensures that the service it provides is of a consistently high standard [1]. The internal quality control of laboratory procedures, equipment and reagents by continual monitoring gives laboratory personnel confidence in the results of tests performed routinely on clinical specimens; however, only EQA, by providing specimens of known content to be tested blind by participating laboratories, can give the laboratory manager an independent means of ensuring that his routine quality control is adequate and effective. EQA also provides a means to monitor the overall standard of diagnostic laboratory performance.

The National External Quality Assessment Scheme (NEQAS) for virology forms part of UKNEQAS for microbiology, which is administered by the Public Health Laboratory Service (PHLS) from the Central Public Health Laboratory. The scheme is managed by the organiser, supported by two sources of expert advice and comment, the PHLS NEQAS Steering Committee and the National Quality Assurance Advisory Panel for Microbiology; the major function of the panel is to offer advice to laboratories whose performance is persistently below average. The virology scheme has been developed over a number of years to provide EQA in the major disciplines of diagnostic virology, and a brief history is outlined below.

1975: The first distributions of simulated specimens for virus isolation and serological tests were made [2].

1980: Distributions of specimens in five separate categories had been developed, comprising virus isolation, rubella serology (IgG detection), hepatitis B surface antigen (HBsAg) detection, electron microscopy and general virus serology [3].

1983: Specimen production and quality control methods were improved. Methods of scoring participants' results had been established, and the analysis of participants' performance begun. The storage and analysis of results and the production of individual laboratory reports were computerised. Simulated specimens for the detection of rubella IgM were distributed on a trial basis [3].

1986: Simulated specimens for the detection of rubella IgM were being distributed on a regular basis and had become a formal part of the scheme. Simulated specimens for the detection of antibody to human immunodeficiency virus (anti-HIV) were introduced.

1988: The scheme had expanded to provide the maximum possible service with the staff and resources available. Work had commenced on the development of simulated specimens for the detection of *Chlamydia trachomatis*, and on improving and rationalising existing distribution types. Distributions of sera for anti-HIV detection were well established.

Table 1 shows the numbers of laboratories currently participating in the scheme including Public Health, National Health Service, Blood Transfusion, private and armed forces laboratories, and those of some commercial firms. All private laboratories known to the scheme organiser are offered recruitment, but as there is no official register of such laboratories, it is likely that some operate without any independent assessment of their standard of service. The recruitment of further foreign participants has been restricted to those who intend to develop a quality assessment scheme for their own country: limited resources dictate that we must concentrate on the UK scheme if further improvements are to be made. Table 2 shows the numbers of distributions made during the period 1984–88.

The operation of the scheme

Full details of the operation of the virology scheme are described elsewhere [3,4], but a brief resumé seems appropriate here, together

Table 1 Participating laboratories

Distribution types	UK laboratories	Foreign laboratories	Total
Hepatitis B serology	262	40	302
HIV serology	200	50	250
Rubella serology (IgG)	196	44	240
Rubella serology (IgM)	101	38	139
General virus serology	95	30	125
Virus isolation	88	16	104
Electron microscopy	61	12	73

Table 2   Numbers of distributions, 1984–88

Distribution type	1984	1985	1986	1987	1988
Hepatitis B serology	2(10)	2(10)	2(11)	2(12)	2(10)
HIV serology	–	–	1(6)	3(18)	3(16)
Rubella serology (IgG)	2(8)	2(8)	2(10)	1(6)	1(4)
Rubella serology (IgM)	–	1(3)	1(3)	2(6)	2(6)
General virus serology	2(8)	2(7)	2(8)	1(4)	1(4)
Virus isolation	2(7)	2(6)	2(7)	3(16)	2(10)
Electron microscopy	1(4)	–	2(10)	2(9)	1(4)
Total	9(37)	9(34)	12(55)	14(71)	12(54)

Numbers in parentheses are numbers of specimens examined

with details of the characterisation of specimens for the two distribution types recently developed.

SEQUENCE OF EVENTS

1   Specimen preparation, with detailed characterisation and quality control.
2   Distribution of specimens to participants together with request/ report forms.
3   Post-distribution quality control of specimens.
4   Issue of intended results so that participants can take investigative action quickly if their results are discrepant.
5   Computer entry of participants' results and method details, allocation of scores and analysis of this data.
6   Production of individual computer reports, including analysis of individual performance compared with that of other laboratories.
7   Production of distribution summary, including the overall results of all UK participants, methodological data and educational comment where appropriate.
8   Six-monthly analysis of participants' performance over a retrospective twelve-month period. Results which fall significantly below average are presented to the National Quality Assurance Advisory Panel for Microbiology, who can offer confidential advice to laboratories concerned.

SPECIMEN CHARACTERISATION FOR RUBELLA IgM AND ANTI-HIV DETECTION

Pre-distribution characterisation of these two types of specimen has been particularly extensive because the current lack of national reference preparations has precluded precise calibration, which has been possible with specimens for rubella IgG and HBsAg detection. For rubella IgM, work is in progress on the production and standardisation of a national reference preparation; for HIV serology, the PHLS does distribute 'reference' sera, but these are primarily intended for routine inclusion as extra control sera independent of commercial assay controls, and are not stricly reference preparations. Meanwhile, so that participants may have more confidence in the authenticity and validity of the NEQAS specimens, predistribution assays include calibration by expert reference laboratories and testing by the most commonly used commercial assays and by in-house techniques. A summary of tests carried out is shown in Table 3.

How EQA can help participants

Diagnostic virology, especially virological serology, has progressed considerably in recent years, both in the range of tests available and in the complexity of the technology employed. Test sensitivity and specificity have been greatly improved, but it may be that laboratories still pay insufficient attention to the routine quality control of modern assays. Effective quality control procedures, such as those practised by clinical chemistry laboratories, are time-consuming and costly to carry out; but there are several reasons for their desirability in diagnostic virology.

1 The extensive use of enzyme immunoassays: despite their improved sensitivity and specificity, both in-house and commercial immunoassays need continual monitoring of within-batch and between-batch performance by the inclusion of independently-standardised control material.

2 Increased reliance on automation and instrumentation: faults and inaccuracies can go unnoticed unless assays are properly monitored.

3 The greater complexity of the work situation and heavier workloads

Table 3 Pre-distribution tests

	Rubella IgM	anti-HIV
Expert/reference laboratory methods	IgM capture ELISA IgM capture RIA	Competitive RIA IgG capture RIA
Commercial assays	IgM capture ELISA antiglobulin ELISA	Competitive ELISA Indirect ELISA Gel particle agglutination Western Blot
In-house assays	IgM capture ELISA Sucrose density gradient/HI	Competitive ELISA

impose stress on laboratory personnel, increasing the opportunity for error.

Suggested procedures for improving the standard of quality control in diagnostic virology can be found in a quality control manual to be published by the PHLS [5].

By enabling the introduction of well-documented simulated specimens as unknowns into laboratory test systems, participation in EQA can help to identify where quality control procedures may be inadequate. There are certain limitations, two being associated with the nature of the specimens: because they are intended to reflect the type of specimens encountered routinely, they are not, on the whole, 'difficult' enough to expose lack of sensitivity or the absence of sufficiently low positive controls; because they consist mostly of diluted sera, they may not accurately represent clinical specimens for all types of assay.

Another limitation of EQA lies in the attitude of participating laboratories: there is a tendency towards a 'best effort' approach to EQA specimens, to see performing well in the scheme as an end in itself, and not as a means to improving laboratory performance. Only when EQA specimens are treated in exactly the same way as routine specimens, by the same personnel as would normally handle them, can they provide the laboratory manager with any useful information about routine capabilities. In instances reported by laboratory managers, erroneous results with EQA specimens have revealed sporadic or persistent deviation from protocols, inaccurate, inadequate or absent test controls, defective reagents, or, simply, transposition of specimens or results, showing that EQA can be a valuable management tool if employed correctly.

Occasionally, more difficult specimens than usual are distributed as 'educational' material, such as those containing very low levels of HBsAg or anti-HIV. Scores are not normally allocated for this type of specimen, but participants should find it helpful to compare their results with those of their peers. Also, some laboratories welcome the distribution of specimens which provide them with viruses they encounter infrequently, such as some of the respiratory viruses; often they can split the specimen and provide junior staff with a valuable and interesting exercise.

Included in the information participants are asked to give on their request/report forms are details of methods used. Analyses of this information are often included in the distribution summaries, giving the results obtained with various kits and techniques. This information is not intended to be an evaluation of methods, as the small numbers and simulated nature of the specimens and the variability of operator procedure do not permit a valid interpretation of this kind; it is intended to allow individual participants to compare their results with those of others using the same techniques. However, sometimes the evidence strongly suggests that particular kits or techniques have not performed as well as others, and we hope that users take the opportunity to investigate alternatives.

Scoring schemes: function and rationale

The function of EQA scoring schemes can be seen as twofold: firstly, they reinforce the message to participants of the intended results, indicating by a simple numerical grading if errors have occurred and with what degree of severity; secondly, they provide the scheme organisers with a numerical and statistical means of assessing individual and overall laboratory performance. This information is communicated confidentially to participants, who can then compare their own level of performance with that of laboratories examining the same specimens. Overall performance figures can be used for educational purposes, and can also be made available to those interested in the quality of diagnostic virology in the UK. The performance of an individual laboratory may be disclosed to the National Advisory Panel if it falls significantly below average in two consecutive six-monthly periods, the identity of the laboratory remaining confidential. A laboratory's identity is only disclosed to the panel if below-average performance persists over three consecutive six-monthly analyses, in order to allow the chairman to write offering advice. The majority of responses from those laboratories answering offers of help have been positive and participants have voluntarily indicated the probable reasons for their erroneous results. Most of the laboratories identified as potential poor performers do not reappear in subsequent analyses, indicating that the EQA scheme is fulfilling its intended purpose by drawing attention to problem areas and that remedial action has been taken. Despite repeated assurances, some participants still regard the scoring schemes as censorious, and are convinced that errors are revealed to external bodies. It can only be reiterated that an individual laboratory's score is for the manager's confidential information only, and that no laboratory 'league table' exists. Participation in the scheme is voluntary, though strongly encouraged, and the goodwill under which the scheme operates depends upon its strict confidentiality.

The basis of the graded scoring schemes for the various virology distribution types is as follows:

Score 2:   fully correct report
Score 1:   partially correct report, eg partial identification of a virus, or an equivocal serological result, referred for further testing
Score 0:   erroneous result which would not have serious clinical consequences, eg false negative rubella IgG

Score −1:   erroneous result which would have serious clinical consequences, eg false negative HBsAg or anti-HIV.

The scoring schemes were initially approved by the steering committee
of NEQAS for microbiology, and were based on the level of performance achievable by good diagnostic laboratories. For HBsAg and anti-
HIV detection, common practice is reflected in that stated referral of
positive results to a more expert laboratory is accepted as an alternative
to in-house confirmation. The scoring scheme is more tolerant of errors
with more difficult specimens (see Table 4).

Specimen scoring categories for HBsAg detection are currently based
on detectability by reverse passive haemagglutination (RPHA), and, for
anti-HIV detection, on classification by competitive radioimmunoassay
(COMPRIA). Now that more sensitive techniques are increasingly available, the average diagnostic laboratory is able to perform many such
serological tests to a very high standard, and these scoring schemes have
recently been reviewed to take such factors into account. For example,
new immunoassays employing recombinant DNA-derived antigens and
monoclonal antibodies enables the detection of anti-HIV at much lower
levels than designated by the current 'low level' positive category, and
the implication of the scoring scheme that a report of 'equivocal,
referred' is satisfactory for this category of specimen is probably out-
dated.

Table 4   Scores allocated for various categories of specimen

Specimen category	HBsAg detection*			Anti-HIV detection†		
	RPHA pos	RPHA weak	HBsAg neg	pos	pos low level	neg
Participant's report						
Positive, confirmed or referred	2	2	0	2	2	0
Positive, not confirmed or referred	1	1	−1	1	1	−1
Equivocal, referred	1	2	1	1	2	1
Equivocal, not referred	0	1	0	0	1	0
Negative	−1	0	2	−1	0	2

*Specimen categories are based on RPHA (reverse passive haemagglutination)
titre: RPHA positive = >32, RPHA weak positive = <16
†Specimen categories are based on percentage inhibition in COMPRIA (competitive RIA): positive = 75–100% inhibition, positive low level = 50–75% inhibition, negative = 0–25% inhibition

Occasionally specimens for HBsAg detection have been distributed in which the antigen concentration has been below the level of detection by RPHA. Scoring for this type of specimen was attempted in 1986, with those participants using sensitive techniques such as enzyme-linked immunosorbent assay (ELISA) or RIA being expected to detect the antigen, and those finding negative results by RPHA alone being allocated a full score. The rationale when this scoring scheme was devised was that laboratories using RPHA alone were not 'wrong' in their inability to detect very low levels of antigen, and that such laboratories might have had justifiable reasons for this practice. Since then, the increasing availability of sensitive and rapid ELISA techniques has generated further discussion on the advisability of using RPHA alone as a screening test for HBsAg. Some workers feel that in certain situations when the HBsAg levels undetectable by RPHA would be expected to be very rare, it is still acceptable in the interests of economy and convenience. This is disputed by others, who feel that there is little justification for the use of RPHA alone unless the time taken to obtain a result is an overriding factor, and then repeat testing by the more sensitive ELISA or RIA should follow within 24 hours. Some evidence obtained from EQA results in favour of including a sensitive assay will be presented later in this chapter. The PHLS hepatitis sub-committee is proposing to publish a recommendation that assays capable of detecting at least 1 IU/ml HBsAg should be used routinely in diagnostic laboratories. From early 1991 this will be reflected in the scoring scheme for NEQAS specimens in that participants will be expected to detect this level of antigen.

The scoring scheme for rubella serology (IgG detection) was based upon the effectiveness of the most commonly-used screening test, single radial haemolysis (SRH); the availability of a well-standardised cut-off control helped to ensure that a high rate of success was possible [4]. In recent years further refinements have been made in rubella screening: laboratories are able to confirm the presence of lower levels of rubella IgG than the recommended cut-off point for immunity of 15 IU/ml [6] by using sensitive techniques such as ELISA and latex agglutination. Again, specimen categories and scores allocated have recently been reviewed to come into line with this increased sensitivity.

The scoring scheme for rubella IgM detection is shown in Table 5. There is no 'low positive' specimen category, as this term is not clinically relevant; any antibody detectable above the test cut-off is considered significant, depending on an analysis of clinical details. The lowest antibody content in a distributed specimen so far has been approximately 7.5 arbitrary units (au)/ml; most test kits incorporate a positive cut-off control of 3au, but it is not certain how comparable these controls are. The score of -1 for false positive and false negative reports stands because of their potentially serious clinical consequences. No

Table 5 Scoring scheme for rubella IgM detection

	Specimen categories	
	rubella IgM positive	rubella IgM negative
Participant's report		
Positive	2	−1
Equivocal	1	1
Negative	−1	2

allowance is made for the referral of specimens giving equivocal results, as it is thought to be good laboratory practice to have alternative tests available to clarify such results and also to confirm positive results.

Scoring schemes for virus identification and electron microscopy have been successful and will probably remain unchanged for the time being from those originally described [4]. Participants are expected to make a full identification of viruses isolated, which is not unreasonable considering the wide availability of suitable reagents and the epidemiological importance of serotyping many viruses. This may seem unfair to smaller 'satellite' laboratories who normally refer their partially-identified isolates to a larger laboratory for full identification; but if it were possible to obtain a full score under such circumstances, perfectly capable laboratories could be tempted to opt out of the exercise during times of stress. It is interesting to note that for the lower serotype adenoviruses, where type identification is of epidemiological significance, particularly in cases of ocular cross-infection, the percentage of laboratories reporting the serotype is lower than for any other virus group. It is justifiable for the scoring scheme to discriminate in favour of those laboratories reporting the serotype, as the majority of those failing to do so indicate that they do not intend to refer the isolate.

Scoring for general virus serology has been described in detail elsewhere [4]. Briefly, participants are requested to record the complement fixing antibody titres of two sera, not representing a clinical pair, to one or more virus antigens. To obtain a full score the ratio of these titres must fall within a statistically-derived range. The purpose is not to determine whether laboratories are capable of detecting a clinically-significant difference between the two serum titres, but rather to assess their precision in performing such tests.

Catering for different types of laboratory
Many different types of laboratory participate in the virology scheme, from expert laboratories which provide a full range of viral diagnostic

services and some reference facilities, to non-virology laboratories which may carry out only rubella IgG or HBsAg screening. Laboratories can choose which distribution types are appropriate to the service they offer, and can also opt out of certain exercises within those distribution types by indicating that they do not examine that type of specimen.

Confirmation of positive results and referral of specimens have already been described in the section on scoring schemes; in serological tests, the capabilities and routine practices of all types of laboratory are taken into account, and concessions are made to less expert or less well-equipped laboratories in the allocation of a full score for equivocal results on more weakly positive specimens, if referred. The assumption is made that referral would lead to a successful outcome. In general, a high standard is expected of all types of laboratory, and repeated failure within a particular diagnostic area should encourage a participant to reassess his capability of providing a satisfactory diagnostic service. There have been several instances of smaller laboratories finding, through failure with EQA specimens, that a small workload has not provided them with sufficient routine experience to be competent, or that the test system to which they are restricted by a limited budget is not adequately sensitive.

In virus identification, no allowance is made for referral for the reason previously stated. Most laboratories attempt full identification of viruses isolated, apart from the adenoviruses, where less importance seems to be attached to serotyping by some laboratories. One problem which may increase with progressive financial constraints is that, for simulated specimens such as nasopharyngeal aspirates from young children, more laboratories may simply screen for respiratory syncytial virus by immunofluorescence, and may not proceed to culture, thereby over-looking other respiratory viruses which may be present.

In hepatitis B serology, attempts to cater for all types of laboratory with the same specimens has led to some deficiencies in the scheme. Blood Transfusion laboratories require an extremely high level of sensitivity in their screening tests for HBsAg, and a greater input of very low positive EQA material would serve their needs better; clinical laboratories testing for a range of hepatitis B markers need specimens representative of all stages of infection, but it is often difficult to provide an adequate volume of interesting sera because of the large numbers of laboratories participating in this distribution. At present, valuable, well-characterised sera are distributed to many laboratories that only screen for HBsAg, and there is a case for making a subdivision within the hepatitis B distribution type: all participating laboratories would receive specimens intended to assess their ability to detect HBsAg, and additional distributions including more interesting specimens could be made available to those laboratories wishing to monitor their tests for other markers. In this way it might be possible to distribute original sera without dilution or mixing, and laboratories could be scored on the results of their tests for other markers, particularly hepatitis B e antigen (HBeAg) and its antibody HBeAb.

What does the scheme test?

The question of what precisely the scheme is assessing has often arisen in discussions with participants. Over the years, request/report forms, particularly for serological tests, have been simplified to exclude most of the specimen and clinical detail, so that requests are appropriate to the work of the maximum number of laboratories. This simplification underlines the fact that the scheme is testing only the laboratory's ability to detect and report correctly the required analyte in the specimen supplied, under the testing conditions prevailing at that time. The whole laboratory diagnostic process cannot be tested because it includes elements of interpretation which can require information from the clinician, from further specimens or from additional tests, and the scheme cannot take such factors into account.

Some participants have blamed poor EQA results on commercial test kits; now that their use is so widespread it must be accepted that the quality of such kits does have a great influence on a laboratory's success. However, in EQA it is not the performance of the kits that is being assessed, but that of the laboratory, whose responsibility it is to investigate the reason for any failure with EQA specimens, whether it be the fault of the manufacturer, such as inadequate initial evaluation of the kit or the supply of controls incompatible with nationally-acceptable standards, or the fault of the laboratory, such as failure to monitor batch quality or deviation from kit protocols. Adequate product evaluation and effective quality control measures should indicate the unsuitability of a particular kit, batch or assay run, and should prevent the reporting of erroneous results.

A generally high standard of performance for UK laboratories

Over the past five years, the overall performance of UK laboratories in virological EQA exercises has been maintained at an acceptably high level. Success rates, as judged by the current scoring schemes, have been especially encouraging in serological disciplines. This reflects the greater possibility for standardisation in serological methods, with the widespread availability of serological reference and control material, compared with techniques for virus identification. The availability of more sensitive, well-evaluated serological assays and the elimination of subjectivity in reading the results of such assays further increase the opportunity for accuracy.

Another reason for this high standard in serology is that many participants employ more than one type of assay routinely, which is commendable, but the application of multiple testing to negative EQA specimens, involving three or more types of assay, must be seen as paying excessive attention to such specimens, as this is unlikely to be routine practice in screening tests for which the majority of results are negative.

The success rates for the four main branches of serodiagnosis covered by the scheme, expressed as the percentage of the total number of participants' reports for which the maximum score was obtained, are shown in Tables 6–9.

The false positivity rate for HBsAg detection (Table 6) is very low; the negative specimen distributed in 1986 and showing only 92% success produced some problems of non-specificity with RPHA and with some new ELISA kits employing monoclonal antibodies, and 3% of reports were positive and 5% equivocal. The false negativity rate is of more concern, but many of those participants who apparently failed with strongly positive specimens would have issued an equivocal report and would have referred the specimen for further testing. For the weakly positive specimens (RPHA titre <16), the scoring scheme allows full marks for a report of 'equivocal, referred', so the 1–10% of participants failing with these specimens were issuing negative reports; the majority of these were using insensitive techniques such as RPHA.

Up to 43 of 233 UK participants examine specimens for markers of hepatitis B infection other than HBsAg, including antibody to HBsAg, antibody to hepatitis B core antigen, IgM antibody to HBcAg, HBeAg, and anti-HBe. No scores are allocated for participants' performance in these tests, but overall results are published in the distribution summary for information, and these usually correlate well with the expected results.

The success rates for anti-HIV detection (Table 7) show an improvement after the initial distribution in 1986, due partly to the availability of

Table 6 Hepatitis B surface antigen detection: success rates

Specimen category	Strong positive (RPHA >32)		Weak positive (RPHA <16)		Negative (RPHA <8)	
	n	% correct	n	% correct	n	% correct
1984	790(6)	96	–	–	405(3)	99
1985	1083(6)	94	186(1)	91	551(3)	98
1986	1491(8)	94	187(1)	90	182(1)	92
1987	1341(7)	96	387(2)	91	591(3)	97
1988	862(4)	96	217(1)	99	1076(5)	98

n = total number of reports
Numbers in parentheses are numbers of specimens examined

Table 7 Anti-HIV detection: success rates

Specimen category	Positive		Low positive		Negative	
	n	% correct	n	% correct	n	% correct
1986	282(3)	95	93(1)	92	188(2)	98
1987	672(6)	99	562(5)	97	782(7)	98
1988	129(1)	98	922(7)	98	792(6)	99

n = total number of reports
Numbers in parentheses are numbers of specimens examined

more sensitive assays and partly to the increasing expertise of laboratories who had only just begun testing in 1986. An additional specimen which gave equivocal results by COMPRIA (inhibition 25–50%) and for which scores were not allocated was distributed in late 1988 and a success rate of 99% was obtained. The few errors in recent distributions have rarely been due to incorrect technical results; the transposition of specimens and failure to confirm positive results were more common.

For rubella IgG detection (Table 8), the apparently high success rates with two of the very weakly positive specimens (15–25 IU/ml) are partly accounted for by the increased leniency of the scoring of this type of specimen. The specimen distributed in 1985 for which only 83% of participants recorded a correct result contained 15 IU/ml of rubella antibody, equivalent to the recommended cut-off level for immunity [6], and many participants were penalised for ignoring or failing to detect the low amount of antibody present and reporting a negative result.

For rubella IgM detection (Table 9), overall performance is affected by poorer performance with more weakly positive specimens. In gener-

Table 8 Rubella IgG detection: success rates

Specimen category	Strong positive (<50 IU/ml)		Weak positive (25–50 IU/ml)		Very weak positive (15–20 IU/ml)		Negative	
	n	% correct	n	% correct	n	% correct	n	% correct
1984	504(4)	99	–	–	252(2)	95	126(1)	97
1985	580(4)	99	290(2)	91	145(1)	83	145(1)	97
1986	291(2)	99	436(3)	96	291(2)	97	436(3)	94
1987	318(2)	99	317(2)	92	–	–	318(2)	97
1988	162(1)	99	342(2)	96	–	–	162(1)	97

n = total number of reports
Numbers in parentheses are numbers of specimens examined

Table 9   Rubella IgM detection: success rates

Specimen category	Positive		Negative	
	n	% correct	n	% correct
1986	156(2)	96	78(1)	100
1987	346(4)	90	82(1)	100
1988	259(3)	97	250(3)	99

n = total number of reports
Numbers in parentheses are numbers of specimens examined

al, the distribution of specimens with a rubella IgM content greater than 15au/ml results in success rates of 99–100%. The low false positive rate is of limited significance considering the small numbers of negative specimens tested and the fact that the constituent sera are selected for freedom from non-specific reaction.

For general virus serology, overall success rates have ranged from 86% to 92%, which is reasonable considering the lack of standardisation of complement fixation test methods. Participants were occasionally less successful with some antigens such as influenza A and measles, but it is difficult to suggest reasons for such failures.

Overall success rates for electron microscopy and virus identification are shown in Tables 10 and 11, with respect to individual virus types or groups. Performance with some of the electron microscopy specimens has been excellent, but it is not strictly valid to compare performance with different specimens within the same virus group, as it has not been possible to quantify the virus particles present. The lower success rates in virus identification reflect the unreliability of the techniques involved, their lack of standardisation and their dependence on subjective observation. The overall standard of performance is reasonable when these factors are taken into consideration. However, the specimens distributed contained high concentrations of tissue culture adapted viruses, and were subjected to extensive quality control to ensure bacteriological sterility, non-toxicity to cell cultures, stability of virus infectivity and specificity. Of necessity, the specimens consequently presented less of a challenge than most clinical material, and it is probable that performance in this diagnostic field is even less good with clinical specimens.

Despite the fact that overall performance in virology EQA does appear to be acceptable, there should be no room for complacency, as the majority of specimens distributed so far have been relatively straightforward. Although the intention is to reflect the type of clinical specimens encountered routinely by participants, there is a need for the

Table 10 Electron microscopy: overall success rates, 1984–88

Virus type or group	n	Mean % correct	Range of % correct
Herpesvirus	166(3)	97	95–98
Adenovirus	211(4)	90	78–98
Papovirus	166(3)	95	91–98
Poxvirus	110(2)	87	81–93
Rotavirus	209(4)	85	73–96
Paramyxovirus	52(1)	83	–
Orthomyxovirus	54(1)	91	–
Parvovirus	50(1)	50	–
Calicivirus	56(1)	46	–
Reovirus	52(1)	29	–
Negative specimens	374(7)	94	88–98

n = total number of reports
Numbers in parentheses are number of specimens examined

Table 11 Virus identification: overall success rates, 1984–88

Virus type or group	n	Mean % correct	Range of % correct
Herpes simplex virus	322(4)	98	95–100
Cytomegalovirus	77(1)	84	–
Varicella zoster	82(1)	72	–
Adenovirus	328(4)	61	53–69
Respiratory syncytial virus	326(4)	82	57–93
Influenza A	164(2)	73	63–82
Influenza B	163(2)	85	83–88
Parainfluenza type 1	238(3)	66	51–73
Parainfluenza type 3	156(2)	89	87–91
Echovirus	416(5)	80	41–92
Coxsackievirus	244(3)	77	59–89
Poliovirus	164(2)	87	82–91
Rhinovirus	78(1)	41	–
Negative specimens	880(11)	96	98–100

n = total number of reports
Numbers in parentheses are number of specimens examined

distribution of more difficult material so that laboratories can assess their true capabilities.

Problem areas revealed by the scheme

The main problem areas revealed by analysis of participants' results involve the choice of method, the combination of methods used or inadequate quality control measures. The use of insensitive methods alone or as secondary tests for EQA specimens suggests that there is a risk of low positive specimens being missed in the clinical situation. Of most concern is the use of RPHA alone as a screening test for HBsAg. Table 12 shows the results recorded by all users of three techniques available for this type of test, both singly and in combination, for various specimens for which the HBsAg content had been accurately determined. The results show the difficulties experienced by users of RPHA at HBsAg levels above 50ng/ml, which were easily detectable by pre- and post-distribution RPHA tests in the NEQAS laboratory. Of 67 participants using RPHA alone, 24% failed to detect HBsAg at 67ng/ml and 34% recorded an equivocal result; of 56 using RPHA alone, 18% failed to detect HBsAg at 106ng/ml and 39% recorded an equivocal result. These failures suggest either poor test kit quality or poor operator technique, both of which could have been detected by the inclusion, in addition to test kit controls, of low positive material calibrated against the British Reference Preparation for HBsAg [7]. Laboratories would normally refer specimens giving equivocal results for further testing, but those giving negative results would have been lost to further investigation. The occurrence of such levels of HBsAg may be relatively infrequent, but laboratories using RPHA alone should realise the limitations of the technique, and that it may be even less sensitive in their hands than expected. The consideration of a more sensitive technique is

Table 12 Participants' results per HBsAg concentration

Specimen content of HBsAg (ng/ml)	Numbers of test users recording these results								
	RPHA			ELISA			RIA		
	pos	equiv	neg	pos	equiv	neg	pos	equiv	neg
51	30	17	69	96	1	0	40	0	0
67	52	44	26	106	0	0	39	0	0
106	65	33	16	95	0	0	40	0	0
229	110	6	0	93	0	0	40	0	0
305	107	9	3	96	0	0	38	0	0
440	117	11	0	105	0	0	38	0	0
1060	117	2	0	96	0	0	38	0	0

of utmost importance if the screening of tissue donors and renal dialysis patients is being undertaken.

RUBELLA IgG DETECTION

Another discipline in which poorer results than expected were obtained with an established technique was rubella IgG detection. Table 13 shows the results recorded by participants using various screening tests for specimens containing low levels of rubella IgG. Users of single radial haemolysis (SRH), the most popular screening test because of its simplicity and efficacy, recorded a surprisingly large number of equivocal and negative results, considering that a cut-off control serum containing 15 IU/ml rubella antibody is available to most laboratories in the UK, and that the second British Standard for rubella IgG is available to those who wish to calibrate additional controls [8]. In pre- and post-distribution tests in the NEQAS laboratory, all the specimens containing 20 IU/ml and above produced zones of haemolysis greater in diameter that that given by the 15 IU/ml control, but many users of SRH, whether through technical inaccuracy or over-cautious interpretation, recorded equivocal results and regarded the immunity conferred by such antibody levels as doubtful. Sera containing low levels of rubella antibody form a small minority of sera tested, but in the interests of accuracy and to avoid confusion when women are tested repeatedly, laboratories should ensure that SRH tests are adequately sensitive and properly standardised, controlled and monitored. A small number of participants using ELISA experienced difficulty with these low level positive specimens. This may be an indication that the control material provided with kits is not compatible with national standards, and adds weight to the argument for the inclusion of independently-standardised control material with every assay run. Haemagglutination inhibition

Table 13 Participants' results per rubella IgG concentration

| Rubella IgG (IU/ml) | Numbers of test users recording these results | | | | | | | | |
| | SRH | | | ELISA | | | HI | | |
	pos	equiv	neg	pos	equiv	neg	pos	equiv	neg
15	34	63	15	48	8	2	8	15	21
20	63	36	2	39	3	0	3	13	3
23	82	21	2	26	2	2	20	12	5
25	70	30	4	40	0	2	6	8	3
27	88	12	1	45	2	1	11	4	1
30	91	8	0	40	2	0	10	5	0
38	94	5	1	38	0	0	8	4	0

(HI) was, as expected, the least sensitive technique with these specimens, but fortunately the number of participants using it alone as a
screening test has reduced considerably in recent years.

One way of improving accuracy is to use an alternative test to confirm
low level positive results. This practice has been recommended for sera
giving antibody levels equal to or less than 15 IU/ml and for some
negative sera [9,10]; both ELISA and latex agglutination have been used
successfully as alternative methods for confirmation. Several participants are still using HI as a secondary test for screen negative and low
level positive specimens, but there is little justification for this because
of the test's lack of sensitivity (Table 13) and tendency to non-specific
reaction [9,10].

RUBELLA IgM DETECTION

In the diagnosis of rubella infection, it is important that the method used
is adequately sensitive, as low levels of rubella IgM can occur in cases of
reinfection and congenital infection [10]. When EQA specimens of
known content were examined, participants using ELISA which incorporated antigen on the solid phase experienced more difficulty in detecting
lower concentrations than those using IgM capture ELISA (Table 14). On
the basis of the results of studies on clinical specimens, techniques
employing IgM capture have now been recommended as preferable for
the diagnosis of rubella because of their greater sensitivity [9]. The
time-consuming and technically-inconvenient sucrose density gradient
with HI (SDG/HI) is now only used by a very few UK laboratories as a
secondary method.

Because of the possibility of false positive results occurring with any
assay, it is advisable to confirm positive results obtained in pregnant
women by an alternative method, unless strong evidence of primary
infection such as seroconversion is available [9,10]. Although no clinical

Table 14 Participants' results per rubella IgM concentration

| Rubella IgM antibody (au/ml)* | Numbers of test users recording these results | | | | | | | | |
| | antiglobulin ELISA | | | IgM capture ELISA | | | IgM capture RIA | | |
	pos	equiv	neg	pos	equiv	neg	pos	equiv	neg
7.5	2	1	18	18	28	10	7	1	0
8	8	9	2	46	2	0	5	0	0
15	15	4	0	47	1	0	5	0	0
30	20	1	0	56	0	0	8	0	0

*au = arbitrary units

detail is supplied with EQA specimens for rubella IgM detection, the majority of laboratories did not indicate the use of a secondary assay to confirm their positive results, and there is some concern that this may reflect their routine practice with samples from pregnant women.

VIRUS IDENTIFICATION

Problems encountered with EQA specimens for virus isolation and identification included failure to make an isolation, misidentification and incomplete identification. Participants were least successful in terms of isolation with specimens containing respiratory viruses, in particular parainfluenzavirus type 1 (Table 15). This would suggest a failure to monitor the sensitivity of cell cultures to respiratory viruses, but as UK laboratories rely mainly on primary baboon kidney cells supplied weekly from a central source for much of their respiratory work, it is difficult for them to confirm the sensitivity of the cells in advance, as is good practice with a continuous or semi-continuous cell line. However, it can be assumed that the majority of laboratories would be using current batches of baboon kidney cells at any one time, and post-distribution tests in the NEQAS laboratory ensure that isolation is easily made in such batches. Information volunteered from several laboratories has indicated that some problems result from deviation from recommended practices, such as failure to use serum-free tissue culture medium or failure to monitor incubator temperature.

Table 15   Failure rates for some virus identification specimens

Virus type	Year of distribution	n	% failing to isolate	% incomplete or misidentification
Parainfluenza	1984	78	21	29
type 1	1987	77	16	13
	1988	83	12	14
Parainfluenza	1986	79	5	4
type 3	1987	77	6	6
Influenza A	1985	82	5	15
	1987	82	27	6
Influenza B	1987	83	7	10
	1988	80	4	9
Respiratory	1987	82	12	0
syncytial virus	1987	81	40	4
	1988	83	4	4
Cytomegalovirus	1987	77	13	3
Varicella zoster	1987	82	13	15

n = total number of reports

It is interesting to note that parainfluenzavirus type 1 presented more difficulty in isolation than parainfluenzavirus type 3, probably due to the lack of cytopathic effect produced by the former or to ineffective haemadsorption procedures. A large percentage of laboratories failed to isolate an influenza type A virus distributed in 1987, which was of the same strain as the more successfully-isolated virus distributed in 1985 (Table 15), but at a slightly lower concentration. The virus was isolated easily in post-distribution tests in the NEQAS laboratory, and it is difficult to explain these sporadic failures, as with the second specimen containing respiratory syncytial virus (RSV) distributed in 1987 (Table 15). Described as a throat swab, it is possible that this type of specimen did not prompt all laboratories to seek RSV. Certainly, the failure to use appropriate cell lines is a contributory factor; success in RSV isolation was less likely when HEp-2 or HeLa cells were not used.

In 1987, simulated specimens containing cytomegalovirus and varicella zoster virus were distributed: 13% of participants failed to isolate these viruses (Table 15). Poor quality of cell line sensitivity is the probable cause of failure here: it is essential to maintain a sensitive fibroblast cell line for this type of work.

The specimens distributed for the investigation of respiratory infection were usually simulated nasopharyngeal aspirates, consisting of infected and uninfected tissue culture cells suspended in a gelatin transport medium. These specimens were suitable not only for isolation but also for rapid detection of viral antigen by immunofluorescence. Unsuccessful participants either did not use this technique or failed to detect specific fluorescence, or misidentified the fluorescence. This suggests poor quality control of fluorescence microscopy and reagents.

Several laboratories have successfully detected viral antigen in simulated specimens by ELISA, and it is hoped that improved techniques for rapid identification such as this will eventually replace much of the relatively insensitive, unreliable and labour-intensive tissue culture work, although isolation will still be necessary for epidemiological purposes.

For some specimens for virus isolation, relatively large percentages of participants lost marks through misidentification or inadequate identification. Some misidentification was evidently due to the transposition of EQA specimens, other errors being due to cross-contamination with routine specimens or to poor quality control of the identification system. As for incomplete identification, many laboratories were unable to proceed beyond a report of 'haemadsorbing agent' for some respiratory viruses, despite adequate time allowed and readily-available reagents. Laboratories with access to a fluorescence microscope would probably find it easier to identify many viruses by immunofluorescence tests on inoculated tissue cultures than by neutralisation.

ELECTRON MICROSCOPY

The major problem encountered by participants examining EQA specimens by electron microscopy (EM) was difficulty in detecting small round viruses (SRV). As can be seen from Table 10, only 46% of participants detected a calicivirus and only 50% a parvovirus, showing no improvement over the success rates obtained for similar specimens distributed in 1982 [11] and 1983 [12]. Similar results had been obtained with specimens containing astrovirus, distributed in 1982 [11] and 1983 [12]. It may be that the resolution of some electron microscopes is not adequate for the detection of all SRV, but whatever the reason for failure, the results emphasise the potential value of such specimens to participants by indicating a deficiency at some stage in their EM procedure.

When a reovirus was distributed in 1984, 79% of participants detected the virus, but only 29% made a correct identification, the remainder confusing the morphology with that of a rotavirus. There was some controversy over the appearance of the particles in this specimen, but it was thought that laboratories should have been able to distinguish them as reoviruses [13].

Simulated specimens for electron microscopy, like all EQA specimens, undergo rigorous quality control to ensure the stability of virus morphology and concentration, but some participants have experienced technical difficulty with those specimens that necessarily contain tissue culture fluid, and sometimes there has been controversy over the adequacy of numbers of virus particles present. Some have felt that specimens for electron microscopy tended to be difficult compared with those for other distribution types and were less representative of clinical specimens. In future it will be necessary to improve the authenticity of these specimens and to develop a method of quantifying their content of virus particles.

Other virology schemes

Virology quality assessment schemes operate in a number of other European countries including France, Norway and Italy. The French scheme covers a wide range of disciplines but, from the information currently available, the Norwegian scheme has been limited to rubella IgG detection and some general virus serology [14], and the Italian scheme to rubella IgG and HBsAg detection [15,16]. In the United States, laboratory proficiency testing programs are operated by the College of American Pathologists (CAP) and by the Centers for Disease Control (CDC), the latter being responsible for the extensive HIV serology scheme that has become necessary [17]. The results of their surveys have proved encouraging so far [18]. The US Department of the Army has also set up a worldwide HIV serology scheme with satisfactory results [19]. The CAP organises proficiency testing in HBsAg detection [20,21,22], rubella IgG detection [21,23], viral antigen detection, virus isolation, *Chlamydia trachomatis* detection and some general virus serology [24]. Because of the large scale of these operations, many samples are produced commercially.

In New Zealand, proficiency testing for HBsAg detection has been carried out by the Blood Transfusion Service [25], and a scheme covering some viral serology operates in Australia. Other national and regional schemes probably exist, but without the publication of results in international journals, information is difficult to obtain. To improve the quality of EQA by facilitating the exchange of information and to assist those national organisations wishing to set up their own schemes, it is intended to compile an international directory of microbiological schemes to which interested parties could contribute.

Conclusion

It is important that the virology EQA scheme is continually reviewed and updated to keep pace with developments in the field. Certain improvements to the scheme have already been identified, namely the modification of current scoring schemes to account for the more sensitive methods now widely used, the distribution of more difficult and educational specimens, the modification of the hepatitis B serology subscheme, and the improvement of EM specimens. In addition, the general virus serology subscheme should be updated, with less emphasis placed on CF tests, apart from those which continue to be useful, and with more emphasis on specimens suitable for testing by ELISA and immunofluorescence as these tests become more prominent.

A questionnaire on laboratory tests for *C trachomatis* distributed to all microbiology scheme participants revealed the wide extent of the diagnostic service offered for this organism [26]. All of the techniques for the detection of *C trachomatis* can present difficulties and EQA specimens could help participants to detect any weaknesses in their chosen test systems. To meet this need a trial distribution has recently been made of simulated specimens for antigen detection by ELISA and by immunofluorescence; a preliminary analysis of results looks promising, with the great majority of participants reporting correctly.

Another developmental aspect on which work has commenced is the lyophilisation of specimens for virus identification. The intention is to improve long-term stability so that distribution preparations can be planned in advance more efficiently, and so that more pre-distribution checks can be carried out on the specimens. At present specimens containing viruses as infected cell suspensions in gelatin transport medium must be prepared shortly before distribution to ensure maximum stability of virus infectivity, and there is little opportunity to repeat the pre-distribution checks performed on pilot specimens for identity, purity, sterility and stability. Another advantage of lyophilisation is the possibility of distributing the same specimen on two different occasions in order to assess whether any improvement in isolation rates had occurred.

A final consideration for the future is the introduction of EQA subschemes for the detection of antibodies to cytomegalovirus, hepatitis A virus and hepatitis C virus. As ever, the feasibility of such plans depends on the generosity of those laboratories who collect and donate valuable sera to NEQAS for virology, without whose cooperation the scheme could not function.

References

1   Snell JJS. United Kingdom National External Quality Assessment Scheme for Microbiology. *Eur J Clin Microbiol* 1985, **4**, 464–7.

2   Hart RJC. Quality control in diagnostic virology. *Proc R Soc Med* 1975, **68**, 622–4.

3   Reed SE, Gardner PS, Snell JJS, Chai O. United Kingdom scheme for external quality assessment in virology. Part I. General method of operation. *J Clin Pathol* 1985, **38**, 534–41.

4   Reed SE, Gardner PS, Stanton J. United Kingdom scheme for external quality assessment in virology. Part II. Specimen distribution, performance assessment, and analyses of participants' methods in detection of rubella antibody, hepatitis B markers, general virus serology, virus identification, and electron microscopy. *J Clin Pathol* 1985, **38**, 542–53.

5   Snell JJS, Farrell ID, Roberts C (eds). Quality control: principles and practice in the microbiology laboratory. London: Public Health Laboratory Service, in press.

6   Kurtz JB, Mortimer PP, Mortimer PR, Morgan-Capner P, Shafi MS, White GBB. Rubella antibody measured by radial haemolysis: characteristics and performance of a simple screening method for use in diagnostic laboratories. *J Hyg Camb* 1980, **84**, 213–22.

7   Seagroatt V, Ferguson M, Magrath DI, Schold G, Cameron CH. British Reference Preparation of hepatitis B surface antigen. *Lancet* 1982, **ii**, 391–2.

8   National Institute for Biological Standards and Control. Proposed second British Standard for anti-rubella serum: unpublished data.

9   Morgan-Capner P. Laboratory diagnosis of rubella. Summary of recommendations of the PHLS Working Party. *PHLS Microbiology Digest* 1988, **5**, 49–52.

10 Morgan-Capner P. Rubella serology – 20 years on. *Serodiagn Immunotherapy Infect Dis* 1988, **2**, 85–9.

11 UKNEQAS, 1982. Summary of electron microscopy distribution no 226.

12 UKNEQAS, 1983. Summary of electron microscopy distribution no 257.

13 UKNEQAS, 1984. Summary of electron microscopy distribution no 289.

14 von Kramer M. External quality assessment on serological diagnostic virology in Norway 1982–84. *NIPH Ann* 1986, **9**, 33–9.

15 De Mayo E, Lamanna A, Morgantetti F, Orsi A. Controlli de qualità in microbiologia. *Quad Sclavo Diagn* 1982, **18**, 10–22.

16 Orsi A, De Mayo E, Morgantetti F. The Italian experience of quality control in microbiology. *Quad Sclavo Diagn* 1984, **20**, 203–11.

17 Schalla WO, Hearn TL, Griffin CW, Taylor RN. Role of the Center for Disease Control in monitoring the quality of laboratory testing for human immunodeficiency virus infection. *Clinical Microbiology Newsletter* 1988, **10**, 156–9.

18 Taylor RN, Przybyszewski VA. Summary of the Center for Disease Control human immunodeficiency virus (HIV) performance evaluation surveys for 1985 and 1986. *Am J Clin Pathol* 1988, **89**, 1–13.

19 Damato J, Fipps DR, Redfield RR, Burke DS. The Department of the Army quality assurance program for human immunodeficiency virus antibody testing. *Lab Medicine* 1988, **19**, 577–80.

20 Taylor RN, Fulford KM. Results of the Center for Disease Control proficiency testing program for the detection of hepatitis B surface antigen. *J Clin Microbiol* 1976, **4**, 32–9.

21 Taylor RN, Fulford KM, Przybyszewski VA, Pope V. Center for Disease Control diagnostic immunology proficiency testing program results for 1978. *J Clin Microbiol* 1979, **10**, 805–14.

22 Hanson MR, Polesky HF. Radioimmunoassay and enzyme immunoassay methods for detecting viral hepatitis markers. *Am J Clin Pathol* 1983, **80**, 590–3.

23 Skendzel LP, Wilcox KR, Edson DC. Evaluation of assays for the detection of antibodies to rubella: a report based on data from the College of American Pathologists surveys of 1982. *Am J Clin Pathol* 1983, **80**, 594–8.

24 L. Rosati, personal communication.

25 Anderson RA, Ramirez AM, Steed IW, Woodfield DG. A review of the hepatitis antigen proficiency survey (HAPS). *NZ J Med Lab Technol* 1986, **40**, 8–9.

26 UKNEQAS, 1988. Summary of questionnaire on methods used for the laboratory diagnosis of *Chlamydia trachomatis* infections.